209 Marston
4:30

MATHEMATICAL MODEL BUILDING
AN INTRODUCTION TO ENGINEERING

MATHEMATICAL MODEL BUILDING

AN INTRODUCTION TO ENGINEERING

Charles R. Mischke

THE IOWA STATE UNIVERSITY PRESS, AMES, IOWA

Charles R. Mischke is professor of mechanical engineering at Iowa State University.

© 1980 The Iowa State University Press
All rights reserved

Printed by
The Iowa State University Press
Ames, Iowa 50010

No part of this publication may be reproduced, stored in a retrieval system, or transmitted in any form or by any means—electronic, mechanical, photocopying, recording, or otherwise—without the prior written permission of the publisher.

Second edition, revised printing, 1980
First and second editions © 1973, 1976, Charles R. Mischke, published under the title *An Introduction to Engineering through Mathematical Modelbuilding*

Library of Congress Cataloging in Publication Data

Mischke, Charles R
    Mathematical model building.

    First, 2d ed. published under title: An introduction to engineering through mathematical modelbuilding.
    Includes index.
    1. Engineering mathematics. 2. Engineering. I. Title.
TA330.M49   1980   620'.007'2   79-25436
ISBN 0-8138-1005-1

To my colleagues and all the students
who made this inquiry into the nature of
mathematical models a memorable experience

CONTENTS

Preface	xi
Chapter 1 What is an Engineer?	3
1.1 Preface to the student	3
1.2 Rational decision making	5
1.3 Engineering	7
1.4 Design and the design process	8
1.5 Basic operations in the design process	12
1.6 Attitudes and abilities	14
Problems	17
Chapter 2 Posing the Design Problem	21
2.1 Introduction	21
2.2 Design imperative	22
2.3 Tie-bolt example	22
2.4 Reflections and generalizations	26
Problems	35
Chapter 3 Figures of Merit	39
3.1 Introduction	39
3.2 Design factor as a figure of merit	43
3.3 Cost as a figure of merit	47
3.4 Reliability as a figure of merit	55
3.5 Time as a figure of merit	64
3.6 Trade-off functions	66
3.7 Qualitative parameters in figures of merit	68
Problems	69
Chapter 4 Engineering and the Digital Computer	73
4.1 What is an algorithm?	73
4.2 Engineering design and the digital computer	76
4.3 IOWA CADET: a computer aid to design	78
4.4 Documentation	81
4.5 Error messaging and tests	83
4.6 Some useful numerical methods	86
4.7 Writing computer programs	95
Problems	101

Chapter 5 Empiricism: The Basis of Mathematical Modeling — 103
- 5.1 Introduction — 103
- 5.2 Importance of definition — 107
- 5.3 Graphical communication — 112
- 5.4 Mathematical curve fitting — 125
- 5.5 Reduction of number of variables — 139
- 5.6 Establishing a complete set of dimensionless variables — 143
- 5.7 Use of partial derivatives — 147
- 5.8 Physical models and similitude — 160
- 5.9 International system of units — 164
- 5.10 Recognition of variability — 167
- Problems — 168

Chapter 6 Mathematical Models of Effects — 179
- 6.1 Introduction — 179
- 6.2 System and control space — 179
- 6.3 Heat and work effects — 183
- 6.4 Tractive effects — 185
- 6.5 Surface effects — 187
- 6.6 Charge effects — 187
- 6.7 Magnetic effects — 188
- 6.8 Other effects — 188
- Problems — 190

Chapter 7 Enunciation of First Principle — 195
- 7.1 Introduction to accountability — 195
- 7.2 Conservation of countables — 197
- 7.3 Creation and destruction of countables — 198
- 7.4 Accountability equation for discrete countables — 201
- 7.5 Accountability equation for a continuum of countables — 208
- 7.6 Enunciation of first principle — 211
- 7.7 Delineation of system and control region — 222
- 7.8 Prediction with mathematical models — 228
- 7.9 Checking validity — 238
- 7.10 Assertive model of an epidemic — 240
- 7.11 Probabilistic deductive and simulative models — 247
- 7.12 Generating random numbers of specified distributions — 253
- Problems — 258

Chapter 8 Optimization — 263
- 8.1 Introduction — 263
- 8.2 Maxima and minima — 265
- 8.3 Method of Lagrange — 268
- 8.4 Statement of optimization problem — 274
- 8.5 Direct search methods — 275
- 8.6 Exhaustive search — 280
- 8.7 Interval-halving and golden section searches — 281
- 8.8 Multidimensional searches — 290
- 8.9 Fence-walking, graphical involvement — 297
- Problems — 302

Chapter 9 Decisions Reached through Mathematical Model Building — 307
- 9.1 Introduction — 307
- 9.2 Decision on overloading of a transformer — 309
- 9.3 Decision on routing of a pipeline — 312

Contents   ix

9.4   Decision on parameters for a circular disk cam   316
9.5   Decision on pan angle for a rock crusher   322
9.6   Decision on proportions of a can   325
9.7   Predicting balancing speeds of electric railway equipment 328
9.8   Decision on speed of a wire winder   333
9.9   Decision on a step-down gear ratio   338
9.10 Decision on a measurement   344
9.11 The engineer's notebook   351

Epilogue   352

Appendixes   355
  A   Some utility routines from IOWA CADET   355
  B   International system of units   366
  C   Theorems of random variable algebra   381

Selected Problem Answers   385

Index   389

PREFACE

    This book was written to introduce engineering students to their fields through the vehicle of mathematical model building. This route was chosen in preference to other alternatives because it appears to be more relevant to students, their questions, and their needs.

    First, the author is still convinced that design ideas, methods, and mores are best left largely to the senior year of study, when the student and the instructor can call on a significant body of knowledge upon confronting the demanding task of synthesis.

    Second, students are not usually convinced that all the courses in chemistry, physics, mathematics, statics, dynamics, strength of materials, thermodynamics, heat-mass-momentum transfer, electrical circuits and field theory, instrumentation, and the design of experiments are really necessary to their preparation to innovate. Their interest often lags, partly because of the lack of any coordinated effort to convince them on rational grounds of the necessity of the ordering of their educational events. This volume represents a first step in this direction.

    Third, in disciplinary courses in the engineering sciences, the emphasis is on *subject matter*. The tool used to teach, organize, and utilize it is the mathematical model. Unfortunately, few instructors take the time to talk about the modeling process itself, sans subject matter. The modeling process must be understood in order to interpret the results properly. Too often students are absorbing disciplinary material and trying to learn the modeling techniques concurrently. Faltering in the modeling process leads to poorer understanding of the disciplinary elements. This volume is addressed principally to deterministic, deductive mathematical modeling before the disciplinary material becomes important. Thereafter students can, in later courses, observe modeling techniques in action at the same time they are absorbing disciplinary material.

    The prerequisites to the understanding of the material are high school physics (mechanics and heat) and calculus (simple differentiation and integration). Freshmen who have high school science preparation can benefit. The better the student preparation, the more telling will be the impact of a course embodying the material of this book.

    The material in this volume is suitable to all disciplines of engineering and others that depend on or study the application of mathematical modeling. The material, the course, and the students require the viewpoint and experience of mature staff. When good students are challenged with meaty material that is relevant to their goals, they respond with interest and enthusiasm. Their questions are searching, challenging, and occasionally profound. Answers should be in kind, promptly delivered, and to the point, with examples at their level of understanding.

Freshman (or sophomore, for that matter) students can appreciate and contribute to some carefully chosen larger issues and decisions in the engineering design process. What they cannot do is make meaningful microdecisions. They cannot design a shaft, select a motor or a pump, design an automobile door or an airbag, specify a production process, choose a material, or design a wave guide or a bubble tower. Generally they cannot contribute to the small decisions. If we get too involved with the small tools and technology, the student who likes them and stays in engineering may be staying for the wrong reason. The student who is not intrigued by them or assimilates too much, too fast, in the wrong discipline, and leaves engineering may be leaving for the wrong reason. A profession is chosen because solving its larger problems is the basis for a lifelong motivation and interest. The tools that are required, whatever they happen to be or are later discovered to be needed, are mastered simply to make the larger effort possible.

After many sections the vital points are summarized or identified. These constitute the essential ideas. The author has provided a little more detail than necessary, on the average, and some sections that are surplus for course objectives. When an instructor omits a section, it is still worthwhile to have the students read the summary of the omitted sections for background. Instructors are encouraged to extend some of the ideas presented here even further, augmenting with emphasis tailored exactly to their objectives.

Mathematical model building is based on equalities (and/or inequalities), the authority for which is to be found in empirical evidence. Chapter 1 constitutes an introduction to the nature of engineering and to the engineering designer's problem and the need for predictive ability. Since mathematical models provide this predictive ability, the motivation to examine them is established. Chapter 2 plunges into the use of mathematical models on an intuitive basis. The experience raises questions whose answers must be at hand before mathematical models are formulated. This experience establishes the pattern for the remainder of the book.

Chapter 3 describes figures of merit and gives some clues as to the parameters of which they are constructed. Some figures of merit are developed through examples and the mathematical model building involved is simple and intuitive. The chapter does not detail how to create a figure of merit because considerable background information remains to be developed. The distinction between tactical and formal figure of merit is drawn. The design factor and some of its *raison d'être* is offered. Cost as a figure of merit is introduced as well as some ideas concerning expectation. Reliability as a figure of merit is examined and problems associated with constraints are identified by encounter. A noncontinuous variable example is offered. Time as an element in a figure of merit is examined. The question of trade-off functions is raised as well as how qualitative parameters that should be in a figure of merit can be handled.

Chapter 4 considers how a digital computer fits into the engineering design regimen. An organized approach, IOWA CADET, is explained and its rationale examined. The necessity for documentation, error messaging, and testing is considered. The chapter is substantially *not* programming and devotes most of its attention to considerations at a broader perspective. Students may have had a programming course or be taking one concurrently. For those students, problems are provided at the end of this chapter and in the remainder of the book. For those who have not yet had a programming experience, examples can be read to illustrate the usefulness of the computer and provide motivation for their future programming experience.

Chapter 5 begins the process of building the rationale and empirical basis for mathematical model building. It considers the continuous and discrete conceptions of matter and indicates how models based on the former concept can be useful even when evidence supports the latter viewpoint. Effective empiricism is a carefully structured process and attention must be given to the proper definition of entities and measurable attributes of entities. The reasons for insisting on operational definitions are given. Given proper definition and the opportunity thus afforded for meaningful measurement, the effectiveness of graphical displays is explored and the authority for depicting loci on graphs is examined. As effective as the graph is as a communicative tool, it is awkward in the calculation process; therefore the motivation for fitting equations to experimental data is considered and some simple and reasonable schemes for curve fitting are examined.

The enormous expenditure of time and effort associated with experimental work (so that conclusions may be reached with high degrees of confidence) is noted and the capital contribution of E. Buckingham to the reduction in the number of experimental variables is examined in some detail. A method is presented for the routine establishment of a complete set of dimensionless variables.

The agreement to an international system of units is noted; subsequent to this section the text uses SI units heavily, in sharp contrast to the prior use of the familiar English system. The chapter continues with further simplification and reduction of effort made possible by some properties of partial derivatives. The next topic is the realization that physical models are possible, based on the ideas just examined, and an example is given. The final section indicates the importance of our ability to communicate variability and to be aware of some of its implications.

The structure of empiricism and its achievements have borne fruit. Chapter 6 is concerned with the recognition of reproducibility in cause, effect, and extent. The precise communication afforded by ideas and terminology associated with system and control region concepts is indicated. It becomes possible to speak of heat and work effects; surface effects; charge effects; magnetic effects; and other effects such as chemical, ballistic, and even impossible effects. The mathematical models (*system* statements) of some of these effects are displayed.

Chapter 7 is the prelude to modeling circumstances under which control region definition is the natural approach. Accountability ideas are examined in situations in which countables are discrete and continuous. The enunciation of first principle for conservation of charge, mass, energy, and momentum for systems is made; then corresponding statements for control regions are made. The nonuniqueness of the control region statements means that such statements must be tailor-made by the engineer for each control region of interest. Next the importance of delineating a system or control region for mathematical model building purposes is discussed. The origin of deterministic deductive mathematical models is indicated by example. A six-step procedure is followed, forming a routine yet rigorous plan of attack. The problem of checking results is addressed. Assertive deductive and simulative models and some introductory probabilistic modeling ideas are introduced.

Chapter 8 examines the optimization problem through a frustrating example; then the three kinds of optima are identified. The realization that simple calculus methods (locating places of zero slope) address but one kind of optimum is made. Edelbaum's sufficient conditions for stationary point maxima and minima are introduced. The method of Lagrange incorporates a second kind of optimum, increasing the power of calculus methodology. Sensitivity analysis is examined. The formal statement of an optimization problem is examined.

Direct search methods, exhaustive, interval-halving, and golden section (all one-independent variable methods) are rationally developed. Gradient sensitive multidimensional searches are introduced. Applications follow, including computer implementation.

Chapter 9 is mathematical modeling in action, and the instructor is encouraged to spend as much time as possible, right here where the action is. Many instructors will have their own problems particularly suited to the disciplinary complexion of their classes. The first eight chapters are the background against which the engineering art of mathematical model building can be practiced. Class treatment of the first eight chapters may be subordinated to the final model building experiences. Most of an engineer's subsequent course work will be rich in detail. Strategic understanding tends to be presumed by upper division instructors.

A bonus can be realized at no additional cost. Encourage the students on completion of this course to retain this text. Since many of the topics to be later developed have been touched on, subsequent instructors can assign review reading and not have to repeat the background. A brief oral review and the pursuit of the major point(s) becomes the thrust of the class session. The remainder of the background helps students see where meat is being placed on the bones. Experienced engineering instructors may wish to review these ideas again as a senior elective after most of the engineering courses have been completed. In this context the text becomes the springboard for more specific discussion rich in technical engineering detail.

The technical typing of the manuscript master was competently accomplished by Jene Spurgin.

MATHEMATICAL MODEL BUILDING
AN INTRODUCTION TO ENGINEERING

What is more important,
ideas or things?
S. M. VLAM

Ideas about things.
A. N. WHITEHEAD

Intellectual progress is made
by finding fault with the last
best thought you had.

> A lot of today's frustration is caused by a surplus of simple answers coupled with an acute shortage of simple problems.

CHAPTER 1

WHAT IS AN ENGINEER?

1.1 PREFACE TO THE STUDENT

Times change. People change. Some things change, others do not. Training for the professions takes longer simply because the body of knowledge on which they are based grows larger. Engineering students choose their careers in large measure because of a desire to create. Scientists explain what *is*; engineers create what *never was*. The enthusiastic freshman seems ready (in attitude) to innovate but the engineering college says, "Not now, for first you must have a foundation in physics; chemistry; mathematics; expository English; and in the language of vision, graphics." Not sure that this is really so, the freshmen apply themselves to their studies, which appear at best tangential to their objectives. What has Rolle's theorem to do with heat exchangers? What have energy surfaces to do with power plants? What have crystal structures to do with turbomachinery rotor design? What have themes on the vanishing American home to do with design of pop-top cans?

As a sophomore the student again asks, "Now?" and the college responds, "Not now, for we must now particularize your new knowledge in the areas of mechanics of solids (statics, dynamics, strength of materials), mechanics of fluids, and structure of materials as well as continue teaching you calculus and differential equations." About half the starting group reaches the junior year and the inquiry, "Now?" is a little more feeble, a little more tired. The answer however is consistent. "You must now study thermodynamics, electrical circuit theory, fluid flow and heat transfer, ac and dc machinery, electronics, manufacturing processes, kinematics, theories of failure, and theory of machines. You must not neglect a sequence in social studies and a sequence in the humanities."

When seniors encounter their first design course, is it any surprise that in response to the project assignment they inquire of the instructor, "What do you want me to do?" The structure of the engineering educational experience has subdued or even suppressed the creative desires that led to the selection of engineering in the first place. Perhaps, with the two-thirds of the students that are no longer with us at the senior level has gone some of the creative instinct and desire. They have simply gone to another place where an outlet is provided earlier.

This is not to say that engineering is not creative nor that opportunities for engineers to be creative do not abound. But it is necessary that students be intellectually convinced that the delays they encounter in exercising their creative bent will in the long run enhance their ability to design. Students can be introduced to the designer's dilemma early in their academic career; they can be convinced that tools are necessary for innovation in the engineering framework. This early conviction can keep the inner lights burning so that the response to their initial design problem is more like "Which of these alternative solutions will you permit me time to develop?"

When you enter your engineering classroom building, do you ever consider that you *bet your life* the engineer who designed the building structure knew what he was doing--fifty years ago? How could he design the building framework this well when the knowledge of materials and theories of failure were much more meager than now? Part of this success was due to a philosophical viewpoint and a methodology that incorporated it. Do you believe that this is worth investigating? Now?

The material in this course and those to follow are part of your essential preparation to do *what has not been done before*. The essential purposes of engineering are not necessarily to design bridges, highways, railroads, lathes, automobiles, electric motors, and turboalternators. This we know how to do. Engineers in selected industries are already custodians of these arts. Much of their design is close to routine, slipping away from the mainstream of engineering and entering the field of technology. You will be called on to conceive of processes, systems, and components that will make food out of what is now nonfood, make resource material out of what presently is not, make water out of that which is not usable water, make a resource out of what is now garbage, make clean air out of what now is not, make a mass transportation system out of what is now chaos. In short, you will be called on to create what never was, under pressure of time and shortage of money and brain power and with insufficient information. You will convince the politicians, the public, and the vested interests that the alternatives leading to human survival (with any quality of living) are few in number. You will do it because your "pinky" is in the pencil sharpener too.

The resources available for engineering education are always limited. The delay of design considerations until all background work is completed is an economical approach to the engineering education problem. Thus engineering methods come late in the curriculum. Despite your enthusiasm you are not ready to discuss creative design, much less participate in it. To convince you of this we are going to embark on an overview of engineering, not from a helicopter but through the window of mathematical modeling. You will see that you have much to learn, but in your journey you will learn many things that will ease that journey, help you learn to learn.

The place to begin in understanding the nature of engineering and its contribution to the human experience is to ponder an orthographic projection of the planet earth. These views do not show fine surface detail; nevertheless, a number of inferences can be drawn that form some of the bases for engineering.

First, the planet is finite in extent and therefore encompasses a finite amount of matter. Second, humans can live in the lower part of the air ocean and even move through it, provided they take all they need with them and expend a great deal of energy to maintain flight. Humans can live in the water ocean for just a matter of moments. They can move on the interface between the air and water oceans, if they take all they need with them and expend energy to sustain movement. They can live in the land ocean if they take all they need with them. At the interface between the air ocean and the land ocean they can live and sustain themselves on some parts of the land surface. The available area of habitation is finite.

Third, humans need food, water, clothing, and shelter. The air ocean serves them for respiration only in its lower 20,000 ft. The air ocean is finite and vital. Water is essential to humans and to the food chains in which they participate. The water resources of the planet are finite.

What is an Engineer?

Humans as predatory nomads were so successful that their numbers grew, and the limiting processes controlling population were modified by the discovery of agricultural foodstuffs. This anchored them to the land and they were required to defend it; and to make the transportation problems tractable and sustain their numbers, they utilized the land that was arable near rivers and the sea so that distribution of food and goods was possible on a large scale. This commitment to the land for sustenance and to the waterways for transportation lead to the emergence of nations and defense of their domains against others.

Engineering is concerned with assessing the present human condition, and offering humanity some alternatives to the natural course of events. The alternatives have to be conceived (invented, created) in the human mind, be compatible with the universe of matter and humanity, be viable in the existing human condition, be more good than bad, and be embraced as preferable to other alternatives by human groups affected.

Therefore, those who would practice the engineering art must be knowledgable as to the present and imminent human condition; be able to envision what needs to be done; and, understanding matter and people, be able to conceive viable alternatives to the natural course of events.

1.2 RATIONAL DECISION MAKING

Having conceived an alternative, it has to be sufficiently delineated (specified) so that it can be tested for *suitability* (will it accomplish its purpose), *feasibility* (can it be carried out with present resources of knowledge and the three m's: materials, men, and money) and *acceptability* (are the probable results worth the estimated cost). Then those in charge of the enterprise must be convinced the decision to implement the alternative can be made. The follow-up problems of production, distribution, maintenance, and retirement are also part of the engineering picture, as well as cognate problems of organization of the work force, materials, and facilities (human and other resources).

People are not rational. They are emotional beings and creatures of habit. They have learned that in certain areas of endeavor a rational approach pays dividends. They understand that most decisions are subjective, no matter how complete the window dressing of objectivity. The tests for suitability, feasibility, and acceptability are rational tests and when applied objectively are very useful in avoiding undesirable consequences. Yet their effectiveness in application is reduced by emotion, subjectivity, misinformation, and ignorance.

The tin can in many ways was a blessing to society for it enabled food products to be preserved and distributed over a long time period with small cost. The annoyance of the requirement for a can opener nagged at consumer and designer alike until ideas and technical materials problems were blended to produce the pop-top can. Now let us apply the three tests. Is the pop-top can suitable? It is a container that can do everything expected of an ordinary tin can. It also allows opening without any tool other than a finger. It is a suitable solution to the problem. Is the pop-top can feasible? The design can be executed in a mass production situation and be economically competitive. Is the pop-top can acceptable? Are the probable results worth the estimated cost? The answer is no, yet the product was marketed successfully. The pop-top can has led to a rain of cans along our highways, in our parks, and on isolated private property. The state of Michigan has tried collecting these and found the cost of collection to be about 30 cents per can (just for collection, not for disposal). The cost of recovery of the refuse exceeded the cost of the can and its contents when new and full. Why was it marketed? The corporation making the suitability-feasibility-acceptability tests did not have to consider the cost of collection. They may have been unaware of the current American behavior concerning litter. They may not have been able to foresee all the consequences of their action of offering the alternative to the marketplace. They may have chosen to ignore conclusions they may have reached. Because they visited ugliness and cost burdens on society after using the suitability-feasibility-acceptability test does not indicate that the test is inherently inadequate. It does indicate that its use does not ensure good decisions, and the decisions are no better than the information and values that go into them. The tests are as valid as the user makes them. Engineers must be familiar and skilled with them to obtain the maximum benefit from them.

We have the carcasses of 40 million automobiles in automobile graveyards. We are adding 8.5 million more each year. The birthrate of automobiles is consistently increasing and the death rate is also increasing (the average car is junked six years after birth). The economics of salvage have changed so that it costs money to recirculate the carcass in the materials cycle. Who should bear the cost? Who should design, operate, and maintain the "car eaters" that will be necessary to recycle the metal and prevent our inundation by this ugly solid refuse? When suitability-feasibility-acceptability questions are asked by car-producing corporations of themselves, the matter of the vast expense visited on the public by road construction and maintenance, the matter of 50,000 deaths (and a million hospitalizations) per year, and the matter of half the police effort in the United States being concerned with automotive problems are lost. When the love affairs of individuals for their cars are exploited to the extent that public transportation systems are in trouble and must be subsidized to exist, the matter of the greater good is lost. Is the suitability-feasibility-acceptability test defective? No, just skills of the persons that use it.

The secretary of health, education and welfare stated in 1969 that the average human body has 12 parts of DDT for every million parts of fatty tissue. By the government's own regulations, cattle, sheep, and hogs are unmarketable when their DDT content exceeds 7 parts per million. DDT was a solution to a problem and was adapted to the solution of many other problems. Did the users, promoters, and producers foresee the spread of DDT to all living things on the earth, including fish under the ice cap? Did the acceptability test fail because it was defective or because the persons using it were ignorant, or blind, or possibly dishonest? (Of course we may have struck a blow against cannibalism, since the average American is no longer fit to eat.)

The scalpel is a menace in the hands of a child; is not much more than a knife in the hands of an adult; and often is a source of life in the hands of a surgeon. The tools of any profession or trade are nothing of themselves but are the means to accomplishment in the hands of the skilled. Do not expect more of the tools of the engineer or architect than those of the stonemason. Do expect to study them, practice with them, and eventually apply them to professional work.

The reader should understand the following ideas as a result of reading and studying this section.

--Suitability test: Will this action, if adopted, indeed accomplish the intended purpose?
--Feasibility test: Can the contemplated action be carried out with resources

of knowledge, men, money, and materials or can the necessary resources be assembled in time?
--Acceptability test: Are the probable results of the contemplated action worth the anticipated costs?
--The steps in applying the suitability-feasibility-acceptability test are:
1. State the precise purpose(s) of the contemplated action.
2. Test for suitability.
3. Test for feasibility.
4. Test for acceptability.
5. If the action is not suitable, feasible, and acceptable, discard. If the action barely failed one test, flag for future reference if the situation should change, and set aside. If the action is suitable, feasible, and acceptable, retain for assessment of relative merit among other alternatives.

The suitability-feasibility-acceptability test is used to eliminate possible alternatives from further consideration. When used with skill, it is a fine filter to eliminate dubious solutions. When used without skill, consequences undesirable and unexpected plague the designer.

You will learn other skills here and in other courses. Give them your diligent attention and effort. Try to master them. The difference between a brand-new truck and a worn-out one is only a few pounds of weight. The difference between a skilled engineer and a technician is not detectable on any scale, for engineers' tools are hardly discernible unless you observe engineers in practice or view their works.

1.3 ENGINEERING

The first civilization not based on slavery is the one that resulted from the industrial revolution. Machines were substituted for the muscles of workers and animals. The food for these machines is *energy*. Humankind's appetite for energy seems insatiable. An index to this appetite is the extent of the generation of electrical power, which has doubled every recent decade (i.e., increased exponentially). Energy available where needed, when needed, under control, and in the right form is bedrock of industrial civilization. The energy chain from resource form (fossil fuel, solar radiation, hydraulic potential, tidal flows, etc.) through conversion to transportable form (electric power) to site consumption (electric, hydraulic, pneumatic, and mechanical motors and devices) and the systems that use them are among the concerns of engineers.

In the process of "creating what never was," engineers have to some extent compartmented their activities. Those who design vehicles that operate in fluid or vacuum environment are sometimes called aerospace engineers. Those who design agricultural equipment and systems (machinery and structures) are often identified as agricultural engineers. Those who apply engineering instrumentation and analysis to biological phenomena and are moving toward the design of systems and devices to meet biological and/or medical needs are sometimes categorized as biomedical engineers. Those who are involved in the design of products that are formed from natural and synthetic minerals rendered durable by a process of heat treatment at high temperatures (nonmetallic, inorganic substances, such as glass, porcelain enamels, abrasives, cements, refractories, etc.) are called ceramic engineers. Those who design processes and equipment associated with changes in the chemical state of matter are designated as chemical engineers. Those who are involved with the design, construction, maintenance, and operation of public and private facilities, such as transportation, structures, water supply, waste disposal, irrigation and drainage, and river and harbor facilities are called civil engineers. Those involved in the design, planning, inauguration, and management of systems involving interaction of money, labor, and machines are called industrial engineers. Those who design systems and component machines for processing of energy, processing of material, vehicles of transport, and automating of production techniques are called mechanical engineers. Those involved with the design and specification of metallic materials used by engineers to meet some need are called metallurgical engineers.

The preceding brief categorization represents a coarse description. The words are more likely to be used by educators, professional societies, and licensing boards than by the engineers themselves. In looking over engineers' shoulders you might be in error in any attempt to guess their formal preparation for their careers or the company titles for their positions. An engineer designing an electric motor is more likely to be a mechanical engineer by education than an electrical engineer, with the job title "design engineer, small ac motor division."

Nevertheless, these distinctions are helpful at this point. In them and their amplification is the word *design*. This is one of the things all engineers have in common.

It is useful to the engineering student to draw a distinction between *things* with which engineers are inevitably associated and *disciplines* in which they study and to which they contribute. These distinctions can be made in any field of engineering. The things tend to be different but the disciplines are in large measure shared among engineers. Our example will involve mechanical engineering. Table 1.1 recognizes objects that are associated with mechanical engineering solutions. In their preparation to become mechanical engineers and in their maintenance and improvement of knowledge they do not study these objects specifically. Their attention at school and thereafter is categorized into disciplines that involve the objects as examples. Table 1.1 also indicates the disciplines of interest to mechanical engineers, who share interest in these with many other kinds of engineers.

The reader should understand the following ideas from this section.

--Engineering offers society suitable, feasible, and acceptable, i.e., satisfactory, alternatives to the natural course of events.
--Engineers create what never was.
--As presently constructed, our civilization is in a large measure dependent on abundant food and energy.
--Engineers are categorized for educational and associative purposes by discipline, but employment categories are usually functional.

1.4 DESIGN AND THE DESIGN PROCESS

What is design? It is the central purpose of engineering. It begins with the recognition of a need and the definition of the problem and continues through the conception of an idea to meet this need. It proceeds with a program of analysis and directed research and development and leads to the construction and evaluation of pilots or prototypes. It concludes with effective multiplication and distribution of a product or system so that the original need may be met wherever it exists.

The usual product of an engineering effort is a *service*, embodied in a *template* (plans and specifications) for building an object or for replicating objects that meet a specified need. Mass-produced objects, such as millions

What is an Engineer?

Table 1.1. Objects and Disciplines of Mechanical Engineering

Objects

Mechanical Engineering:

- Machine elements
  - Structures, fasteners
  - Shafts, bearings, seals
  - Gears, brakes, clutches
  - Springs, linkages, cams
- Machines
  - Turbomachinery
    - Fans, blowers, pumps, compressors, turbines
  - Reciprocating machinery
    - Engines, pumps, compressors, hydraulic motors
  - Other combinations of machine elements
- Machine systems
  - Stationary power stations
  - Vehicular power plants
  - Vehicles
  - Production facilities
- Worker-machine systems
  - Biomedical apparatus
  - Machine-operator complexes
  - Environmental controllers
  - Human-and-computer controllers

Disciplines

Mechanical Engineering:

- Design of machines
  - Synthesis, analysis theory
  - Kinematics, dynamics
  - Strength, reliability
  - Energetics
  - Liability, safety
  - Simulation, analog
  - Computer-aided design
- Testing of machines
  - Instrumentation, measurement
  - Inference statistics
  - Test codes, standards
- Manufacture of machines
  - Materials science
  - Manufacturing methods
  - Quality control
- Control of machines
  - Automatic feedback control
  - Vibration
  - Information theory
- Maintenance of machines
  - Reliability
  - Servicing, scheduling

of barrels of detergent or gasoline; thousands of locomotives, planes, tractors, radios; millions of yards of cloth or carpet; or billions of kilowatt-hours of electricity are examples of products replicated from the engineering template.

What skills and knowledge do engineers need to function as designers?

1. Engineers must be able to recognize needs. This awareness has to be tuned to sharpen their perceptions.

2. They must be able to define problems. Problems are crudely defined by telephone calls, letters, conversations with seniors and colleagues. When a prototype tractor broke in two during field-test work, the chief engineer's statement of the problem to the design team leader was a model of understatement. "You have a problem over there--fix it." The project engineer fortunately knew where "over there" was. It took considerable time and effort to define the problem. Then and only then could the engineer devote resources of skill and labor toward its resolution. The refining of the problem statement from "fix it" to "redesign part 6491N so as to avoid fatigue failure in a rib fillet" can be the matter of hours or months, depending on the complexity of the device and the subtlety of the failure cause.

3. They must be able to invent (conceive, innovate, dream up, envision) schemes that are solutions to the problems of need. Creativeness (inventiveness in the engineering sense) is not a common commodity and commands respect when it is recognized by others.

4. They must be able to predict, that is, anticipate, nature's (and/or humanity's) reaction to their proposals prior to any implementation. It would be marvelous if we could predict completely the behavior of material complexes or human organizations on paper. Paper is cheap and engineering time is somewhat expensive, but whatever the cost, interrogations of nature are far more expensive. Engineers spend a good deal of their educational time bettering their ability to foresee the consequences of alternatives, and if they must put questions to the universe, they learn how to put the question and how to interpret the response.

5. They must be able to design experiments and to draw valid inferences from the data.

6. They must be able to test and evaluate prototypes or models of prototypes.

7. They must be able to completely delineate the solution with plans and specifications (the template) so that the complex will function as intended, regardless of the builders or users.

8. They must understand production resources and distribution systems.

9. Above all they must be intellectually honest in understanding the subjective elements of their art--the impact of human organizations, human knowledge, human skills, and human tastes on the engineering function. They must also appreciate the impact of their own characters on their performance. Such elements as energy, perseverance, pursuit of excellence, imagination, inventiveness, decision-making ability, courage, analytical skill, communication skill, luck, ethics, morals, and sense of future steer them more than they will usually admit, and a good designer makes an honest assessment of these matters.

This recitation of skills and attributes may lead you to envision extraordinary individuals as the only people who can become engineers. But different skills and abilities in individuals and different demands of projects are often wedded to common advantage. Likewise, many engineering functions are segmented, often finely, tapping the best skills of different individuals.

Consider item 9 in more detail. The designer is cloistered within four walls with four windows, as depicted in Fig. 1.1. Through the north window one sees the dollar sign and the constraints that go with it. The engineer is imprudent who does not see the human psychology behind it; indeed, human psychology, politics, and the codifications labeled law should be perceived. The engineer who sees only the dollar sign and not human organization will find it difficult to function well. The view from the east window consists of produc-

What is an Engineer? 11

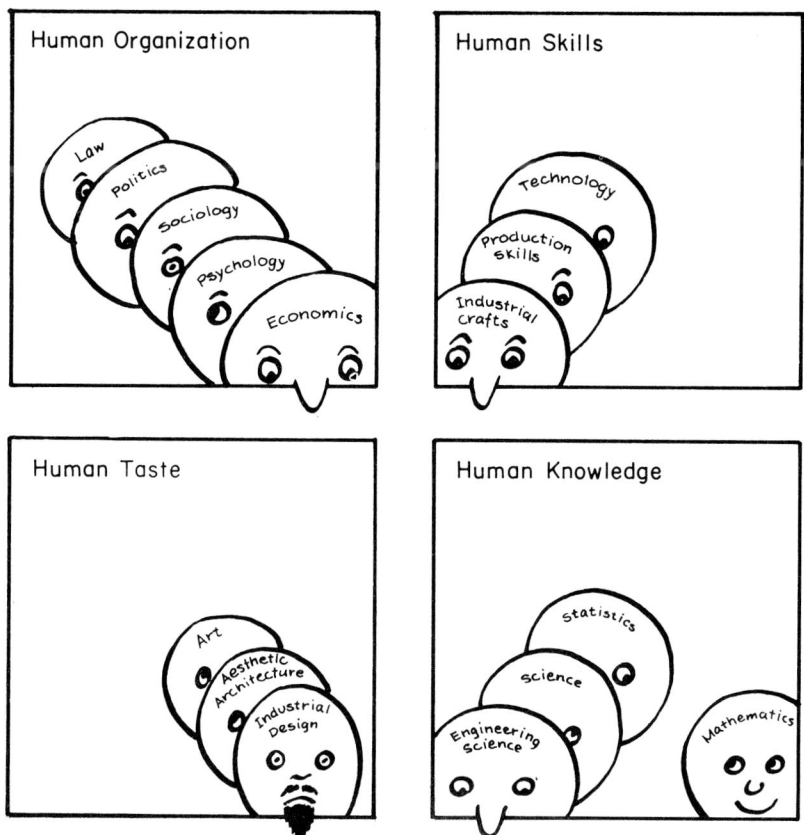

Fig. 1.1. The views from the four windows to the engineering designer's cubical: north, east, south, west.

tion skills, and engineering technology and human skills should be perceived in the background. The view from the south window is the often grim face of the industrial designer who insists on doing things that seem extrafunctional. The engineer who does not view the industrial designer as the speaker for aesthetic architecture, for what humankind regards as good taste, is misunderstanding a function and ignoring opportunity. The west window reveals engineering science, with human knowledge hovering in the background. All these influences on the engineering designer are external. Human organization speaks to the engineer most often in the guise of economics, but economics is not the essence of human organization. Similarly with human skills, human taste, and human knowledge.

There also are internal influences on designers (see Fig. 1.2). They are personal, yet real, often limiting. Their energy and perseverance affect their performance, as do imagination and inventiveness; courage; and skills in decisions, analysis, and communication. Their luck, ethics, and morals have real impact. Certainly their sense of history and the future affect their performance. Successful designers recognize the internal and external influences on them.

In this section the following points have been made:

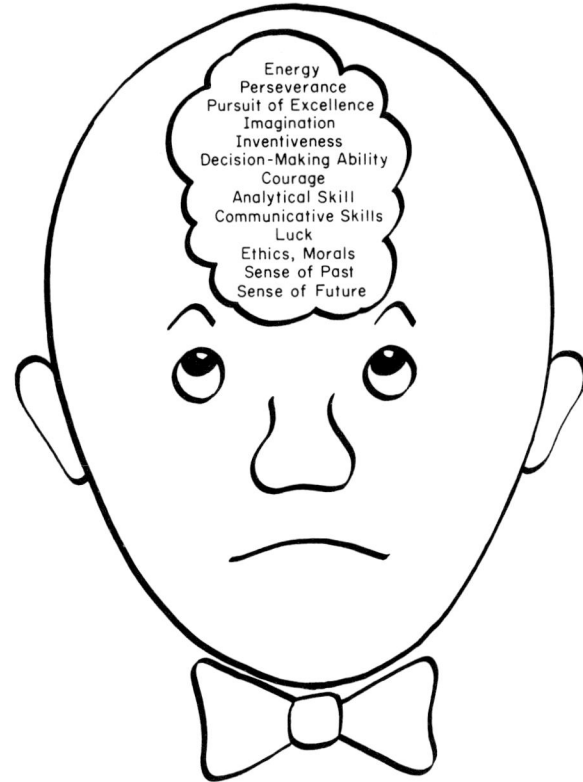

Fig. 1.2. Internal influences on the engineering designer.

--Design is the central purpose of engineering and this role is shared in common by all kinds of engineers.
--Engineers design or support the designer and the design and manufacturing process.
--The usual product of an engineering effort is a template to be followed in order to meet a need.
--An engineer must be able to
  recognize need
  define problems
  conceive alternatives
  predict consequences
  design experiments and draw inferences
  test and evaluate
  delineate solutions
  understand production and distribution
  be intellectually honest

1.5 BASIC OPERATIONS IN THE DESIGN PROCESS

A morphology or structure of design has been attempted by several engineers, based on observation of successful engineering designers. These attempts at structuring the design process are helpful to the beginner but also point up the hopelessness of attempting a fine view of the structure.

The basic operations performed in the design process are as follows:

Identification of the need. An analysis of whether a need exists at all is required. A stated need is not always in harmony with the requirements of the people to be served. The engineer must restate these needs in terms of the true requirements.

Information collection and organization. All factors that relate to the system need to be considered. When necessary, experiments must be devised to obtain data otherwise unavailable.

Identification and statement of system parameters and variables. All factors influencing the system (the so-called boundary conditions) must be identified. Engineering systems can be resolved into simpler elements that, when described or prescribed in appropriate detail and properly synthesized, will constitute the design of the system. *Inputs* are those resources and other environmental factors that are converted (modified) by the system in question. *Outputs* are goods and by-products of the system, both desired and undesired. *Transforming means* are used to obtain the relationship between inputs and outputs. *Constraints* are all elements and factors that express limitation and/or need to be accounted for in the design.

Criteria development for optimal design. These are the rules for judging relative merit. First, develop a value system; then form the criteria relationship among the values.

Synthesis. The process of evolving systems to convert the inputs into desired outputs is synthesis. At this step only the suitability requirement is met. The steps are conception, idealization, prediction, and evaluation by suitability-feasibility-acceptability criteria.

Optimizing and sufficing. Optimizing is maximizing merit according to the criteria relationship; sufficing is the satisfaction of the suitability-feasibility-acceptability test.

Test and evaluation. Prototypes and/or models are tested and evaluated.

Iteration. The preceding three operations are found throughout the design process. Many iterations will be taken around several or all of them. In particular the engineer continually reexamines previous findings and decisions in the light of new information.

In the typical design course the identification of the need is usually done by the instructor. This practice may be one reason why engineers in practice may be too prone to accept the identification of the need by someone else.

The position of the humanities in engineering is unique. It is the introduction of a value system into problems that largely distinguishes the work of the engineer from the work of the scientist. It is generally acknowledged that an engineering problem has unique answers only in terms of a given value system. If the value scales are changed, so must the evolved answers change. We are well aware that different safety factors (presumably measures of the value of human life and limb) are used in the aircraft industry and for railway cars. Compare the safety records.

Until about 50 years ago we were perhaps the most inventive people in the world. From this position we have declined steadily until today we may be one of the least inventive nations, if we measure fecundity by the number of inventions per million dollars of industrial activity. Research efforts are up perhaps tenfold over previous decades, but patent applications have risen only by 17 percent. Why?

Much specialized *knowledge* is requisite to engineering design, along with an understanding of and a familiarity with the physical laws that explain our environment and the mathematical tools that make analysis possible.

A foundation of *skills* is essential to the solution of analytical problems. Facility with graphical expression, essential to the conceptual stage and to the communication of the results, is important to the engineering designer.

1.6 ATTITUDES AND ABILITIES

Many *attitudes* and *abilities* are essential for the successful engineer. A questioning approach, great and insatiable curiosity, flexibility of mind, intellectual integrity, acceptance of responsibility, willingness and ability to make decisions (sometimes very rapidly), reservation of the right to revise decisions when required by new facts, ability to express ideas coherently and effectively, the courage and patience to defend them against superficial condemnation, recognition that judgment based on prior experience is very valuable, and the desire to develop judgmental skill are essential attitudes.

Our species is uniquely creative and it has but one creative instrument: the mind and spirit of the individual. Not much has been created by two people. Only rare examples of good collaborations exist in music, art, poetry, mathematics, and philosophy. Once the miracle of creation takes place the group can build and extend it, but a group rarely invents anything. That preciousness lies in the one mind alone.

Section 1.4 includes an enumeration of the skills and knowledge needed to function well as an engineering designer. Item 3 indicates that designers must be able to conceive, innovate, dream up, envision--in short, invent-- schemes that may be the basis for solutions to problems. Inventiveness (in the engineering sense) is the ability to produce new and useful ideas for accomplishing engineering goals. What are the hallmarks of the results of engineering inventiveness? A quality of newness or uniqueness is present. The results are useful or appreciated or both, purposeful or beautiful or both. There is simplicity where formerly there was complexity. *New* relationships are often a recognizable feature.

Who is inventive in the engineering sense? A prerequisite to inventiveness is a fund of scientific and technical knowledge and understanding. There is no correlation between IQ and inventiveness above the college entrance threshold through gifted. Beyond gifted the correlation appears to be negative. Inventive people value things theoretical and aesthetic more than they value things concrete and pragmatic. Inventive persons are more perceptive than judging--more curious about what it is and how it works than what ought to be; as a result they hear more, see more, remember more, and in short, have more to which to relate. Inventive persons in their judgment are more feeling than thinking.

Engineers as a group are not different from other groups. They prefer specificity, the real and well defined, and are uncomfortable with vagueness. However, they are hired as professional innovators and become involved in new problems that are *always* poorly defined. How do they suppress their very normal human preferences? They know that their enemy in this inventive endeavor is habit or rhythm, called *set* by psychologists, a predisposition to a habitual manner or method of approach to a problem. Since set is your most formidable adversary, think about where it lurks.

--When anxious, we revert to familiar patterns.
--When rushed, we revert to familiar patterns.
--When on familiar ground, we revert to previously successful patterns.

--Some things have to be believed to be seen. People naturally learn more facts that support their predispositions than facts that challenge them. When following our "guiding star(s)," we revert to familiar patterns.
--When conscientiously avoiding set, we revert to familiar patterns for avoiding set. (Of all snobbery, antisnob snobbery may be the worst.)

Know thine enemy.

Your second enemy is lack of perception in the real world and in the world of ideas. A most rewarding experience for an instructor is to place a detailed geodetic survey map before a class of students. Bright ideas abound, whether the discussion involves power dams, irrigation, highways, mass transit routes, microwave tower location, or pipeline right-of-way acquisition. Why? In large measure the students see a view they have never seen before even if they have lived in the area all their lives: contours, watersheds, sidewalk jungles, etc. The view has always been there but the students became aware of it precipitously. Practice awareness--see, catalog, digest, challenge, review. Nothing can stop an idea whose time has come. It happens (is expressed) because of time, place, and *the presence of a prepared mind*. Prepare your mind. Time and place will unfold before you. If an idea is needed immediately, take a prepared mind to the place and remove set.

In the suitability-feasibility-acceptability test, we test ideas in three distinct stages. When we consider suitability sans feasibility and acceptability, we call the event *brainstorming*. A good set of rules: (1) Suppress any thought or comment concerning feasibility or acceptability. Spit out ideas, scores of them. Applaud every new idea because it is there. (2) The name of the game is numbers. A brainstorming session yielding ten new ideas is less valuable than one yielding twenty ideas. (3) Far out is in! The wilder the idea the better. The wilder the idea the less the likelihood that it has ever been expressed in connection with the current problem. After brainstorming, march in with feasibility and acceptability considerations and clean out the deadwood.

Inversion consists of turning an old approach inside out. People get into automobiles. Why cannot a person put on a car, much as one puts on a pair of trousers? Analogy consists of taking a solution from one environment and inserting it into another. Moving sidewalks may have come from thing-conveyors imported into a people-mover environment. Buck Rogers science-fiction flying belts are now a partial reality. There are analogs in nature, in literature, and in other technical disciplines. Empathy is putting yourself in another's place, perhaps even putting yourself in *its* place. The anticipation of the forces on a catapulted jet pilot's body may be related to the design engineer's automotive experience as a passenger as much as to technical knowledge of dynamics.

If only--gravity could be "turned down" at night, controllable by a thermostat, or were smaller on the oceans. If only--liquids would not wet solid surfaces. Thinking wishfully may trigger ideas.

Exhaust the possibilities that you already have. Use systematic procedures to assure this. Do not be above making a grid. If you think combustion is a conceptual answer, list all the fuels along one side of a grid and all the oxidants along the other, then check the nodes of those you understand already and pursue all the blank boxes. A blank may mean not only have you not thought about it before, but perhaps no one else has either in the context of your problem. You have the problem. The time is here. Is your mind as fully prepared as it can be?

All forms of communication--written, oral, and graphic--are vital to the engineering process. Communication may be the critical skill in group engi-

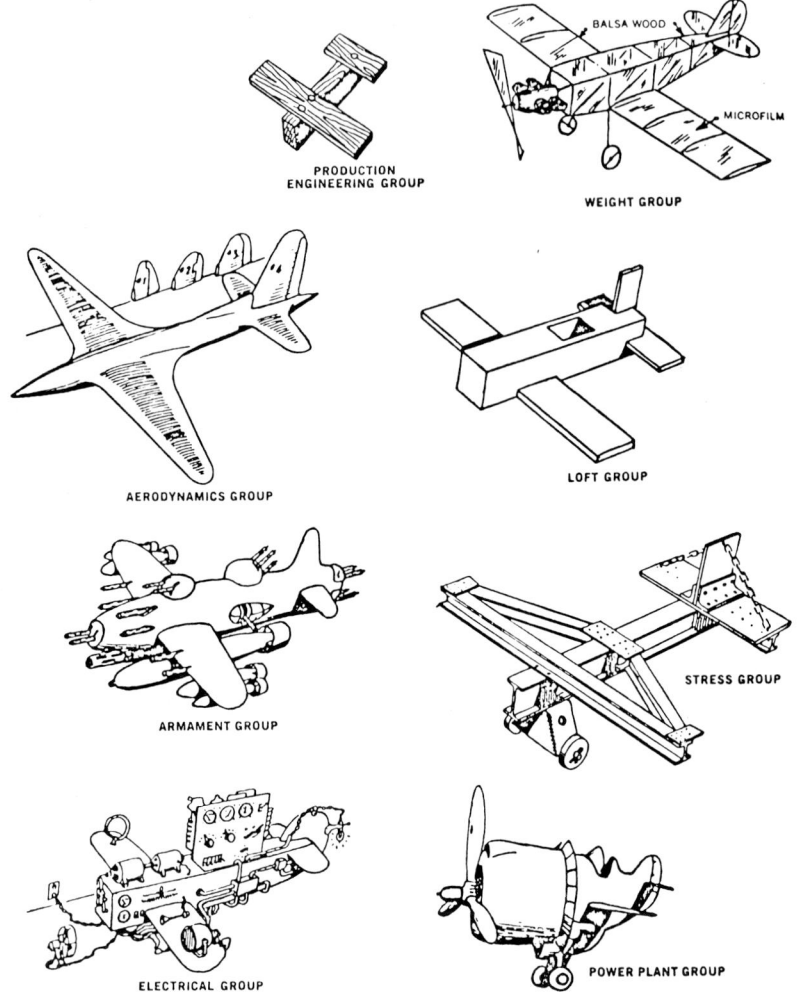

Fig. 1.3. The same military aircraft (it now would be called a "weapons system") as seen in the mind's eye by each of the several engineering groups involved in its preliminary design. The project engineer's task is to change all these conceptual visualizations into one that meets the design specifications. (L. R. Webster, Systems Engineering: The Role of Reliability, *Mech. Eng.* Vol. 91, No. 1, Jan. 1969. Used with permission.)

neering design efforts. The engineer in charge of an aircraft design such as depicted in Fig. 1.3 has just explained all the specifications a proposed military aircraft must possess to the group leaders who work on it. While all have agreed they understand the specifications and have the aircraft designer's concept clearly in mind, the individual mental images that might appear in the cartoonist's balloons above each head would resemble the sketches of Fig. 1.3. The divergence of comprehension of the concept is brought about by differing perceptions of what is important. The engineer in charge must detect the divergence of each individual conception from the whole and bring all of them into congruence without destroying individual contributions and their value to the project.

PROBLEMS

1.1. What is the average annual electrical energy production in the world (Btu/yr)? At what rate is energy from the sun incident on the earth? In view of the difference between these rates and other considerations, comment on the future of society and civilization.

1.2. Who or what permits people to practice engineering? Who or what lets corporations practice engineering?

1.3. Obtain the legal definition of engineering from your state board of engineering examiners.

1.4. What is the suitability-feasibility-acceptability test? What purposes does it serve? What are its strengths? What are its pitfalls?

1.5. The announced purpose of the negative income tax is to eliminate poverty. This negative tax was one of many alternative proposals to reduce or eliminate poverty. Apply the suitability-feasibility-acceptability test and reach a conclusion as to whether the negative income tax represents a satisfactory alternative.

1.6. An explosion on the University of Wisconsin campus involved millions of dollars and loss of life and intellectual property. A suggestion has been made that future insurance on real property on the campus come from an assessment against each student of $100 at the time of enrollment. If there are no claims against the fund, the $100 (less a small service fee) will be refunded to the student at the next enrollment. The student then brings the assessment to $100. Another way to view the operation is to consider the $100 fee as permanent, with a small per semester service charge levied at each enrollment. On graduation (or withdrawal from the university) the $100 is returned if there are no claims against the fund. This suggestion constitutes one of many suggested courses of action. To be retained as a satisfactory alternative, it must pass the suitability-feasibility-acceptability test. Demonstrate your understanding of the test by applying it to this suggestion.

1.7. At Iowa State University the Towers dormitories are a considerable distance from the academic portion of the campus. Complaints have been heard from students who cannot go back for lunch and return again in time for class. A suggested remedy consists of running a shuttle bus between the dorms and the library with a swing around the campus. A nominal fare of 10 cents or 15 cents has been mentioned. Examine this proposal for suitability, feasibility, and acceptability and reach a decision as to whether the suggestion is satisfactory.

1.8. In speaking about noise problems someone said, "It might be a good thing if people's ears would bleed, then people might get aroused." How serious is the noise pollution problem?

1.9. How much water is necessary to produce a pound of meat? A pound of corn?

1.10. Only yesterday, average intelligent people understood their mechanical environment. This is no longer true. What effects on our civilization do you foresee?

1.11. What is the source of oxygen renewal for our atmosphere?

1.12. The following concentrations of lead were found in the Greenland ice sheet:

Year	μg Pb/T snow
800 B.C.	8
A.D. 1750	30
1815	40
1860	48
1920	70
1940	90
1950	130
1960	160

Can you explain this? In the Antarctic the concentration is currently about 8 μg/T. Any ideas why?

1.13. Inventiveness in the engineering sense is defined as the demonstrable ability to obtain new and useful ideas for accomplishing engineering goals. Discuss the influence of heredity, environment, and prior training on inventiveness.

1.14. How can inventive potential be realized?

1.15. Are scientific and technical knowledge and understanding a prerequisite to inventiveness in the engineering sense?

1.16. Engineers in general do not like to discuss the subject of inventiveness. Do you have any thoughts as to why?

1.17. An inventive act has a quality of newness or uniqueness. The embodiment is either useful or appreciated (or purposeful or beautiful) or both. The inventive act brings simplicity where formerly there was complexity (this is sometimes called elegance). New relationships are often a recognizable feature. With these thoughts in mind, comment on whether the following developments were inventive solutions to problems: rear engine jet aircraft, ball typewriter, drafting machine, floating vacuum cleaner, circular slide rule, the torque bolt, ballpoint pen.

1.18. Add ten entries to this list of problems in need of inventive solution: window washing, house building, excessive noise, automobile safety, postal service, waste disposal, mass transportation.

1.19. America cannot feed the world. The hungry nations can be divided into three classifications: (1) those that cannot be saved regardless of the aid given them, (2) those who can survive without aid, and (3) those who can be saved by quick aid. Where would you classify India, Pakistan, Egypt, Nigeria, Bolivia, others? Cite your reasons.

1.20. Do you think scientists outnumber engineers in voicing positions on social, economic, and political issues? Why?

1.21. Are scientists responsible for what they discover? Are engineers responsible for what they create?

1.22. Do you think engineers are conservative in the matter of dress, politics, and social behavior? Can you shed some light as to why this may be so?

1.23. What is the U.S. energy consumption of oil, gas, coal, and nuclear sources? What fraction is consumed in transporting goods and people, in heating and cooling homes and industrial buildings, as raw material, or for processing in manufacture?

1.24. What are the energy costs in Btu/T·mi for moving freight over inland waterways, pipeline, railroad, truck, and airplane? What are the costs in cents per ton mile?

1.25. What are the energy costs of a person moving (measured in Btu per passenger mile) in intercity travel via bus, automobile, train, and air-

plane? What are the costs per passenger mile?
1.26. What are the energy costs of a person moving (measured in Btu per passenger mile) in intracity travel via bus, automobile, train, bicycle, and walking? What are the costs per passenger mile?
1.27. Freight often moves by rail or barge between the same two points. By what fraction, on the average, do barge miles differ from rail miles?

> Some view the fan dancer as the problem; others the fan.

CHAPTER 2
POSING THE DESIGN PROBLEM

2.1 INTRODUCTION

It is axiomatic that a problem must be posed before it can be solved. Occasionally we see solutions running around in search of a problem. The information plazas on Iowa interstate highways, electric toothbrushes, inertia nutcrackers, self-cooking meals, and thermostatically controlled hair curlers are examples. Often the need is for a service and immediately a machine is designed to provide the service. Service is what the customer wants, but does it require a machine? The automobile was conceived as an answer to the farm-to-town and intertown transportation problem. Since that solution (say 1930) the automobile has changed, not in concept but in style and refinements and in the minds of people. In the city it is a solution looking for a problem. It has been adopted for intracity transportation, but it is a forced fit. It takes about 12 lb of vehicle and 1/2 hp prime mover to move a person at traffic speeds of 25 mph. A 2 T automobile is an inept solution, adapted to the task by individuals, aided and abetted by municipalities that have come to know better but are nevertheless intransigent.

The first step in solving a problem is in ascertaining the problem. The pursuit of the solution and the skills involved are different for sundry problems. Figure 2.1 depicts a distinction that is not well understood in contemporary America. The black box depicts a system or component that accepts an input, provides an output, and in the transformation is obedient to the laws of nature. If you are given the system, input, and laws of nature and are required to find the output, the solution is arrived at by a process called *analysis*, in which the dominant intellectual skill is *deduction*. Similarly if the input is required, given everything else, the process is called *inverse analysis* and the dominant intellectual skill remains *deduction*. Given all but the laws of nature, with the laws themselves being sought, the problem-solving regimen is called *science* and the dominant intellectual skill required is *induction*. While analysis plays a role in science, by itself it is not science, and the essential scientific act is the induction.

When we are given the laws of nature, the input and output required, then we are seeking the system (or component or process). The problem-solving regimen is called *engineering* and the dominant intellectual skill is called *synthesis*. While analysis and prior science play a role in engineering, science and analysis alone are not engineering; the essential engineering act is the synthesis. Synthesis is different from induction; neither is better or worse, harder or easier, more or less rewarding, or more important. Science explains what *is* and the creative act is an induction. Engineering creates what *never was* and the creative act is a synthesis. To engineer is to synthesize; to engineer is to design; to synthesize is to design.

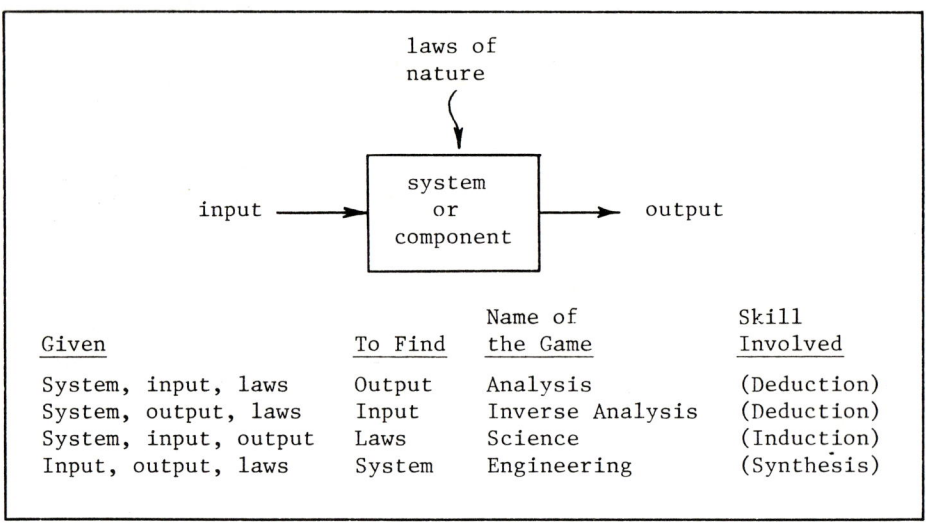

Fig. 2.1. The name(s) of the game(s).

2.2 DESIGN IMPERATIVE

All design problem statements should be posed in the imperative mood and the following model is better than most.

> Design (subject to certain problem-solving constraints) a component, system, or process that will perform a specified task (subject to certain solution constraints) optimally.

The parenthetical expressions refer to qualifications placed on the solution. Immediately, two vital amplifications are needed.

    1. State exactly how you will recognize a satisfactory alternative.
    2. State how you will distinguish between two satisfactory alternatives to identify the better.

Then the task becomes to

    3. Invent alternative solutions.
    4. Through analysis and test, simulate and predict the performance of each alternative, retain satisfactory alternatives, and discard unsatisfactory alternatives.
    5. Choose the best satisfactory alternative as an approximation to optimality.
    6. Implement your decision by delineating the chosen alternative.

2.3 TIE-BOLT EXAMPLE

An example, first handled intuitively and then more formally, will serve as initial illumination of the ideas just advanced. The phone rings. It is your project leader. "One of the tie-bolts just failed in the field tests. I think we can suppress the failure by making the bolts absorb more energy before individual failure. Get on it right away. Can you have new specs down

Can we do this by making holes in the bolt.

Posing the Design Problem 23

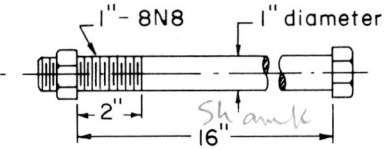

Fig. 2.2. A 1 in. diameter bolt, 8 threads/in. (From *Machine Design* by J. E. Shigley, Fig. 3-10a. Copyright 1956, McGraw-Hill. Used with permission of the publisher.)

to the experimental shop first thing tomorrow morning?"

You would expect the problem to be posed more specifically by the project leader but final form of the problem definition is the individual engineer's responsibility, inasmuch as each individual does it somewhat differently in some detail. Anyway, this is part of what you are retained to do, so let's get at the problem the only way we know how at the moment: we'll wade in and see what comes to mind. Let's look at the bolt, depicted in Fig. 2.2. The root diameter of the threads is 0.850 in. The allowable working stress for the material in this application was previously established as 40,000 psi. Now to define the problem a bit better: "Redesign the tie-bolt so that it will absorb more energy in tension before rupture."

We are not sure what the word "optimally" means just yet. How will I recognize a satisfactory solution? At the moment, a satisfactory solution is one that absorbs more energy before tensile failure. This solution could be brought about by improving the bolt material by processing, or by changing the composition. This would require that new bolts be made from a more expensive material that may not be at hand. Is there something else we can do? Perhaps an understanding of what contributes favorably and unfavorably to energy absorption would suggest a course of action. Consider the tensile deformation of a steel cylinder as depicted in Fig. 2.3. The work done by the force F during elastic deformation is

W.D = Force × distance

$$W = F\delta/2$$

where δ is the increase in bolt length l.

For a prismatic body, Hooke's equation relating axial normal stress and axial strain is

$$\varepsilon_x = s_x/E$$

Stress / Strain = Modulus of Elasticity so $E = \frac{S_x}{E_x}$

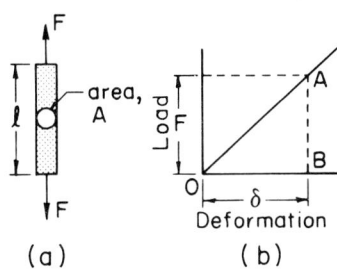

Fig. 2.3. The relation of one-dimensional deformation to applied load in an elastic prismatic body. (*Machine Design*, Shigley, Fig. 3-9. Used with permission.)

where: ε_x = strain in the x-direction, in./in. $= \ell/L$
    s_x = normal stress in the x-direction, psi $= F/A$
    E = material modulus of elasticity, psi

Uniform strain is defined in terms of the elongation of the bar and the original length of the bar and can be expressed as

$$\varepsilon_x = \delta_x/\ell$$

E_x = strain
δx = change in length
ℓ = original length
strain = ℓ/L or $\varepsilon_x = \delta x/\ell$
$\delta_x = \varepsilon_x \ell$

where: δ_x = elongation in the x-direction, in.
    ℓ = unloaded (original) cylinder length, in.

The uniform normal stress is defined as

$$s_x = F/A$$

S_x = stress $F = S \times A$

where: F = external load whose line of action contains the centroid of the cross section of the prismatic bar, lbf
    A = cross-sectional area, in.^{2}

In terms of the preceding, we can write

$$W = \frac{F\delta}{2} = \frac{(s_x A)(\varepsilon_x \ell)}{2} = \frac{(s_x A)}{2} \frac{(s_x \ell)}{E} = \frac{s_x^2 \ell A}{2E}$$

By inspection of this equation we observe that the work absorption of the cylinder in bringing the stress up to some value s_x varies directly with the volume of the bar ℓA and inversely with the modulus of elasticity E. To improve the energy absorption capability of the bar we may:

1. Change the steel so that the working stress s_x may be raised.
2. Increase the length of the bar.
3. Increase the cross section of the bar.
4. Change the material so that s_x^2/E may be increased.

Are the above alternatives open to us in this problem, and will they prove effective? Before we decide, let us model the bolt more realistically as a cylinder of two diameters, as sketched in Fig. 2.4.

The area of the root cylinder is $(0.850)^2 \pi/4 = 0.565$ in.^{2}. In the bar the tension is the same throughout, so

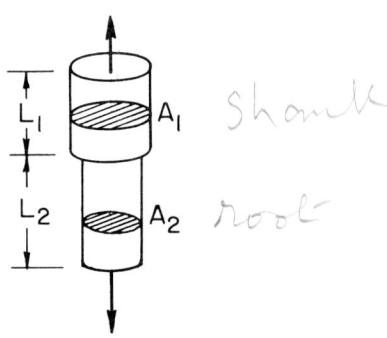

Fig. 2.4. A more realistic mathematical model of the bolt, acknowledging the two active diameters in the bolt.

Posing the Design Problem

$$F = s_1 A_1 = s_2 A_2 \quad \text{or} \quad s_1 = s_2 A_2 / A_1$$

The energy absorbed by the bolt is

$$W = \frac{s_1^2 l_1 A_1}{2E} + \frac{s_2^2 A_2 l_2}{2E} \quad \text{or} \quad W = \frac{s_2^2 A_2^2 l_1 / A_1^2}{2E} + \frac{s_2^2 A_2 l_2}{2E}$$

$$= \frac{s_2^2 A_2 l_1}{2E}\left(\frac{A_2}{A_1} + \frac{l_2}{l_1}\right)$$

Substitution for s_1 yields

$$W = \frac{s_2^2 A_2 l_1}{2E}\left(\frac{A_2}{A_1} + \frac{l_2}{l_1}\right)$$

Inspection of this equation, which is a more realistic model of the bolt, shows that energy absorption of the bolt may be increased by increasing s_1 to the working stress limit of 40,000 psi [s_1 is now $(0.565)(40,000)/0.785 = 28,000$ psi]. This can be brought about by *decreasing* A_1 (the simpler model indicated that an increase in cross-sectional area would improve energy absorption). How about that?

Evaluating W for the present bolt gives

$$W = \frac{40,000(0.565)(14)}{(2)(30,000,000)}\left(\frac{0.565}{0.785} + \frac{2}{14}\right) = 182.2 \text{ lbf·in.}$$

The limit on the reduction of A_1 is A_2, for at that point s_1 and s_2 become 40,000 psi. This math model shows our alternatives to be:

1. Change the steel so that the working stress of the bolt may be raised.
2. Decrease A_1 so that the energy absorption is increased.
3. Change the material so that s_2^2/E may be increased.

Resolve the discrepancy between the previous model suggesting an increase in A_1 and the above model suggesting a decrease in A_1. What is the lesson here?
How will I recognize a satisfactory solution? It will be the alternative that raises the energy absorption capacity of the bolt and, in view of the press of time, allows modification of an existing (now unsatisfactory) bolt.
How will I distinguish between two satisfactory solutions? I will choose the solution that results in the greater energy absorption.
What will optimality mean in this case? The optimal solution is the solution that affords the greatest energy absorption by modifying existing bolts.
Invent alternative solutions:

1. Reduce the unthreaded shank diameter by turning down the shank in a lathe.
2. Reduce the unthreaded shank cross-sectional area by drilling a hole from the bolt head end, along the centerline, 14 in. deep.

Actually, an infinite number of solutions of both classes exist. Solutions of class 2 will be rejected out of hand because drilling a hole of small diameter 14 in. deep is a very specialized operation. Of the infinite number of solutions of class 1, the best will be the one where W takes on the largest possible value and s_1 does not exceed 40,000 psi. This occurs when $A_1 = A_2$. The energy that is absorbed is

$$W = \frac{s_2^2 A_2 l_1}{2E}\left(\frac{A_2}{A_1} + \frac{l_2}{l_1}\right) = \frac{(40,000)^2(0.565)(14)}{(2)(30,000,000)}\left(\frac{0.565}{0.565} + \frac{2}{14}\right) = 241 \text{ lbf·in.}$$

The important change in the specifications is sketched in Fig. 2.5.

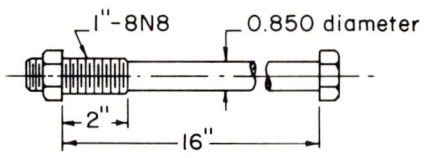

Fig. 2.5. The revised bolt specification. (*Machine Design*, Shigley, Fig. 3-10b. Used with permission.)

2.4 REFLECTIONS AND GENERALIZATIONS

We have done a number of things, mostly instinctively. Do you realize what we did? We

--constructed mathematical models
--adapted Hooke's equations to this particular circumstance
--used definitions of stress, strain, and work and equated work to energy change
--acted on the premise that steel is an elastic material
--made deductions
--drew inferences
--made judgments
--made a decision (and a few things not mentioned)

Is there any counterpart in nature to these scribblings on paper? Dare we risk our position by sending a job order to the shop to modify the existing bolts? Will the project engineer authorize the job order? Does it make sense that the part (bolt) will absorb more energy if it is trimmed down? Can I convince the project engineer that this is sound? Have I used any general methods without realizing it? Am I aware of the implicit assumptions (presumptions) in my intuitive approach? How are they limiting?

In response to the bulk of the questions, we must admit that the authority for doing some of the things we did is presently somewhat hazy. One objective of this book is to supply the understanding that will give you more confidence in what you are doing, expose the specific underlying assumptions, and allow you to judge or test the relevance to reality. Indeed, much of your future engineering course work continues in this vein.

Let us now view the same problem as a special case of a more general viewpoint. In any design problem in which an engineer is expected to make one or more decisions, there must be at least one design variable. Usually there are several. Let us denote the set $\{x_n\}$ as the set of n design variables. Another way to display the set is by roster as $(x_1, x_2, \ldots, x_n)$.

If the set is specified (each x_i is given a value), the set defines a design alternative. Two different sets define two different design alternatives, i.e., two different designs. One alternative will, in general, be preferable to the other. How can we distinguish between them and select the better? One way is to establish a *merit function* (objective function, criterion function, etc.). This merit function must be a function of the set $\{x_n\}$ or $M = M(x_1, x_2, \ldots, x_n)$, and by convention the designer seeks to maximize it.

Posing the Design Problem

If the merit function M can be created (invented, discovered, cobbled up, found, borrowed), the designer's task is to maximize M. Can M be created? In some cases, yes. For instance, if a least cost design is sought the merit function M can be

$$M = -\text{cost}(x_1, x_2, \ldots, x_n) \quad \text{or} \quad M = 1/\text{cost}(x_1, x_2, \ldots, x_n)$$

The minus sign or reciprocal will allow the most desirable design to be associated with the largest M even though cost is to be minimized.

One may seek to minimize or maximize a time interval, energy transfer, weight, ground periphery of a cam, error in a function generator, reliability, or a margin of safety. A figure of merit might be a blend of cost, capacity, safety, and reliability. The only requirement is that the figure of merit function allow discrimination between alternatives with certainty. The marketplace must respond to the same figure of merit or we will have a solution on our hands that will not serve a need in a free marketplace. The valuable ability of the figure of merit function to point the way toward superior alternatives is exploited later.

The engineer cannot simply substitute numbers for $(x_1, x_2, \ldots, x_n)$ to maximize M, for in each case the design variables are differently related. The mathematical models that the designer uses to constrain $(x_1, x_2, \ldots, x_n)$ realistically are called *functional constraints*. They can be symbolically displayed as

$$g_1(x_1, x_2, \ldots, x_2) = 0$$

$$g_2(x_1, x_2, \ldots, x_n) = 0$$

$$g_m(x_1, x_2, \ldots, x_n) = 0 \quad (m < n)$$

representing m functional constraints. If there are m of these and there are n design variables (n > m), the designer has some latitude.

Also, certain values of x_i are prohibited. For example, if x_i is the diameter of a shaft, it cannot be negative. For practical reasons it cannot be 1000 in. in diameter, either. Thus statements such as $1/2 \leq d \leq 3$ or symbolically,

$$z_1 \leq d \leq Z_1$$

are called *regional constraints*, since they define regions of feasibility and infeasibility. These regions are sketched in Fig. 2.6. Symbolically we may write

$$z_1 \leq f_1(x_1, x_2, \ldots, x_n) \leq Z_1$$

$$z_2 \leq f_2(x_1, x_2, \ldots, x_n) \leq Z_2$$

·

$$z_\lambda \leq f_\lambda(x_1, x_2, \ldots, x_n) \leq Z_\lambda$$

Examples of parameters such as $z_1, z_2, \ldots, z_\lambda$ are stress levels, size variables, pressure angles, temperatures, etc.

The bolt problem can be viewed now in the fashion outlined above. We can argue ourselves into the position that the design variable is the cross-sec-

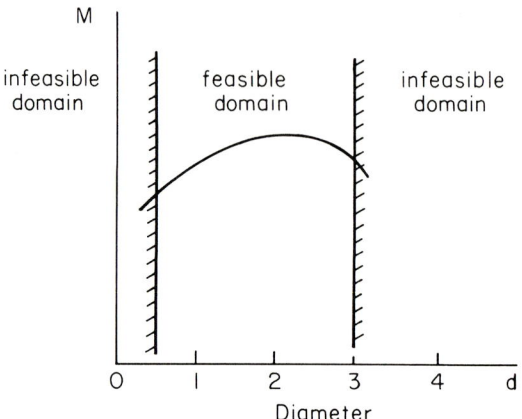

Fig. 2.6. The feasible domain of a shaft problem with an arbitrary figure of merit function displayed.

tional area A of the bolt shank, or

$$M = M(A)$$

What shall the figure of merit be? Since we seek to increase the strain energy absorption of the bolt above the present amount, we designate the strain energy of distortion to be the figure of merit. Using our previous arguments

$$M(A) = W$$

The functional constraints include Hooke's equation for one-dimensional strain, which has been incorporated into the strain energy relations for the cylindrical models. The tension is presumed to be the same throughout the length of the bolt model; thus,

$$W = W_{root} + W_{shank}$$

$$W_{root} = \frac{s_2^2 A_2 l_2}{2E}$$

$$W_{shank} = \frac{s^2 A l_1}{2E}$$

$$A = s_2 A_2 / s$$

$$s_2 = 40{,}000 \text{ psi}$$

This problem is simple enough that the functional constraints may be combined into an explicit expression for merit,

$$M(A) = \frac{s_2^2 A_2 l_1}{2E} \left(\frac{A_2}{A} + \frac{l_2}{l_1} \right)$$

The regional constraint on A is that A cannot be larger than A_1, since we cannot conveniently "glue" on material to the existing bolt. It cannot be smaller than A_2 because the stress level cannot exceed 40,000 psi in the root or

Posing the Design Problem   29

shank cylinder. The regional constraint may be expressed as $A_2 \leq A \leq A_1$. An alternative expression would be $28,800 \leq s \leq 40,000$. The designer's problem is to maximize the figure of merit, subject to the functional and regional constraints. A plot of A as abscissa and M as the ordinate, such as shown in Fig. 2.7, is helpful. The best design is where A is least within the feasible domain, i.e., where $A = A_2$.

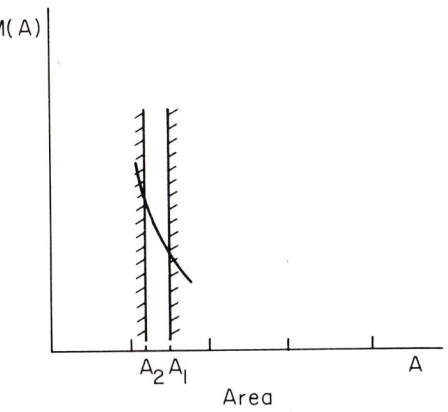

Fig. 2.7. The feasible domain of the bolt problem with a figure of merit that is the energy absorption capacity of the bolt.

The viewpoint taken here is different from that for the intuitive approach. This viewpoint has the advantage of reminding the engineer to make important engineering decisions.

  1. How will I recognize the best alternative? (Largest figure of merit.)
  2. How will I simulate nature? (Functional constraints and figure of merit model reality.)
  3. What parameters can I control? (The set of design variables.)
  4. In what domain may I search for alternative designs? (Within regional constraints.)
  5. What search domain is fruitless? (Outside regional constraints.)

An important feature of this approach is that the "trend" of the merit surface will suggest designs exhibiting superior merit to the one(s) at hand, a valuable contribution to an optimization procedure.

As another brief example, in an electrical vein, consider the problem of a farmer during an exceptionally cold winter whose oil-fired hen house heaters cannot hold the poultry house temperature at sufficiently high level at night. The nearest electrical source is the barn some 800 ft away. Available there is 240 V at 60 Hz ac. The farmer has on hand 4000 ft of heavy wire with a resistance of 0.1 Ω/100 ft. He is considering adding a resistance heater to augment the oil-fired heater and is wondering what size electrical heater would be needed to save the chickens. He gives the problem to you, an engineering freshman home on winter recess. Figure 2.8 depicts the essential geometry of the farmstead and the equivalent circuit. The impedance of the line is denoted Z_1 and the impedance of the heater is Z_2. The current I through the heater, using Ohm's resistive effect, is

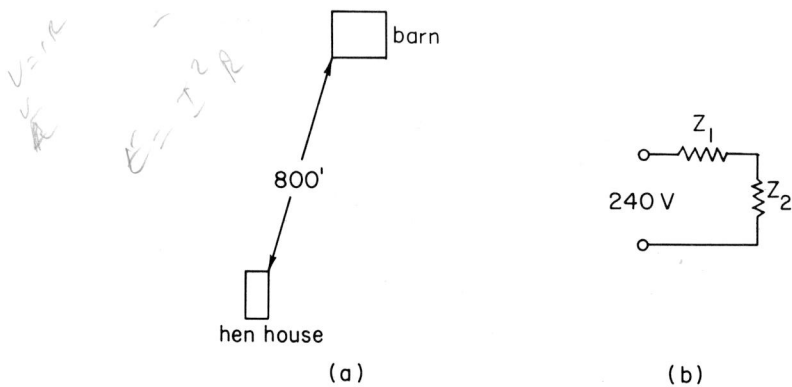

Fig. 2.8. (a) The location of the hen house with respect to the barn; (b) the equivalent heater circuit incorporating the line impedance and the heater impedance.

$$I = E/(Z_1 + Z_2)$$

The power dissipated in the hen house heater is taken as the figure of merit and M is expressed as

$$M = I^2 Z_2 = \left[\frac{E}{(Z_1 + Z_2)}\right]^2 Z_2 = E^2 Z_2 / (Z_1 + Z_2)^2$$

A plot of M is depicted in Fig. 2.9. The maximum merit is to be sought, and using the available wire for the supply line leaves the designer with one design decision, the heater impedance Z_2. The choice of Z_2 resulting in the maximum merit will give an upper bound on capability of the electrical augmentation. The maximum is found by differentiating M with respect to Z_2 and equating the derivative to zero. This is a heuristic move based on the geometric interpretation of the nature of the M function as depicted in Fig. 2.9. The result of this process is that $Z_2 = Z_1$. About 1700 ft of wire will be needed to connect the two current paths between barn and hen house (800 ft of

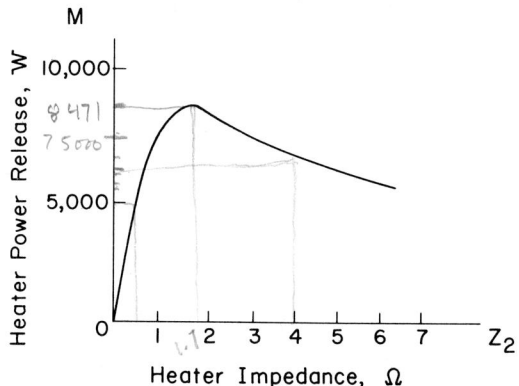

Fig. 2.9. The figure of merit, heater power release, as a function of the single decision variable, the heater impedance.

Posing the Design Problem

two conductors plus lead-ins), involving an impedance of (17.0)(0.1) = 1.7 Ω. The current with $Z_1 = Z_2 = 1.7$ Ω is

$$I = E/Z = E/(Z_1 + Z_2) = 240/(1.7 + 1.7) = 70.59 \text{ A}$$

The power dissipated in the hen house is the figure of merit,

$$M = E^2 Z_2/(Z_1 + Z_2)^2 = 240^2(1.7)/(1.7 + 1.7)^2 = 8471 \text{ W}$$

You expect that a 2 Ω resistance heater might be available. You inform your father that a 240 V, 7200 W, 2 Ω resistor will do and you will go into town tomorrow and see what you can find.

Your father, of course, cannot wait and connects both ends of his 4000 ft coil of wire across the line, strings the wire to the hen house, and hangs the remainder on the wall.

--What are the advantages of your father's solution to the problem?
--What are the disadvantages?
--What are the advantages of your solution to the problem?
--What are the disadvantages of your solution?
--What did your father's mumbling about matching impedances and "what are colleges coming to" mean?

As another example, consider that two pad eyes on a wall and a structural frame are required to support a pin on which is to be imposed a load of 100,000 lbf, 10 in. from the pad eyes, as depicted in Fig. 2.10(a). The design imperative for this problem might read:

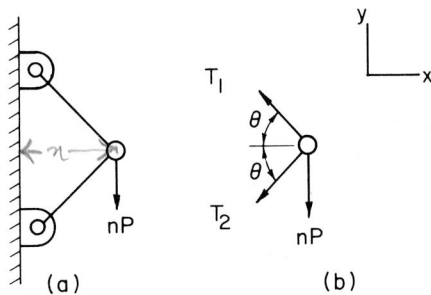

Fig. 2.10. (a) The essential geometry of the link-and-pin structure; (b) the geometry of the link tensions and the load vector.

> Design a (link-and-pin) structure that will support a load of 100,000 lbf at a distance of 10 in. from the pad eyes on a vertical wall (using identical links in an isosceles triangle configuration), the structure being of least weight.

The parenthetical expression "link-and-pin" is a problem-solving constraint and the designer cannot use a gusset, web, or truss as a solution specie. The parenthetical expression "using identical links in an isosceles triangle configuration" is a solution constraint ruling out nonidentical links and non-isosceles triangle configurations. The solution constraint does not restrict materials or cost.

The first move is to set down in writing just how to recognize a satisfactory alternative, should one be encountered. In this case the designer writes, "Any isosceles triangle link-and-pin structure that will statically support 200,000 lbf or more will be satisfactory if it is suitable, feasible, and acceptable." This statement implies a design factor (factor of safety) of two or greater. This design factor is traditionally used to account for unexpected conditions in material, its processing, or its loading (more about this in Sec. 3.2). The next move is to write exactly how to distinguish between satisfactory alternatives. "I will choose a figure of merit equal to the negative of the weight of the links." In this statement the engineer, drawing on experience, predicts that the major portion of the structure weight will be in the links. The design statement involves weight minimization. The engineer is willing to be judged professionally in using this approximation to the optimal structure. As yet there is no clear picture of the geometry of the links. Again, the engineer is confident from past experience, that the link weights will be roughly proportional to the pin-to-pin length. Again, as an approximation to the original merit function, $M = -(\text{weight})$,

$$M \doteq -(2A\rho l)$$

(where A is the link cross-sectional area, ρ is material specific weight, and l is link pin-to-pin length) is chosen as the working approximation unless or until the design proves the approximation too coarse a criterion.

Figure 2.10(b) indicates the load pin as a free body (system). Since the pin is at rest and will remain at rest, summation of forces in the x-direction and y-direction is zero:

$$\Sigma F_x = 0 = -T_1 \cos\theta - T_2 \cos\theta = 0$$

$$\Sigma F_y = 0 = T_1 \sin\theta - T_2 \sin\theta - nP = 0$$

It follows that simultaneous algebraic equations for the link tensions T_1 and T_2 are

$$T_1 \cos\theta + T_2 \cos\theta = 0$$

$$T_1 \sin\theta - T_2 \sin\theta = nP$$

Therefore $T_1 = nP/(2\sin\theta)$ and $T_2 = -nP/(2\sin\theta)$. If the material to be used exhibits the same ultimate strength in tension and compression (which is not usually true but the assumption simplifies the case for readers who have not yet studied strength of materials) and if A is the link cross-sectional area, then

$$A = T_1/S = nP/(2 S \sin\theta)$$

if column action is ignored (compressive link is a strut). The figure of merit can now be expressed as

$$M = -(\text{weight}) = -2A\rho l = -2 \cdot \frac{nP}{2 S \sin\theta} \cdot \rho \cdot \frac{x}{\cos\theta}$$

recognizing that $l = x/\cos\theta$. The figure of merit terms can be grouped as follows:

$$M = -2nPx \left(\frac{\rho}{S}\right)\left(\frac{1}{\sin 2\theta}\right)$$

The parameters, n, P, and x are not available to manipulation by the designer. The quantity ρ/S, a property of the material that can be called the material selection variable, is a design variable; let us assign the name *material se-*

Posing the Design Problem

lection design variable. The quantity $1/\sin 2\theta$, a property of the solution or structure geometry, also is a design variable; let us call it the *configuration design variable*. Mathematically, the figure of merit function is of the form

$$M(x_1, x_2) = -2nPx\left(\frac{\rho}{S}\right)\left(\frac{1}{\sin 2\theta}\right)$$

where $x_1 = \rho/S$ and $x_2 = 1/\sin 2\theta$ are the material design variable and configuration design variable, respectively. Optimality requires that the figure of merit be maximized. Since the merit function is of the form of (constant) $\times$ $(x_1) \times (x_2)$, it is intuitively clear that the extreme of M occurs simultaneously with the extreme of x_1 and the extreme of x_2. We can find the extreme of x_1 independently of the extreme of x_2.

The desired extreme of x_2 is a stationary point (a minimum exhibiting horizontal slope) and may be found by differentiation.

$$\frac{\partial M}{\partial \theta} = 0 = \frac{\partial}{\partial \theta}(\sin 2\theta)^{-1}$$

It follows that $\cos 2\theta = 0$ and that $\theta = 45°$ *for any material*.

The best material selection from available materials (see Table 2.1) is that which produces the minimum extreme in design variable x_1 (see Fig. 2.11).

For 2340 steel: $\frac{\rho}{S} = \frac{0.283}{282,000} = 1(10^{-6})$ in.$^{-1}$

For one alloy titanium: $\frac{\rho}{S} = \frac{0.163}{150,000} = 1.085(10^{-6})$ in.$^{-1}$

For 7075-T4 aluminum: $\frac{\rho}{S} = \frac{0.100}{82,000} = 1.23(10^{-6})$ in.$^{-1}$

Of these three, the nickel steel is best. Using the 2340 steel for the material and $\theta = 45°$ for geometry, the cross-sectional area of the links should be

$$A = \frac{nP}{2\,S\,\sin\theta} = \frac{2(100,000)}{2(282,000)(0.707)} = 0.501 \text{ in.}^2$$

Table 2.1. Some Characteristics of Engineering Materials

	w, Weight Density (lbf/in^3)	c, Unit Cost ($/lbf)	S_y, Tensile Yield (kpsi)	S_u, Tensile Ultimate (kpsi)	Notes
AISI 1018	0.283	0.17	32	58	Hot rolled
AISI 1035	.283	0.21	81	110	OQ&T 2hr 1200°F
AISI 2340	.283	0.38	174	282	OQ&T 2hr 400°F
AISI 4130	.283	0.37	133	146	OQ&T 2hr 1000°F
AISI 4340	.283	0.38	234	260	OQ&T 2hr 600°F
Nitralloy 135	.283		165	181	OQ&T 2hr 1100°F
304 Stainless	.290	0.75	33	75	
416 Stainless	.278	0.54	40	75	
446 Stainless	.273	0.71	45	75	
Ti-5Al-2.5Sn Titanium Alloy	.163	12.00	120	130	
2024-T4 Aluminum	.100	0.68	46	68	
7075-T6 Aluminum	.101	0.72	72	82	
AM-C585 Magnesium	.066	0.81	32	46	
AM-65S Magnesium	0.067	0.78	28	40	

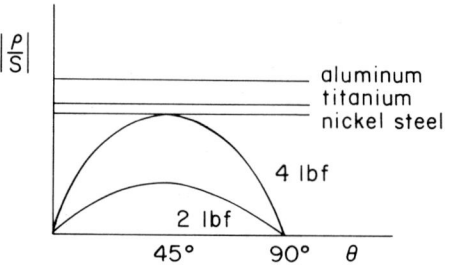

Fig. 2.11. Contours of weight on a plot of the two design variables as abscissas. The choice of θ is continuous whereas the choice of ρ/S is discrete. The smallest ρ/S that is tangent to a merit contour represents the optimal design decision.

$M = -2(2)(100,000)(10)(10^{-6})(1) = -4 \text{ lbf}$

$l = x/\cos\theta = 10/0.707 = 14.14 \text{ in.}$

$T_1 = \dfrac{nP}{2 \sin \theta} = \dfrac{2(100,000)}{2(0.707)} = 141,400 \text{ lbf at rupture}$

Since x_1 and x_2 are mathematically independent, there are no functional constraints. The regional constraints on θ are $0° \leq \theta \leq 90°$. Regional constraints on ρ/S are unmentioned but constitute the positive domain.

Suppose the figure of merit is required to be sensitive to material cost. The merit function would take the form $M = -(\text{cost}) = -2cA\rho l$, where c is the material cost per unit volume. Using the same approach as previously, the figure of merit becomes

$M = -2nPx\left(\dfrac{c\rho}{S}\right)\left(\dfrac{1}{\sin 2\theta}\right)$

A different material selection variable appears. The configuration design variable is the same, so the minimum extreme still occurs at $\theta = 45°$.

What changes in the solution occur as the solution constraint of identical links is relaxed? What changes in the solution occur as the vertical wall condition is changed to an inclined wall?

Another example is in order to indicate that a figure of merit is not necessarily a continuous function. Let us consider a transportation routing problem involving finding the best path from city 1 to city 6. A matrix can be formed to display the mileages between various cities along the way from city 1 to city 6. Infeasible or undesirable paths between cities are simply indicated as blank entries under "Distance" in Table 2.2. The table also indicates the driving times between cities, accounting for traffic congestion

Table 2.2. Distances and Driving Times between Cities

	Distance, mi					Driving Time, hr				
	2	3	4	5	6	2	3	4	5	6
1	128	155	178			3:00	3:15	3:20		
2		175	132	141			3:40	2:50	3:00	
3				142					3:40	
4					294					7:30
5					152					3:00

Posing the Design Problem

Table 2.3. Figure of Merit Matrix for Truck Routing Problem

	\multicolumn{5}{c}{Cost, $}
	2	3	4	5	6
1	158.00	187.50	211.33		
2		211.70	160.33	171.00	
3				178.70	
4					369.00
5					182.00

and grades. If the primary consideration as to route is the cost of moving freight without intermediate stops, a cost matrix can be formed. The total cost of wages for drivers ($10/hr on the highway) and tractors and trailers ($1/mile) are shown in Table 2.3. The cost entry from city 1 to city 2 can be computed from (128)($1) + 3($10) = $158. This cost matrix (the negative of each entry) can be considered to be a figure of merit and the objective would be to find the route that maximizes the merit. The detail of the procedure for maximizing merit is not the object here; we simply wish to indicate that a figure of merit function may be discrete rather than continuous.

PROBLEMS

2.1. Experience with highway tunnel traffic indicates that the *average* spacing between vehicles increases at a faster rate than does speed. Data from a New York tunnel indicate that between 15 and 35 mph the space x between vehicles in miles can be expressed as $x = 0.324/(42.1 - v)$, where v is the vehicle speed in mph. Ignoring the length of the individual vehicles, establish the speed limit that will allow the tunnel to handle the largest number of cars per hour.

2.2. A tire manufacturing company is able to make x grade A tires and y grade B tires per day simultaneously. If it makes only grade A tires its capacity is 400 tires per day. If it makes only grade B tires its capacity is 800 per day. The net profit on grade A tires is twice the net profit on grade B tires. When its production is mixed, its capacity is given by

$$y = (40 - 10x)/(5 - x) \quad (0 \le x \le 4)$$

where x and y are expressed in hundreds of tires. What production schedule will maximize the net profit?

2.3. It costs a manufacturer a + bx dollars to produce x units per week. The price that can be commanded in the marketplace to move x units per week is c - ex. What level of production maximizes profits? What is the corresponding price? What is the level of profit at optimal production rate? If a tax of t dollars is imposed on this item and the manufacturer still wishes to maximize profits, at what price should the item be sold?

2.4. Table 2.1 gives typical characteristics of some engineering materials, including cost. Suppose we are specifying the geometry and material of a tensile member (tie-rod). The cost ($) of the material is $ = cwLA, where c is the unit cost of material in dollars/lb, w is the weight density in lbf/in.^{3}, L is the length of the bar, and A is the cross-sectional area. The tensile stress in the rod is P/A. Substituting P/S_u for A in the cost equation, we obtain $ = $cwLP/S_u$. With different materials, c_w/S_u varies. What is the most economical material that can be selected from Table 2.1 (cost of forming and heat treating not considered)?

2.5. In Prob. 2.4, what material would make the lightest tie-rod?

2.6. If the endurance strength of the materials is significant in lieu of the ultimate strength, how would the answers to Probs. 2.4 and 2.5 change?

2.7. Engineers sometimes face situations wherein the required reliability of a system must exceed the reliability of the individual parts! When equipment functions permit, reliability can be improved by providing multiple redundancy, i.e., providing two or more items for the same purpose. The electric utility industry, which must provide a highly reliable system, accomplishes this in part by providing standby equipment such as two or three feedwater pumps with only one on-the-line at any given time. Your chief engineer is confronted with a situation that has two functions, both of which must be successful for the task to be successfully accomplished. The equipment cost is $1000 for task A and $1500 for task B. The reliabilities exhibited are 0.5 and 0.8, respectively. (These reliabilities have been set low for computational ease.) To meet the requirement that the reliability of the complex must be high, the chief engineer plans to use redundancies of equipment for task A and task B. The chief engineer details you to determine what it will cost to maintain reliabilities between 0.6 and 0.9 on the complex. Submit to your instructor what you would submit to your chief engineer.

2.8. A vertical load P must be supported at a distance x from a vertical wall. Two pinned links are arranged as depicted in Fig. P2.8 to support the load. The members need not be identical. If no column action is encountered in the compressive member, what is the configuration for least material? (Specify α, β.)

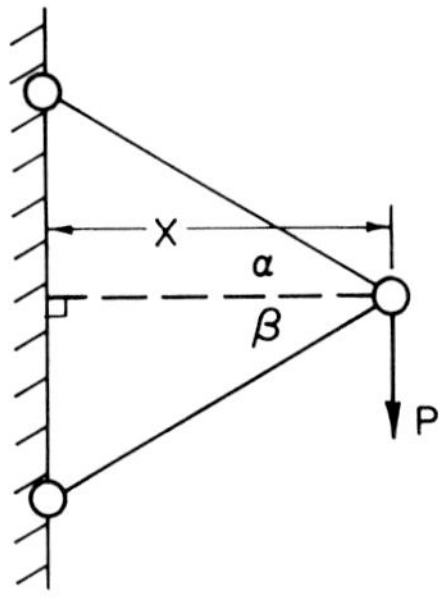

Fig. P2.8

2.9. The wall of Prob. 2.8 is inclined at an angle ξ from the vertical, as shown in Fig. P2.9. If the configuration is symmetrical with the load P still at a distance x perpendicular from the wall, what is the configuration for the least material (if the two members are identical and $\xi = 30°$)? How does the structure weight compare to the vertical wall problem?

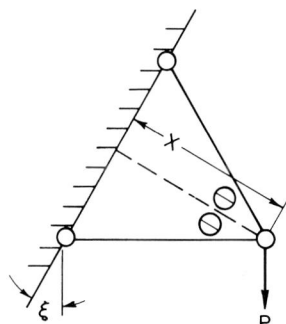

Fig. P2.9

2.10. A vertical load P must be supported at a distance x from a vertical wall. It is planned to use two pinned links to support the load. Column action is to be ignored. The tensile link may fail by fracture at its ultimate load and the compressive strut may fail at its yielding load. What is the configuration for least weight structure? The two links need not be identical in either geometry or material.

> I often say that when you can measure what you are speaking about and express it in numbers, you know something about it; but when you cannot express it in numbers your knowledge is of a meager and unsatisfactory kind.
>
> **Lord Kelvin**

CHAPTER 3

FIGURES OF MERIT

3.1 INTRODUCTION

In this chapter the purpose is to describe more fully what a figure of merit is and of what it may be a function. Later chapters address the detail of how figure of merit functions are constructed (mathematical model building) and how information such as design decisions are educed from them.

The words *figure of merit, utility function, objective function,* and *criterion function* all have similar import and represent different descriptive phrases for similar ideas. *Objective function* is used in formal optimization jargon (by convention) for the function to be minimized. *Utility function* has its roots in economics as a measure of usefulness. All phrases refer to a function that expresses worthiness of a design set. Such functions could be maximized or minimized as appropriate.

The design imperative admonishes the designer to seek an optimal solution. A useful tool manifests itself in a figure of merit because maximizing the result of the test for merit is a most efficient way to maximize solution merit. To be useful, a figure of merit needs the following attributes: (1) In comparing figures of merit for two design alternatives, the set of design variables yielding the higher figure of merit is (with certainty) the superior of the two alternatives. Solutions need not be "twice as good" to exhibit double the figure of merit; it is merely that a better solution results in a larger figure of merit. The figure of merit, then, is a number whose magnitude is an index to the merit or desirability of a solution to a problem. When a multiplicity of factors vie for dominance, the fabrication of a valid and meaningful figure of merit may range from easy through difficult to impossible (in terms of one's current knowledge). (2) The figure of merit must be valid, i.e., the marketplace in which the design competes must be responsive to the same value system.

The fact that a figure of merit is elusive is in itself instructive of the vagueness that can be present in the design situation. The need for competitive solutions to engineering problems is nevertheless imposed on the designer, who must perform creditably (as good or better than the competition) in the face of many uncertainties, substituting skillful judgments for unavailable rigor.

If a figure of merit is a function of a single independent variable (the designer has but one decision to be made), the figure of merit may be expressed mathematically as $M = M(x_1)$ and interpreted graphically as a plane curve such as in Fig. 3.1(a). For instance, suppose that

$$M = M(x_1) = x_1 - x_1^2$$

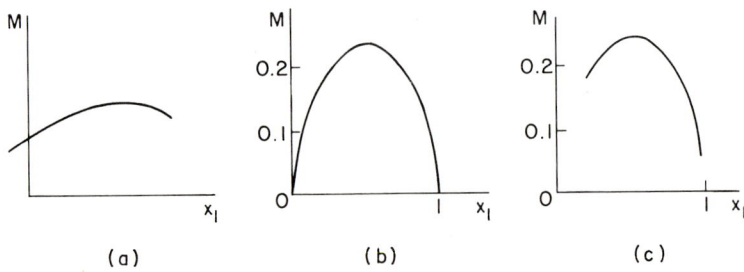

Fig. 3.1. Figures of merit as functions of a single design variable.

The geometric interpretation appears in Fig. 3.1(b). There may be regional constraints such as

$$x_1 \geq 0.25 \qquad x_1 \leq 0.95$$

and the solution constrained to lie in the region $0.25 \leq x_1 \leq 0.95$ as suggested by Fig. 3.1(c). A figure of merit may be a function of two independent variables (the designer has two design decisions to be made) and may be expressed as

$$M = M(x_1, x_2)$$

If the design variables x_1 and x_2 are not mathematically independent, the engineer will discover that they are related. The expression of this relationship is called a functional constraint and it might be expressed as

$$g(x_1, x_2) = 0$$

Such a constraint is also known as an *equality constraint*. For instance, if

$$M = M(x_1, x_2) = 2x_1^2 + 3x_1 x_2$$

and the functional constraint is

$$g(x_1, x_2) = x_1^2 + 1 - x_2 = 0$$

then back-substitution of the functional constraint into the merit function gives

$$M = 2x_1^2 + 3x_1 x_2 = 2x_1^2 + 3x_1(x_1^2 + 1)$$

which allows the merit function to be expressed as a function of a single variable x_1, and it is recognized that there is actually only one design decision to be made. Back-substitution of functional constraints into the figure of merit function reduces the dimensionality of the optimization problem (which follows later), and all the simplifications that then occur accrue to the designer. Sometimes this cannot be done. For instance, when explicit solutions of one design variable in terms of another cannot be accomplished, the penalty of increased dimensionality of the problem is accepted. For example, suppose

$$M = M(x_1, x_2) = x_1 + 6x_1 x_2$$

Subject to $x_1 \ln x_1 = x_2 \tan x_2$

An explicit expression for x_1 as a function of x_2 or vice versa is not obvi-

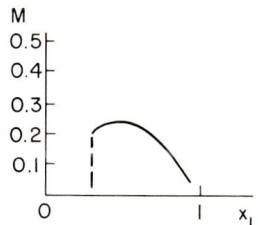

Fig. 3.2. A figure of merit function exhibiting zero merit in the infeasible domain.

ous; the figure of merit must be viewed as a surface above the $x_1 x_2$ plane that contains a locus (curve) of admissible solutions, and the designer must understand this.

A figure of merit is often "tuned" to give zero merit (Fig. 3.2), negative merit, or drastically reduced merit in domains wherein the regional constraints are violated. When a design decision $\{x_n\}$ is inserted in the merit function, examination of the figure of merit shows whether the decision is within the feasible domain or in the infeasible domain. For example, if

$$M = x_1 - x_1^2 \quad (0.25 \leq x_1 \leq 0.95)$$
$$= 0 \quad \text{(elsewhere)}$$

then M may be evaluated with any x_1 whatsoever and it will be evident by inspection that x_1 is feasible or infeasible. Another way to accomplish this is by declaring

$$M = x_1 - x_1^2 \quad (0.25 \leq x_1 \leq 0.95)$$
$$= -\left| x_1 - \frac{0.25 + 0.95}{2} \right| \quad \text{(elsewhere)}$$

Infeasible solutions display negative merit. An expression such as the one displayed above is called a *formal figure of merit function*. See Fig. 3.3. No errors in merit are incurred in blind substitution for x_1. An expression such as

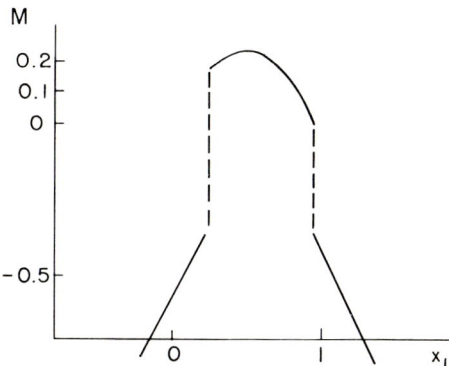

Fig. 3.3. A figure of merit function exhibiting negative merit in the infeasible domain, decreasing as the distance from the feasible domain increases.

$$M = x_1 - x_1^2$$

is called a *tactical figure of merit function*, since the burden is on the user to avoid substitution of an infeasible x_1. For example, $M(0.1) = 0.1 - 0.1^2 = 0.09$ and clearly the decision to make $x_1 = 0.1$ is without merit, since x_1 is infeasible; yet this merit exceeds $M(0.94) = 0.94 - 0.94^2 = 0.0564$, which is feasible.

This distinction between formal and tactical figure of merit may seem arbitrary, even artificial, at this point, but when you use the digital computer to optimize (maximize a figure of merit function), you will frequently avail yourself of a coded search algorithm that has a free hand to evaluate $M(x_1)$ anywhere the stratagem demands, and the result of the interrogation of the merit function must not be confusing to the search program. Tactical figure of merit programs are easier to write but the user is responsible for keeping the search program evaluations within feasible space. Formal figure of merit functions are used whenever the user cannot assure search containment within feasible space.

Consider a fuse link that must be provided for an electrical circuit. Two constraints are imposed on the solution: (1) Up to 10 A must be carried indefinitely. (2) The circuit must be broken by currents of 11 A or more. A fuse link is either acceptable or unacceptable under these conditions. In formulating a figure of merit, a u-function that has the properties

$$u(x) = 0 \quad (x < 0)$$
$$= 1 \quad (x \geq 0)$$

is useful. If I is the lowest steady current ending in a broken circuit, a yes-no figure of merit can be expressed as

$$M = [u(I - 10) - u(I - 11)]$$

Six fuse links are available with properties displayed in Table 3.1. The merit of fuse links B, C, D, and E is unity and A and F display a merit of ze-

Table 3.1. Properties of Available Fuse Links

Fuse Link	Lowest Steady-State Current, A	Cost, dollars
A	9.7	0.0194
B	10.0	.0200
C	10.3	.0206
D	10.6	.0212
E	10.9	.0218
F	11.2	0.0224

ro. The fuse links B, C, D, and E are admitted to the status of feasible alternatives. However, the alternatives cannot be ordered by a figure of merit sensitive only to feasibility-infeasibility. If a third constraint--the chosen link must have the smallest possible cost--is added, a merit function that will allow the ordering of the alternatives can be constructed. Since many computer optimization programs are written to maximize *or* minimize consistently, we will adopt the convention of always maximizing a figure of merit function. Two possible merit functions come to mind:

Figures of Merit

Table 3.2. Comparative Figures of Merit

Fuse Link	M	M_1	M_2
A	0	0	0
B	1	-0.0200	50
C	1	-0.0206	48.5
D	1	-0.0212	47.3
E	1	-0.0218	45.9
F	0	0	0

$$M_1 = -[u(I - 10) - u(I - 11)]\text{cost}(I)$$

$$M_2 = [u(I - 10) - u(I - 11)]/[\text{cost}(I)]$$

Table 3.2 displays the figures of merit of the six links. Clearly link B meets all requirements. The advantage of the M_1 formulation of the figure of merit is that the difference in the figures of merit for links B and C is the increase in cost attributable to selecting alternative C over B, i.e., the incremental cost of violating the merit ordering. The advantages of the M_2 formulation are that it is positive and geometric thinking is more convenient.

If the approach to this problem solution is to consider every possible solution, evaluate each solution figure of merit and order them (as in Table 3.2); then a formal figure of merit is recommended. If the approach is to screen out infeasible alternatives before evaluating the figure of merit function, a tactical figure of merit function may be utilized. The decision as to which approach to use in a particular circumstance is influenced by the difficulty in evaluating the formal figure of merit function in every case compared to the difficulty in screening out infeasible decision sets prior to evaluating a tactical figure of merit.

The concept of an abrupt "chop" to zero merit whenever a regional constraint is violated is helpful in understanding the geometric interpretation of the formal figure of merit function. For reasons associated with search strategies, it is useful to shape the penalized merit function in the infeasible domain to avoid frustrating or defeating the search strategy used in a computer solution. These considerations must be deferred to a later time.

The reader should understand the following ideas as a result of study of this section.

--A figure of merit is a number whose magnitude is an index to the desirability of a solution (decision) set.
--The figure of merit function must be valid in the marketplace in which the design competes, i.e., it must respond to the same value system.
--A formal figure of merit function can be evaluated anywhere and feasibility or infeasibility is determinable by inspection of the figure of merit.
--A tactical figure of merit function must have interrogation confined to feasible space.

3.2 DESIGN FACTOR AS A FIGURE OF MERIT

When insufficient strength leads to failure of a design configuration, engineers speak in terms of a deterministic *design factor* (or conventional *factor of safety*). At the critical location

$$\text{design factor} = \frac{\text{ultimate load}}{\text{imposed load}} = \frac{\text{strength}}{\text{load-induced stress}}$$

where the strength may be the tensile yield strength or endurance strength, etc., as appropriate, to the usage of the machine element concerned. The stress is the load-induced stress at the same location. The design factor n is a simple quotient and can be algebraically expressed as

$$n = S/s$$

where S is strength and s is stress. In a crude way a design factor of 2 means that the element is twice as strong as it needs to be. This statement is misleading without considerable qualification, and your subsequent courses in strength of materials and design applications thereof will treat this idea of design factor in great detail.

Why is a design factor needed? Many unknowns and uncertainties are associated with the design problem.

--Uncertainty as to material properties within a part, within a bar of stock, and within a load of material (as steel). Properties used in design may not come from actual test but from historical experience, since parts are often designed before the steel from which they will be manufactured is even made.
--Uncertainty due to discrepancy between the size of the designed part and the size (necessarily small) of the test specimen. The influence of size on strength is such that smaller parts tend to exhibit larger strengths.
--Uncertainty as to the actual effects of the manufacturing process on the strength at critical locations in the part. Processes such as upsetting, cold or hot forming, thermal treatment, and surface treatment influence strength.
--Uncertainty as to the true effect of peripheral assembly operations on strength. Nearby weldments, mechanical fasteners, force fits have influence on strength that is difficult to assess with precision.
--Uncertainty as to the influence of elapsed time on strength. Aging in steels and aluminums (not exclusively) occurs; in fact, some strengthening mechanisms are time dependent. Corrosion is another enemy of surface that is very important in fatigue strength. Uncertainty as to the operating environment is also present.
--Uncertainty as to the validity (precision) of the mathematical model employed in reaching a decision on a geometric specification of a part.
--Uncertainty as to the intensity and dispersion in loads that may or will be applied to the machine element and the approximations inherent in impact load assessment.
--Uncertainty as to stress concentrations actually present in a manufactured part picked at random. Changes in tool radius due to wear or tool regrinding or tool replacement can have a dramatic impact on stress levels in working parts.

Prior experience or legal code (constraints on design) influences the selection of an appropriate design factor. The decision on design factor may have an upper bound placed on it, not by economics alone but by attainability, as the following example will demonstrate.

In hoisting operations using wire rope, stresses in the rope arise from

1. tension due to the load and the load carrier

Figures of Merit

2. tension due to the weight of the rope itself
3. bending about sheaves in fairleads or winch drums

An endurance tension for wire rope running a million times over a sheave is given by

$$F_n = S_u D d / 2000$$

where: F_n = allowable tension in a wire rope for 1 million flexures over sheave, i.e., rope endurance strength in tension, lbf
S_u = ultimate tensile strength of individual wires within the rope, lbf/in.^{2}
D = sheave diameter, in.
d = wire rope nominal diameter, in.

The strength of the rope repeatedly running over a sheave is the endurance strength in tension less the tensile equivalent of the flexural (bending) load. In keeping with our definition, the design factor for a wire rope in tension and bending about a sheave is the strength in tension divided by the working rope tension, or in symbols,

$$n = (F_n - F_b)/F_t$$

where F_b is the tension that would induce stress equal to that caused by flexing around the sheaves, and F_t is the rope tension. For a 6 × 19 mild plow steel rope,

$$F_b = \frac{321,600 d^3}{D} \text{ lbf} \quad \text{and} \quad F_n = 100 D d \text{ lbf}$$

and the weight of the rope itself is approximately $1.6 d^2$ lbf/ft.

Consider the following problem. As depicted in Fig. 3.4, a mine hoist 480 ft deep lifts a load and cage weighing 1 T with a maximum acceleration of 2 ft/s^2. The hoisting sheave has a 72 in. diameter. Determine an expression

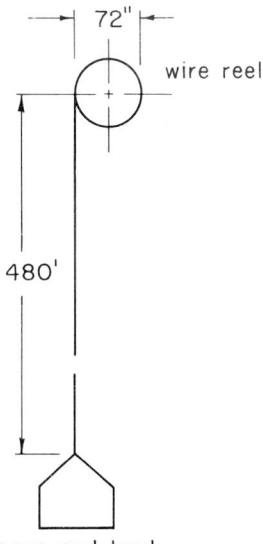

Fig. 3.4. The essential geometry of a mine hoist and its wire reel.

for the design factor that is a function of the nominal rope diameter d and ascertain how the design factor varies with the rope size. We choose the figure of merit as the design factor (as a crude measure of safety) and desire to maximize it. The constraints on the rope diameter are $1/4 \le d \le 2\ 3/4$ in the available diameters.

The first task is to ascertain the elements of the expression for the design factor n. The rope endurance strength in tension is

$$F_n = 100Dd = 100(72)(d) = 7200d \text{ lbf}$$

The second task is to ascertain the tension that would induce the same extreme stress as the flexure.

$$F_b = \frac{321,600d^3}{D} = \frac{321,600d^3}{72} = 4467d^3 \text{ lbf}$$

The working tension is greatest just below the hoisting drum. The rope tension is the weight of the hoist and cage plus the weight of the rope itself magnified by the acceleration of 2 ft/s². If W is the weight of the hoist and cage and wl is the weight of the pendent rope, the force F_t is given by

$$F_t = (W + wl)(1 + a/g) = (2000 + 1.6d^2 l)(1 + 2/32.174) = 2124 + 1.7d^2 l$$

Particularizing this expression for the 480 ft of rope yields

$$F_t = 2124 + 1.7(480)d^2 = 2124 + 816d^2 \text{ lbf}$$

Our figure of merit is defined as

$$M = n = \frac{F_n - F_b}{F_t} = \frac{7200d - 4467d^3}{2124 + 816d^2} \tag{3.1}$$

The tactical figure of merit function is zero at $d = 0$ and at $d = 1.27$ in., values that cause the numerator of Eq. (3.1) to vanish. This puts bounds on the diameter, for we know it must be larger than zero and smaller than 1 1/4 in. Within this range we evaluate the merit function at available wire rope diameters (see Fig. 3.5).

d	M
0.250	0.795
0.375	1.101
0.500	1.307
0.625	1.396
0.750	1.361
0.875	1.203
1.000	0.930
1.125	0.557
1.250	0.081

Values of the merit function less than unity imply no margin of safety at all. The design factor exhibits a maximum for 5/8 in. diameter wire rope. Here is an instance of "if a little bit is good, more isn't necessarily better." Suppose a design factor of 1.4 is simply unacceptable and it must be larger. What can be done? One possibility is to multistrand the ropes, i.e., use more than one rope to support the load. Then each of m ropes would support W/m of the load of cage and mineral. All other elements in Eq. (3.1) remain the same,

Figures of Merit

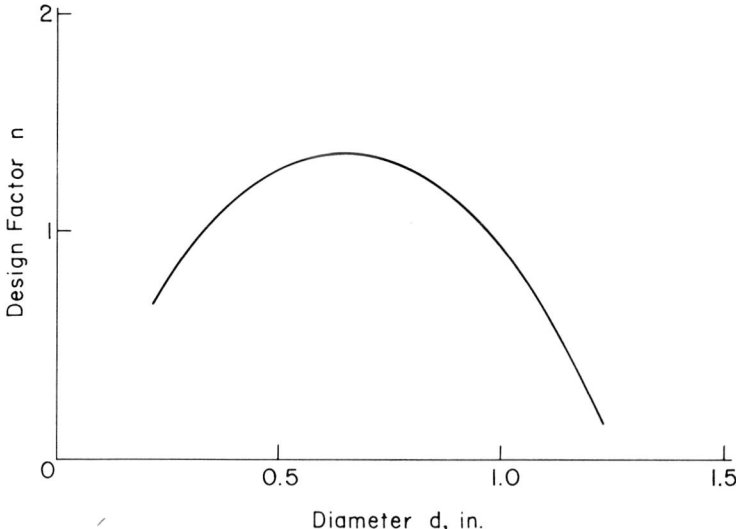

Fig. 3.5. The figure of merit function for the mine hoist problem as a function of the single design decision (choice of nominal wire rope size) treating the design variable as continuous.

so for multistranding with m wire ropes,

$$M = \frac{7200d - 4467d^3}{2124/m + 816d^2}$$

For two wires the rope diameter for maximum safety does not change appreciably, so we evaluate the figure of merit at $d = 0.625$ and obtain $M = n = 2.47$, which does not quite double the design factor. Can you explain why?

The reader should understand the following ideas as a result of reading and studying this section.

--A design factor is used in establishing geometric specifications and its presence is due to many uncertainties associated with strength, stress, load, and geometry.
--A design factor may not be arbitrarily assigned, for it may not even be attainable.
--The design factor may be the figure of merit or an element in a figure of merit.
--A discrete problem (no continuum of wire rope sizes is available) sometimes can be handled as a continuous problem; then discreteness is taken into account. This is not always possible.

3.3 COST AS A FIGURE OF MERIT

Cost is almost always an element in a merit function, usually very influential and occasionally the exclusive constituent. It is not a simple element, given the complexity of economics. Money has a time value, since human nature is such that most people would rather have a dollar now than the promise of a dollar at some future date. To lend money (and exchange a dollar now for more than a dollar later) is to postpone satisfactions that money can provide. The charge for this postponement is called *interest*. Interest is usually computed as a fraction of principal for a time interval, and this fraction varies with the scarcity of money and the future outlook.

A corporation earning 15 percent on its capitalization and considering a new company project must demand that the new endeavor return to the company at least 15 percent for comparable risk; otherwise it will invest money in other areas of the company where the demonstrated rate of return is 15 percent.

An element in cost is taxation and the bases for taxing are manifold. A profit may be taxed, as may the throughput of a pipeline, a ton-mile of transportation, or a payroll. The tax situation must enter the cost element of the figure of merit.

The merit of an enterprise may be viewed as the total payroll; total of production, gross income, or profit (before or after taxes); or the unit cost of production. Accounting practices sometimes obscure the cost of an alternative. A person considers driving to the airport either directly or with a short detour for some purpose not connected with the trip. What is the cost of the detour? The accounting approach is to establish the *expected* cost, which is established by the historical costs of driving a car one mile and multiplying the excess miles involved in the detour. What is the cost if this is done once? Is not the cost of doing it once the difference between expenses incurred detouring and not detouring? The cost can be established only by doing it both ways, something we are normally unwilling or unable to do. This difference in cost is called *incremental cost* and equals the cost of fuel, oil, rubber, and any repair or maintenance occasioned by the detour leg of the journey. It is probable that this cost is much less than the accounting cost.

A classroom in a private school has 40 seats and a class of 20 is conducted in it. The course tuition is $100. What is the cost·of adding another student? Accounting costs run: instructor, $1000; room rental, $200; and student materials, supplied by the student. The $1200 costs (overhead) are prorated per student and amount to $60 per enrollee. If one becomes accustomed to thinking in the usual accounting terms, it is easy to say that of an additional tuition income of $100, $60 is used against overhead and $40 accrues to the institution. The differential cost of having and not having student 21 may even be zero. The costs of registration and intramural paperwork are zero if the registrar and departmental offices are not working to full capacity. If the incremental cost of adding a student is zero, let us add more students! The problem arises when the class becomes too large and another classroom and a second instructor are needed. The incremental cost associated with student 41 is very large indeed; the institution could afford to send that student tuition prepaid to another school (for the course). Incremental costs are important when considering alternatives, and care must be taken to ascertain whether incremental or accounting costs are the better model of reality.

A figure of merit function can (in addition to expressing relative worth of an alternative) also identify which of a constellation of alternatives is optimal at a point in design space. The next example illustrates this capability of a figure of merit function.

A manufacturer of widget machines purchases relays from a vendor and without inspection installs them in machines. Upon assembly a defective relay can be spotted by its obvious malfunction. If a relay's contacts do not mechanically close it is detected on the test stand immediately; if insufficient magnetism is induced to hold the contacts closed without chatter it is detected from the sound plus circuit symptoms; if the coil grounds fully or partially when it first achieves operating temperature or hot coil magnetism is insufficient, it is detected late in the test. The detection of the malfunction

Figures of Merit

is without tangible cost, but the removal from the machine and replacement by another relay has been established at $10. Experience with the last 5000 relays has been as follows: 20 mechanically open circuits (bent contacts), 10 coil chatters (weak fields), and 5 hot coil defects. Alternatives confronting the production engineering staff are

(a) Continue status quo, i.e., do not inspect relays.
(b) Install an inspection procedure that will detect bent contacts, i.e., detect $20/(20 + 10 + 5)$ or $4/7$ of the bad relays. The estimated cost of this inspection is $0.11 per relay.
(c) Install an inspection procedure that will detect bent contacts and weak field, i.e., detect $(20 + 10)/(20 + 10 + 5)$ or $6/7$ of the bad relays. The estimated cost of this inspection is $0.15 per relay.
(d) Install complete inspection to detect bent contacts, weak coil fields, and bad hot coils, i.e., detect all the bad relays. The estimated cost of this inspection is $0.90 per relay.

An analysis is required to provide a rational basis for a decision.

The chief manufacturing engineer has asked you to look into the matter. You decide that the basis for any recommendation you make will be in terms of incremental cost per machine inspected, tested, and shipped. The status quo is costing approximately $10 per bad relay with $35/5000$ of the relays defective. This amounts to $10(35)/3000 = \$0.07$ (approximately) per machine tested and shipped. The production department indicates that it can supply inspection costs once the inspection specifications are drawn. Let p = probability of a bad relay in vendor's delivery and f = fraction of bad relays detected by inspection station. Figure 3.6 indicates the sequence of events. Vendor's

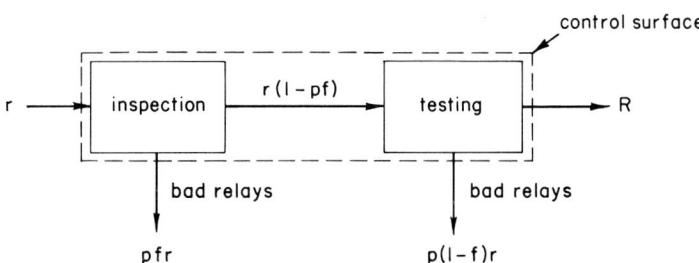

Fig. 3.6. The widget inspection and test facility, showing the movement of relays.

relays move to the inspection station. The inspection procedure removes fraction f of the bad relays from the stream; the stream of inspection-detected bad relays is pfr. The test station removes the remainder of the bad relays that number $p(1 - f)r$. A control surface is defined such that the inspection station and the test station are within, as depicted in Fig. 3.6. Relays are neither created nor destroyed within the control region; consequently all the relays that enter the inspection process emerge as either good or bad relays. The number of influxing relays is r. The effluxing relays appear in three streams numbering pfr from the inspection station, $p(1 - f)r$ from the test station, and R to the shipping room. Thus a conservation of relays equation is

$$r = pfr + p(1 - f)r + R$$

which yields $R = r(1 - p)$. If we set $R = 1$ relay, then

$$r = R/(1 - p) = 1/(1 - p)$$

establishing r as the number of relays inspected for each good widget machine shipped.

The cost of inspection is x dollars per relay inspected or xr dollars per machine shipped. The cost of bad relays detected on the test stand is W = 10 dollars for each bad relay encountered or Wp(1 - f)r dollars per machine shipped. The incremental cost of alternatives per machine shipped is

$$\$ = xr + Wp(1 - f)r = [x + (1 - f)Wp]/(1 - p)$$

If p is small, say 0.01, the above equation is approximated by $\$ = x + (1 - f)Wp$. The probability of a bad relay in the vendor's delivery is historically p = (20 + 10 + 5)/5000 = 0.007.

In alternative (a), the noninspection cost is x = 0 and the fraction detected by this inspection is f = 0. The incremental cost of alternative (a) per machine shipped is

$$\$_a = 0 + 10(0.007)(1 - 0) = 0.07 \text{ dollars}$$

In alternative (b), the estimated cost of inspection is x = 0.11 dollars per relay inspected and the alternative will detect f = 4/7 of the bad relays; hence the incremental cost per machine shipped is

$$\$_b = 0.11 + 10(0.007)(1 - 4/7) = 0.14 \text{ dollars}$$

In alternative (c), the estimated cost of inspections is x = 0.15 dollars per relay inspected and the alternative will detect f = 6/7 of the bad relays; hence the incremental cost per machine shipped is

$$\$_c = 0.15 + 10(0.007)(1 - 6/7) = 0.16 \text{ dollars}$$

Alternative (d), the estimated cost of inspections, is x = 0.90 dollars per relay inspected and the alternative will detect f = 1 (all) the bad relays; hence the incremental cost per machine shipped is

$$\$_d = 0.90 + 10(0.007)(1 - 1) = 0.90 \text{ dollars}$$

The alternative incremental costs are listed in ascending order of cost. The least cost alternative, (a), is the status quo.

The equations for the alternatives expressed as a function of p for small values of p are

$$\$_a = 10p$$

$$\$_b = 0.11 + 4.29p$$

$$\$_c = 0.15 + 1.43p$$

$$\$_d = 0.90$$

$$M(p) = -[\min(\$_a, \$_b, \$_c, \$_d)]$$

$$M(p) = -[\min(10p, 0.11 + 4.29p, 0.15 + 1.43p, 0.90)]$$

Figure 3.7 shows the negative of the costs of each alternative as well as the merit function M(p) in a heavier line. The lowest cost alternative is (a) until the probability of a bad relay from the vendor rises to p = 0.0175; the next least cost alternative is inspection procedure (c).

Figures of Merit

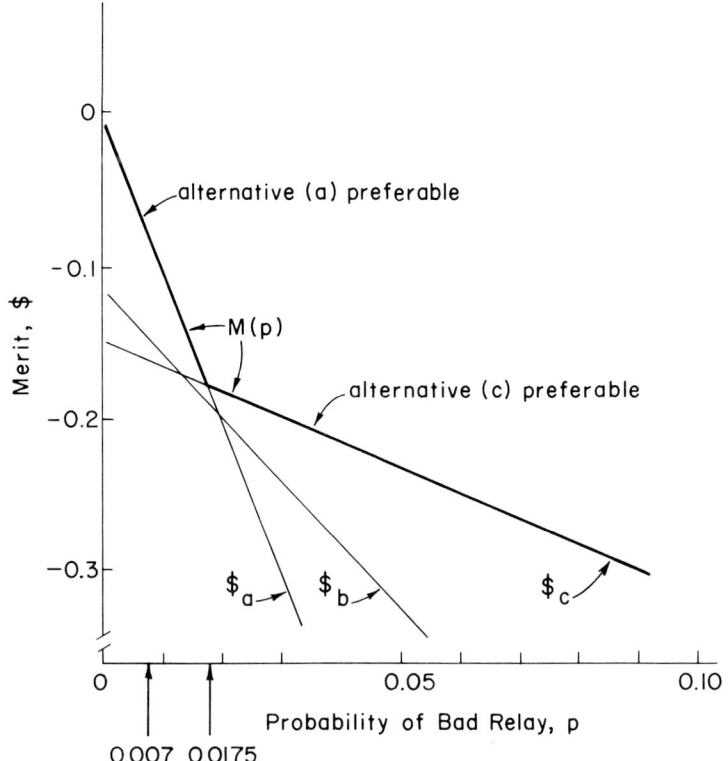

Fig. 3.7. The figure of merit of the widget testing problem displayed as a function of the probability of encountering a bad relay; also, display of alternative inspection procedures.

You report that the best alternative at the present time is not to inspect (and complain to the vendor). When the failure probability reaches 0.0175 or the number of defective relays reaches 17 1/2 per 1000 (and the probability of the failure modes retain the same relative proportions), it is time to institute inspection procedure (c) if the vendor cannot be expected to respond to encouragement to improve.

Earlier in this section we use the term *expected value* in relation to the accounting method of estimating cost of driving the detour miles on the trip to the airport. Expected value has a more technical meaning completely consistent with our previous usage. If one were to measure the extreme width of every leaf on a tree, all the widths x would lie between zero and (say) 5 in. If we classified the leaves by width into arbitrary intervals of 0.1 in. range and plotted frequency of observation (number of leaves with a width in a particular range), we would obtain a histogram such as depicted in Fig. 3.8(a). If we divide every class frequency by $N_0 \Delta x$ [(total leaf population) × (width of class interval)], we would have a diagram of similar shape as in Fig. 3.8(b); but now the area of each bar is $[N/(N_0 \Delta x)](\Delta x) = N/N_0$, the probability of encountering a leaf of the width in the class range. The ordinate $N_x/(N_0 \Delta x)$ is called the probability density; and as the leaves increase without upper bound, the limit

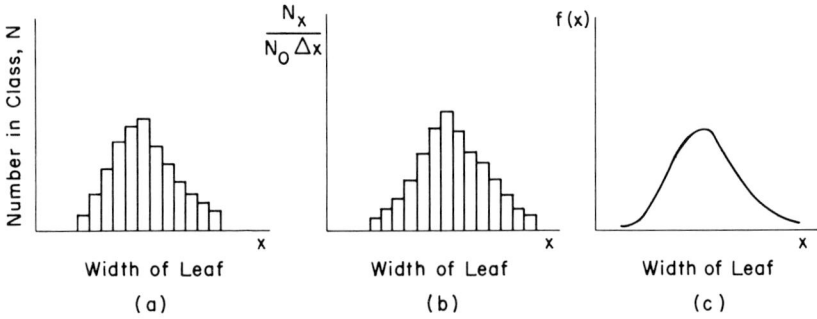

Fig. 3.8. (a) A histogram of leaf widths from a tree; (b) the same histogram with probability as the ordinate; (c) the probability density function f(x), representing the limit as the class interval, leaf width, approaches zero.

$$\lim_{\substack{N_0 \to \infty \\ \Delta x \to 0}} [N_x/(N_0 \Delta x)] = f(x)$$

is still called the probability density but it becomes a continuous function such as depicted in Fig. 3.8(c). The area under the curve is unity, since the probability of a leaf picked at random exhibiting an extreme width between $-\infty$ and $+\infty$ inches is one (certainty). This observation is a property of the *probability density function* (pdf) and is expressed as

$$\int_{-\infty}^{\infty} f(x) \, dx = 1$$

The expected value of x is the weighted average of all possible x's.

(probability of width lying between x and x + dx) = f(x) dx

The weight assigned to an observation of the size x is f(x) dx; consequently,

$$(\text{expected value of } x) \equiv E(x) = \int_{-\infty}^{\infty} x f(x) \, dx = \mu_x$$

This measure of the central tendency, usually denoted μ_x, is the population mean. When the population is finite, the expected value of x is

$$E(x) = \sum_{i=1}^{N} X_i/N = \mu_x \qquad (3.2)$$

which is also denoted μ_x. In the case of random samples drawn from a parent population, the mean of the sample is an unbiased estimator of the population mean,

$$\hat{\mu}_x = \sum_{i=1}^{n} x_i/n \qquad (3.3)$$

where n is the sample census.

In describing variates, it is necessary to add to the description represented by the mean some measure of the dispersion (variability) of the quanti-

Figures of Merit

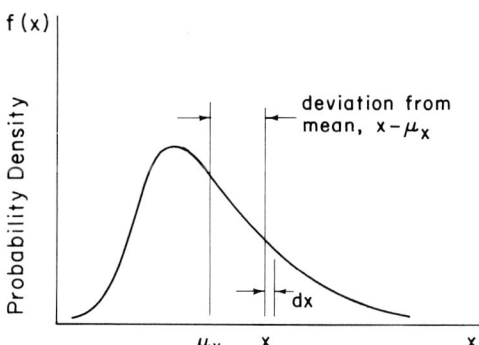

Fig. 3.9. A probability density function showing a deviation from the mean.

ty. One measure of the dispersion is called the *variance*, which is defined as the average squared deviation from the mean. In Fig. 3.9 the probability of a deviation (displacement) from the mean of size $x - \mu_x$ is the area $f(x)\,dx$. Following the same weighting procedure as before,

$$\text{variance} = \sigma_x^2 = \int_{-\infty}^{\infty} (x - \mu_x)^2 f(x)\,dx \tag{3.4}$$

Since the variance has the units of the variate squared, the more common term of description is the *standard deviation*, which is the square root of the variance.

The integral representing the definition of the variance may be simplified by carrying out the squaring operation.

$$\int_{-\infty}^{\infty} (x - \mu_x)^2 f(x)\,dx = \int_{-\infty}^{\infty} x^2 f(x)\,dx - 2\mu_x \int_{-\infty}^{\infty} x f(x)\,dx + \mu_x^2 \int_{-\infty}^{\infty} f(x)\,dx$$

The right-hand integral has a value of unity as has been shown before. The next integral to the left is the definition of the mean value of x. The integral just to the right of the equal sign is the definition of mean value of x^2 (as distinguished from the squared mean value of x. Consequently,

$$\sigma_x^2 = \mu_{x^2} - 2\mu_x(\mu_x) + \mu_x^2 = \mu_{x^2} - \mu_x^2 \tag{3.5}$$

which is a useful relationship. For example, suppose the mean value of the diameter μ_d of right circular cylindrical bars made by an automatic screw machine is 1.001 in. with a standard deviation of $\sigma_d = 0.001$ in. What is the mean cross-sectional area of bars picked at random? Solve the above relationship for μ_{d^2}.

$$\mu_{d^2} = \mu_x^2 + \sigma_d^2 = 1.001^2 + 0.001^2 = 1.002002$$

$$\mu_A = (\Pi/4)\mu_{d^2} = (3.14159/4)(1.002002) = 0.786971$$

This answer is very close to the one obtained by ignoring the contribution of the variance because it is so small. This is not always the case; if the standard deviation is 0.1 rather than 0.001, then $\mu_A = 0.794824$.

When sampling randomly with replacement from an infinite or a finite population, the unbiased estimator of the parent population standard deviation from the sample parameters is

$$\hat{\sigma}_x = \left[\Sigma \frac{(x - \hat{\mu}_x)^2}{n - 1} \right]^{1/2} \qquad (3.6)$$

where $\hat{\mu}_x$ is $\Sigma x/n$, the sample mean and unbiased estimator of the population mean. When sampling randomly without replacement from a finite population, the unbiased estimator of the population standard deviation is given by

$$\hat{\sigma}_x = \left[\frac{N - n}{N - 1} \Sigma \frac{(x - \hat{\mu}_x)^2}{n - 1} \right]^{1/2} \qquad (3.7)$$

The term *risk* is usually used when the odds (probabilities) of the outcome are predictable. Underwriting life insurance policies involves risk because mortality tables are available (probabilities of death may be well estimated from history). The term *uncertainty* is usually used when the odds are not predictable. Uncertainty occurs in areas in which we have had no experience or understanding. The outcome of a war or the first manned spaceflight involves uncertainty.

When risk is involved the expected value of a parameter can be useful, but this is not always the case. Let an event have k possible outcomes and let the probability of each outcome be known.

| probability | p_1 | p_2 | ... | p_k |
| outcome | x_1 | x_2 | ... | x_k |

The expected outcome is the expected value of x that is

$$E(x) = p_1 x_1 + p_2 x_2 + \ldots + p_k x_k = \sum_{i=1}^{k} p_i x_i$$

As an example, let an event have two possible outcomes. The probability of winning is p_w and the probability of losing is p_l. The event is a wager that is known to pay twice its value on winning and nothing on losing; the expected return is the expected value of x:

$$E(x) = \Sigma p_i x_i = p_w (2x) + p_l (0) = 2 p_w x$$

If the event is the flip of an honest coin, $p_w = 1/2$ and the expected value of x is x. In an environment of alternative investments yielding a higher rate of return, this investment has little merit. If a company investment of x will yield 1.4x with a 65 percent chance of success and will yield 0.1x in salvage if unsuccessful, what is the expected return?

$$E(x) = \Sigma p_i x_i = p_w x_w + p_l x_l = 0.65(1.4x) + 0.35(0.1x) = 0.95x$$

The yield on the average would not recover investment. The predicted 5 percent loss is observed in the history of a large number of investments. Any *one* investment can succeed or fail. Suppose on the flip of a coin the payoff is 5x on winning and nothing on losing. The expected return is

$$E(x) = \Sigma p_i x_i = p_w x_w + p_l x_l = (1/2)(5x) + (1/2)(0) = 2.5x$$

In an environment of alternatives yielding 1.1x the opportunity might seem attractive, but if the amount of money risked was most of or all the resources

Figures of Merit   55

of the investor, should the risk be assumed? The expected value, $E(x) = 2.5x$, stands mute on this point. Thoughtful persons shy away from investing nearly all their resources. Can you risk ruin?

If f is the current rate of return on company investment, $(1 + f)x$ return can be expected. Let

$$E(x) = n(1 + f)x$$

where n is the number of times company investment is expected from a contemplated new investment. A merit function might be the excess of return above company investment, say,

$$M(x) = E(x) - (1 + f)x = n(1 + f)x - (1 + f)x = (n - 1)(1 + f)x$$

The merit function should be also sensitive to extent of resources committed. The term $(1 - x/R)$ vanishes when $x = R$ and assigns small penalty when x is a trivial fraction of R. We write

$$M(x) = (1 - x/R)(n - 1)(1 + f)x$$

The merit function exhibits a maximum at $x/R = 1/2$, independent of company rate of return f, magnification n, or the extent of resources R. There is nothing magic here. If the fraction of resources term is $(1 - x/R)^k$, show that maximum merit corresponds to investment of resource fraction $x/R = 1/(k + 1)$.

The reader of this section should have identified the following points:

--Cost may be a figure of merit or an element in it.
--Incremental cost and expected cost are different.
--The figure of merit function can be used to identify alternative decisions.
--Expected value (mean) and standard deviation of populations have specific technical definitions.
--Risk is the term used when probabilities of outcomes are available a priori or can be estimated.
--Uncertainty is the term used when probabilities of outcomes are not available a priori.
--While risk associated with an event may be favorable, can you afford to risk all on a single venture?

3.4 RELIABILITY AS A FIGURE OF MERIT

The reliability of a system is simply the probability that the system will function as required during a stipulated time interval. If the life of automobile mufflers exhibits a probability density function as depicted in Fig. 3.10, the reliability exhibited by mufflers at x miles of driving is the probability of observing a life of x or greater.

$$R(x) = \int_x^\infty f(x)\,dx \qquad (3.8)$$

Some relationships can be established a priori. Most are determined a posteriori, if at all. Reliability is a function of the age of the component. The reliability of a complex or system of components can be inferred from the reliability of the individual parts.

The *cumulative density function* $F(x)$ is defined as the probability of observing a life of x or less (sometimes called the unreliability) and is given by

$$F(x) = \int_{-\infty}^x f(x)\,dx \qquad (3.9)$$

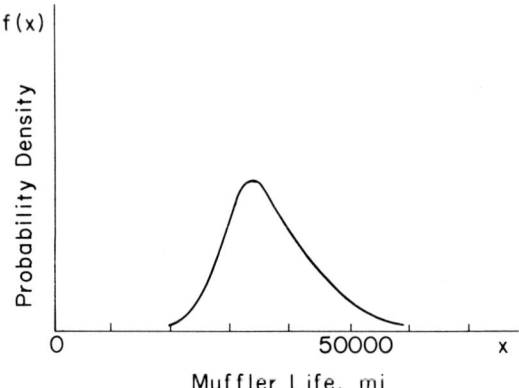

Fig. 3.10. The probability density function for muffler life.

The first derivative of F(x) with respect to x is obtained by differentiating the previous equation with respect to x, yielding

$$\frac{dF(x)}{dx} = f(x) \qquad (3.10)$$

Differentiation of Eq. (3.8) with respect to x yields

$$\frac{dR(x)}{dx} = -f(x) \qquad (3.11)$$

and as suspected, the relationship between R(x) and F(x) is

$$R(x) = 1 - F(x) \qquad (3.12)$$

and between the derivatives is

$$\frac{dR(x)}{dx} = -\frac{dF(x)}{dx}$$

With a population of components of n_0 at life-measure zero and a population of n(x) at life-measure x, from the definition of reliability we may declare

$$R(x) = n(x)/n_0$$

Differentiating both sides with respect to x yields

$$\frac{dR(x)}{dx} = \frac{dn(x)}{n_0 dx}$$

If we define hazard as the failure fraction per unit life measure, then

$$\text{average hazard} = \Delta n / [n(x) \Delta x]$$

where Δn is the number of failures in the life interval Δx. It follows that instantaneous hazard Z(x) is

$$Z(x) = \lim_{\Delta x \to 0} -\left(\frac{\Delta n}{n(x)\Delta x}\right) = -\frac{dn(x)}{n(x)dx} \qquad (3.13)$$

Substituting previous results, we obtain

$$Z(x) = -\frac{dn(x)}{n(x)dx} = -\frac{dn(x)}{(n/n_0)n_0 dx} = -\frac{dn(x)}{R(x)n_0 dx} = -\frac{dR(x)/dx}{R(x)} \qquad (3.14)$$

Figures of Merit

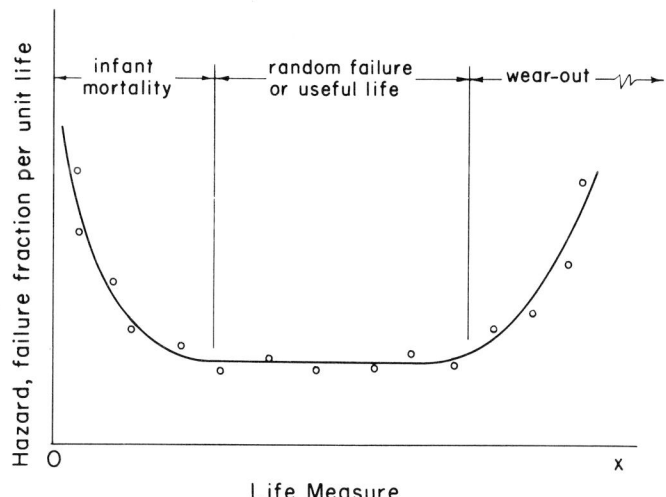

Fig. 3.11. The "bath-tub" curve, showing the infant mortality interval, the useful life interval, and the wear-out interval associated with manufactured products of some complexity.

$$Z(x) = \frac{f(x)}{R(x)} \tag{3.15}$$

This equation is important, since the hazard is observable, $Z_{avg} = -\Delta n/(n_{avg}\Delta x)$, where n_{avg} is the average census of survivors during the life interval Δx. Experiments with complete machines (complex machinery) have yielded the stereotype in Fig. 3.11, when the hazard function is plotted against the life measure--the "bathtub" curve. The first zone or interval is referred to as the *infant mortality* interval. The second domain exhibits substantially constant hazard and this interval is called the *random failure* or *useful life* interval. Since the failure fraction per unit life interval is independent of history in this interval, failure incidence is completely random. This observation is the basis for a good deal of the simpler reliability theory. In the final or *wear-out* interval, cumulative effects of wear begin to show and failure increases to intolerable levels. The component is either repaired, rebuilt, or retired. For the useful life of machines, electronic devices, and other complicated aggregates of components, the reliability can be well predicted by the models that express the constant hazard phenomenon.

A general useful result is now developed. Beginning with Eq.(3.14), $Z(x) = -[dR(x)/dx]/R(x)$, we separate the variables:

$$\frac{dR(x)}{R(x)} = -Z(x)dx$$

Integrating the left side of the above equation between $R = 1$ and $R = R$ and the right side between $x = -\infty$ and $x = x$, we have

$$\int_1^{R(x)} \frac{dR(x)}{R(x)} = -\int_{-\infty}^x Z(x)\, dx$$

$$\ln R(x) \Big|_1^{R(x)} = -\int_{-\infty}^x Z(x)\, dx$$

$$R(x) = \exp[-\int_{-\infty}^{x} Z(x)\, dx] \tag{3.16}$$

which is a useful general equation, since the hazard is measurable experimentally. If, as in the case of the useful life interval of complex devices, $Z(x) = k$ = constant, then measuring life after break-in, i.e., after infant mortality interval, is

$$R(x) = \exp(-\int_{0}^{x} k\, dx) = e^{-kx} \tag{3.17}$$

This equation is called the *survival* equation for constant hazard failures.

The reliability of a single bearing in particular circumstances might be R_1. The probability of failure of the bearing is $(1 - R_1)$ or p_1; therefore, $R_1 = 1 - p_1$. Similarly, for the bearing at the other end of the shaft, $R_2 = 1 - p_2$. If either bearing fails or if both fail, the shaft is incapacitated. The reliability of the bearings as a pair is the product of the individual reliabilities, or $R = R_1 R_2 = (1 - p_1)(1 - p_2)$. If the bearings are identical with identical loadings, $R = (1 - p_1)^2$. This arrangement, not necessarily geometrically expressible, where the failure of *either* or *both* constitutes system failure, is called a *series* system and is diagramatically depicted in Fig. 3.12. In general for k elements in series

$$R = R_1 R_2 \cdots R_k \tag{3.18}$$

and for k *identical* elements

$$R = R_1^k \tag{3.19}$$

It is possible for elements to be arranged on an either/or basis, such as boiler feedwater pumps (one active, one standby), and the situation can be depicted as in Fig. 3.13. The probability that both fail simultaneously (a si-

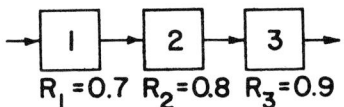

Fig. 3.12. A complex with components in reliability series. (From C. R. Mischke, 1968, *An Introduction to Computer-Aided Design*, Prentice-Hall, Englewood Cliffs, N.J., p. 28. Used with permission.)

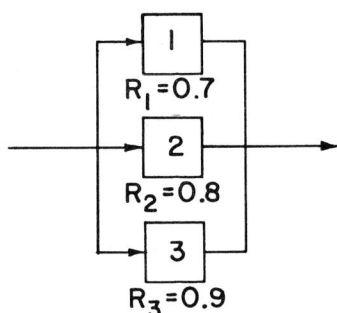

Fig. 3.13. A complex with three components in reliability parallel. (From Mischke, *Computer-Aided Design*, p. 28. Used with permission.)

Figures of Merit   59

multaneous failure is necessary to the failure of a parallel complex) is the product of the failure probabilities,

$$p = p_1 p_2 = (1 - R_1)(1 - R_2) = (1 - R)$$

or $R = 1 - (1 - R_1)(1 - R_2)$. If the two components are identical, $R = 1 - (1 - R_1)^2$. In general for a complex of parallel components,

$$R = 1 - (1 - R_1)(1 - R_2)\ldots(1 - R_k) \tag{3.20}$$

and if the k elements are identical,

$$R = 1 - (1 - R_1)^k \tag{3.21}$$

If, in Fig. 3.13, the reliabilities of the components are $R_1 = 0.7$, $R_2 = 0.8$, and $R_3 = 0.9$, then from Eq. (3.20), $R = 1 - (1 - 0.7)(1 - 0.8)(1 - 0.9) = 0.994$.

Note that the reliability of the series complex is inferior to that of the individual components and the reliability of the parallel complex is superior to that of the components. Systems can be built with overall reliability higher than that of the individual components by using parallel (redundant) arrangements. In high reliability complexes (electrical power utilities, military and hospital units), redundancy has been used for a long time. Although costly, it is often the only alternative to a long and expensive development program.

As an example of the use of reliability as a figure of merit, consider the case of establishing the highest possible reliability for $550 as depicted in Fig. 3.14, wherein component 1 costs $100 and exhibits a reliability of $R_1 = 0.5$ for the life measure in question and component 2 costs $150 and exhibits a reliability of $R_2 = 0.8$. In Fig. 3.15, complex A consists of parallel arrangements of components of type 1 and complex B consists of parallel arrangements of components of type 2. From Eq. (3.21),

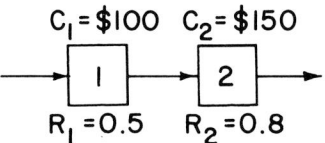

Fig. 3.14. A complex of two components in series. (From Mischke, *Computer-Aided Design*, p. 29. Used with permission.)

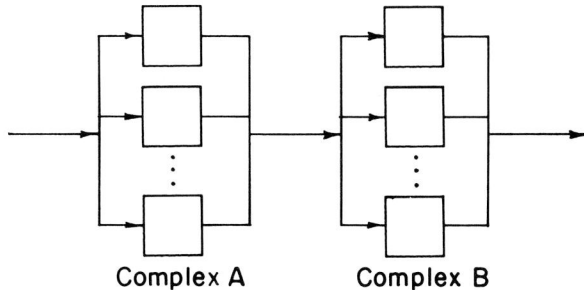

Fig. 3.15. Two complexes in series whose individual components are in parallel. (From Mischke, *Computer-Aided Design*, p. 29. Used with permission.)

$$R_A = 1 - (1 - R_1)^j \qquad R_B = 1 - (1 - R_2)^k$$

where j is the number of components of type 1 in A and k is the number of components of type 2 in B. As a figure of merit we choose the overall reliability. The designer has just two decisions to make, the number of components in each complex.

$$M(j, k) = R_A R_B = \left[1 - (1 - R_1)^j\right]\left[1 - (1 - R_2)^k\right]$$

We would seek to maximize M subject to a regional constraint (a relationship between j and k through cost). We write

$$jC_1 + kC_2 \leq 550 \qquad (j = 0, 1, 2, \ldots; \; k = 0, 1, 2, \ldots)$$

or $100j + 150k \leq 550$, remembering that j and k are integers. *One* design decision is involved if we are against the regional constraint and *two* if we are not. We will view the design space as having two independent variables, j and k, and Fig. 3.16 depicts the feasible area of solution space. Our design decisions are discrete and are depicted as nodes.

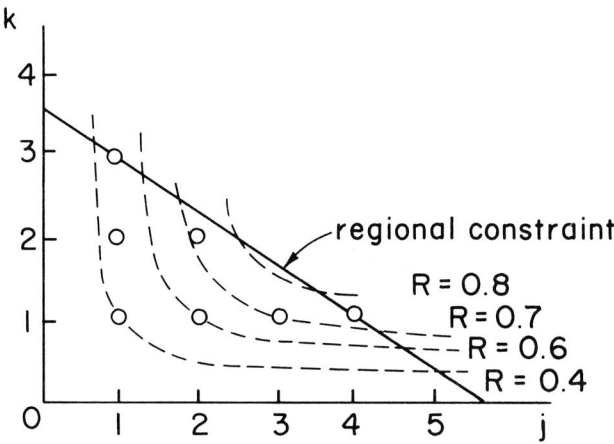

Fig. 3.16. Reliability contours on a graph with the redundancy decisions, j and k, as axes.

Our first approach to a solution will be exhaustive. We write the figure of merit specifically as

$$M(j, k) = [1 - (1 - 0.5)^j][1 - (1 - 0.8)^k] = (1 - 0.5^j)(1 - 0.2^k)$$

and using the inequality, $j \leq (550 - 150k)/100$, we evaluate

$$k = 1 \quad j \leq [550 - 150(1)]/100 = 400/100 = 4$$
$$k = 2 \quad j \leq [550 - 150(2)]/100 = 250/100 = 2.5$$
$$k = 3 \quad j \leq [550 - 150(3)]/100 = 100/100 = 1$$

It follows that

$$M(1, 1) = (1 - 0.5^1)(1 - 0.2^1) = 0.400$$
$$M(2, 1) = (1 - 0.5^2)(1 - 0.2^1) = 0.600$$

Figures of Merit

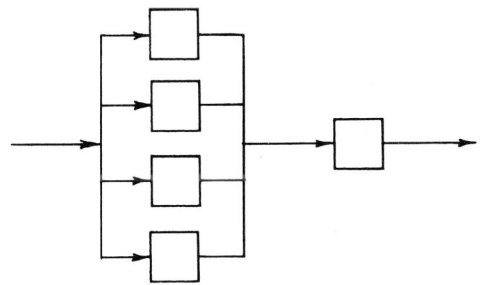

Fig. 3.17. Redundancy decisions with the highest reliability obtainable for a cost of $550 or less. (From Mischke, *Computer-Aided Design*, p. 33. Used with permission.)

$M(3, 1) = (1 - 0.5^3)(1 - 0.2^1) = 0.700$

$M(4, 1) = (1 - 0.5^4)(1 - 0.2^1) = 0.750$

$M(2, 2) = (1 - 0.5^2)(1 - 0.2^2) = 0.720$

$M(1, 2) = (1 - 0.5^1)(1 - 0.2^2) = 0.480$

$M(1, 3) = (1 - 0.5^1)(1 - 0.2^3) = 0.496$

Best alternative: $j = 4$, $k = 1$; $\$ = 100j + 150k = 100(4) + 150(1) = 550$; $R = 0.750$; and Fig. 3.17 depicts the optimal system.

An approach to this problem that begins to show a strategic plan is to map the contours of cost onto Fig. 3.16. We note that cost is $\$ = 100j + 150k$ and solving for k yields

$k = [\$ - 100j]/150 = \$/150 - (2/3)j$

For various values of the parameter cost, $, we observe a family of parallel straight lines of slope -2/3 and an ordinate intercept of $/150. If we took a feasible decision, say $j = 1$, $k = 1$, we would be on the $\$ = 250$ contour. Improvement in reliability would occur if we were on the $\$ = 550$ contour (the regional constraint). The best position on the regional constraint depends on how the merit (reliability) contours cross the regional constraint. The next move is to "walk the constraint" but it is dangerous when the design variables are discrete, since a node off the constraint that would be ignored could be just barely off and be the optimal solution.

As another example of using reliability as a figure of merit, consider the situation depicted in Fig. 3.18 with two components in reliability series. The survival equations of the components are

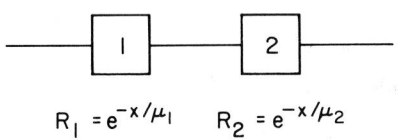

Fig. 3.18. Two components in reliability series with their reliability functions declared.

$$R_1 = e^{-x/\mu_1} \quad R_2 = e^{-x/\mu_2}$$

where μ_1 and μ_2 are mean times to failure. At a life measure of 1000 hr, the combined reliability is to be as large as possible. The estimated cost of building component 1 is $100R_1$ dollars and the estimated cost of building component 2 is $200R_2$ dollars. The total cost for the component pair is not to exceed $100. The problem is to establish the mean time to failure for component 1 and component 2. We will use the system reliability as the figure of merit, $M = R_1R_2$, which must be maximized subject to the following regional constraints:

$$100R_1 + 200R_2 \leq 100 \quad 0 \leq R_1 \leq 1 \quad 0 \leq R_2 \leq 1$$

A map of the solution space is depicted in Fig. 3.19. Since R_1 and R_2 are continuous, the maximum merit is represented by the contour tangent to the diagonal regional constraint. We use the equality $100R_1 + 200R_2 = 100$, since we know we will be on that extremity of constraint.

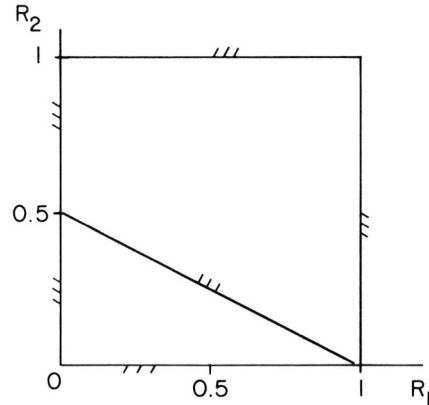

Fig. 3.19. The feasible domain using R_1 and R_2 as axes.

$$M = R_1R_2 = R_1[100 - 100R_1]/200 = 0.5R_1(1 - R_1) = 0.5R_1 - 0.5R_1^2$$

Recognizing there is but one decision variable if we insist on maximum merit, we differentiate the above equation with respect to R_1 and set it equal to zero in the manner of the calculus.

$$\frac{dM}{dR_1} = 0.5 - 2(0.5R_1) = 0$$

from which $R_1 = 0.5$, $R_2 = 0.25$, and we note that $d^2M/dR_1^2 = -1$, denoting a stationary point maximum. From the survival equations we may write

$$\ln R_1 = -x/\mu_1 \quad \ln R_2 = -x/\mu_2$$

and solving for μ_1 and μ_2, we obtain

$$\mu_1 = -x/\ln R_1 = -1000/\ln 0.5 = 1443 \text{ hr}$$

$$\mu_2 = -x/\ln R_2 = -1000/\ln 0.25 = 721 \text{ hr}$$

Some hidden dangers exist due to our methodology; they should be identi-

Figures of Merit

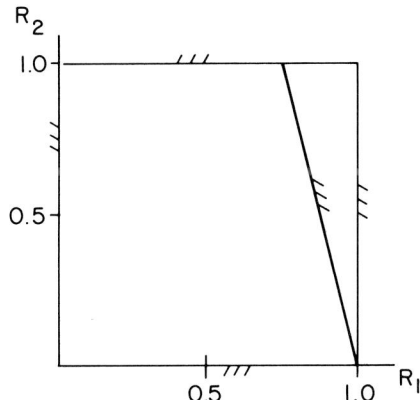

Fig. 3.20. The feasible domain using R_1 and R_2 as axes.

fied lest glibness lead us astray. Suppose the constraint had been $100R_1 + 25R_2 \leq 100$. The corresponding solution space is depicted in Fig. 3.20. If we use the calculus as before,

$$M = R_1 R_2 = R_1(100 - 100R_1)/25 = 4R_1 - 4R_1^2$$

$$\frac{dM}{dR_1} = 0 = 4 - 8R_1$$

from which $R_1 = 0.5$ and $R_2 = 2$, clearly impossible! By inspection the maximum merit occurs at the junction of *two* constraints, and $R_2 = 1$.

$$R_1 = (100 - 25R_2)/100 = [100 - 25(1)]/100 = 0.75$$

The mean time to failure specifications are

$$\mu_1 = -x/\ln R_1 = -1000/\ln 0.75 = 3476 \text{ hr}$$

$$\mu_2 = -x/\ln R_2 = -(1000/\ln 1) \to \infty$$

As a practical interpretation, $\mu_2 \gg \mu_1 = 3476$ hr.

The lesson here is that calculus methods of maximization and minimization as usually taught are not dependable in the presence of constraints. To suppress some of these difficulties, the work of Lagrange should be studied. More on this point is found in Sec. 8.3.

As a result of reading and studying the material of this section, note that:

--The figure of merit function need not be continuous.
--Two design decisions are made unless we are up against a constraint.
--We used an exhaustive search strategy because the problem was small and because we have not learned any strategies of search yet.
--The maximum merit was not found by calculus methodology. A derivative is a less than meaningful concept in an integer solution space.
--Reliability can be a figure of merit or it can be used as an element in a figure of merit function.
--Reliability, probability density function, and hazard function have specific technical definitions.

--When elements of a complex are in reliability series, the overall reliability is less than the reliability of any of the components.
--When elements of a complex are in reliability parallel (redundant), the reliability of the complex is superior to the reliability of any component.

3.5 TIME AS A FIGURE OF MERIT

Time can appear in a function of merit either implicitly or explicitly. Time assumes engineering importance as either absolute time (calendar time, as in a completion date) or relative time (interval time, as in a process time constant).

Absolute time is important when a market exists in the future until a specified date. A corporation or governmental agency might accept bids, until a specified date, to provide a system to meet definite specifications. Alternatives of superior quality will be discarded if the calendar time available to do the requisite engineering work is insufficient.

Relative time is important when a time interval in a system is critical. A company marketing equipment with a response time of 0.001 s, faced with future competition providing time constants of 0.0002 s, will make its merit function sensitive to response time. If only one way to accomplish a task for a specific application exists and response time is important, a minimum of response time may be sought; and the response time, expressed negatively or reciprocally, can be the figure of merit.

Times other than cutoff dates can often be related to costs, and the time factors in a function of merit may appear as a cost. For example, if the processing time of a production element must be less than 4 days, a Boolean variable can assign zero merit to all alternatives exhibiting a processing time in excess of 4 days, and the cost of effecting the alternative (which involves process time) can be encoded into the merit function.

Consider the problem of specifying the gear ratio between the traction motors and the wheels of a subway or commuter train. If the system is being built with equipment having a gear ratio tailor-made for the service and the distance between stations is S, what is the optimal gear ratio to minimize time between stations, i.e., the start-to-stop time?

The dc series traction motor has torque-speed characteristics as displayed in Fig. 3.21. At low speeds, armature current is limited to permissible values by resistor banks and series and parallel electrical arrangements of motor wiring. During this controlled current period, the motor torque varies in a somewhat sawtoothed fashion, but its average is shown as the initial horizontal line in the diagram. The running characteristic is exhibited when full line voltage is across the traction motor.

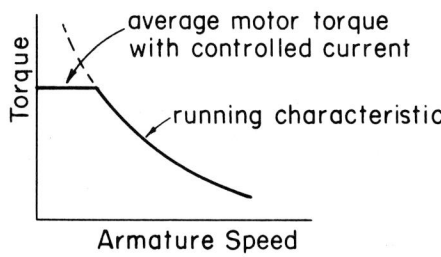

Fig. 3.21. Torque-speed characteristics of a typical dc traction motor. (From Mischke, *Computer-Aided Design*, p. 33. Used with permission.)

Figures of Merit

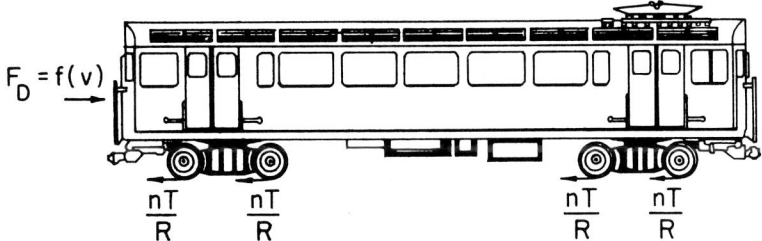

Fig. 3.22. Free body diagram of an electrically propelled rapid transit car with four traction motors. (Copyright 1965, Central Electric Railfans' Association, Bull. 108, p. 86. Used with permission.)

A free body of the railway car is represented in Fig. 3.22. The principal forces that affect train speed are the tractive efforts of the wheels and the drag forces, bearing and rolling friction and aerodynamic drag, which are functions of vehicle velocity. Presuming negligible inertia in the rotating parts (wheels, gears, and armatures), from Newton's law

$$\Sigma F = (W/g)a \qquad \frac{4nT}{R} - f(v) = \frac{W}{g}\frac{dv}{dt} \qquad dt = \frac{W\,dv}{g[4nT/R - f(v)]}$$

where W is the weight of the car and its passengers, a is the acceleration of the car, v is the velocity of the car, n is the step-down gear ratio, T is the motor torque, R is the radius of the wheels, and g is the gravitational constant. Figure 3.23 shows the train speed-time curve asymptotic to an ordinate

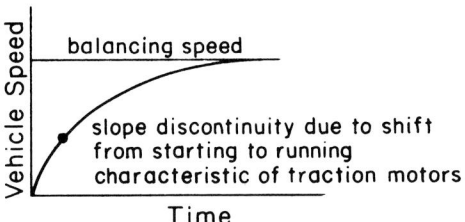

Fig. 3.23. The train's speed-time curve is asymptotic to the balancing speed. (From Mischke, *Computer-Aided Design*, p. 33. Used with permission.)

called the balancing speed, which is the speed a rapid transit car will obtain on level track in still air. Several similar cars in train will exhibit a higher balancing speed than a single car. Other balancing speeds exist under other conditions of grade, curvature, and line voltage.

The train will brake for the approaching station before attaining balancing speed. The braking in recently built equipment is initially dynamic braking, wherein the traction motors as electrical generators dissipate energy across resistor banks or deliver energy back to the third rail (overhead catenary). As the train slows, this braking effort fades and must be augmented by mechanical air brakes of conventional types. The distance from the station at which braking commences is a function of train speed, as is the duration of the braking phase. A superposition of the braking and accelerating curves is shown in Fig. 3.24. The area under the v-t curve is the distance traversed start to stop. Different gear ratios lead to different acceleration curves on

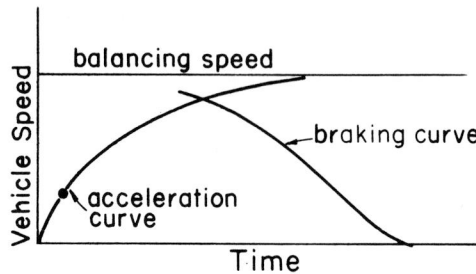

Fig. 3.24. The distance between stations can be shown to be the area under the speed-time curve. (From Mischke, *Computer-Aided Design*, p. 33. Used with permission.)

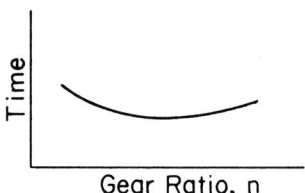

Fig. 3.25. The time start-to-stop will exhibit a minimum for some gear ratio n between the traction motor shaft and the axle of the car wheels. (From Mischke, *Computer-Aided Design*, p. 34. Used with permission.)

the v-t plot, and different train speeds move the deceleration curve for service braking to the left or to the right, affecting the start-to-stop time. The designer must relate start-to-stop times to the gear ratio. Clearly, a very high reduction ratio involves a low balancing speed and a long traveling time between stations. A very low reduction ratio involves poor acceleration and attainment of low speed on a short run and a long time elapsed between stations. At some intermediate gear ratio, a minimum start-to-stop time is exhibited. Figure 3.25 suggests this.

This simple problem involves complicated integrations. It is admirably suited to computer solution. The possibility of simulating an entire route with variable station distances is open, and the minimum schedule time for an entire run might be the quantity sought. The effects of grade, curvature, speed limits at switches, junctions, and special track work can easily be introduced. If the equipment is to be used interchangeably in local and express (skip-stop) service, the optimum gear ratio associated with each service will be different. The use of trailers (nonpowered cars) affects the ratios also. The cost of power differs and a compromise (trade-off) between cost of service and speed of service must be made.

3.6 TRADE-OFF FUNCTIONS

The engineer, constantly confronted with conflicting objectives, tries to keep costs low; safety as well as capacity high; and meet a host of other specifications and gross size, compatibility with other equipment, etc. Clearly, a larger safety margin increases cost and larger capacity increases cost. For a given budget, how shall safety and capacity be distributed? A marketplace appears wherein a certain amount of safety can be traded for a

Figures of Merit

different amount of capacity. How will safety and capacity be measured? What is the legal tender of this marketplace? How can one establish values?

One approach is to think in terms of money. If, in a simple problem, cost is a function of design factor n and capacity C, then $ = f(n, C)$ and

$$d\$ = \frac{\partial \$}{\partial n} dn + \frac{\partial \$}{\partial C} dC \qquad (3.22)$$

The partial derivatives $\partial\$/\partial n$ and $\partial\$/\partial C$ must be examined. If it is more costly to increase the design factor when capacity is high and cost is directly proportional to design factor, the derivative is $\partial\$/\partial n = aC$. When the design factor is held constant and the capacity is varied, the increase in cost is proportional to the magnitude of the design factor, i.e., $\partial\$/\partial C = bn$. Substituting in Eq. (3.22), we have

$$d\$ = aC\,dn + bn\,dC$$

Integrating the partial derivative, $\partial\$/\partial n = aC$, we obtain

$$\$ = aCn + f_1(C)$$

Integrating the partial derivative, $\partial\$/\partial C = bn$, we obtain

$$\$ = bnC + f_2(n)$$

The cost function determined either way is identical, i.e.,

$$aCn + f_1(C) \equiv bnC + f_2(n)$$

It follows from the property of an identity that $f_1(C) = 0$, $f_2(n) = 0$, and $a = b$. Therefore, setting $a = K$,

$$\$ = KnC$$

If the designer decides that the merit of the design is to be measured by the reciprocal of cost, the tactical figure of merit becomes

$$m = 1/\$ = 1/KnC$$

If limitations are imposed (such as $n \geq 5$ and $C \geq 5$ T) and K is evaluated from the fact that the cost of one combination of design factor and capacity is known, say the cost ($n = 2.5$, $C = 4$) of a crane hook is \$40, then $K = 4$ and

$$M = [B_1(n)B_2(C)]/(4nC)$$

where $B_1(n)$ and $B_2(C)$ are Boolean functions of n and C that are zero when $n < 5$ and $C < 5$. A plot of this merit surface is shown in Fig. 3.26.

For this example, the alternative with the highest figure of merit would be the one where $n = 5$ and $C = 5$. The extremum of the merit surface need not be a stationary point. A maximum or minimum found by a differentiation process is only occasionally the merit extremum sought.

The previously posed question of how much change in a design factor is equal in value to a change in capacity can be answered from the merit function. The change in n for a change in C at constant merit M is

$$n = B_1(n)B_2(C)/(4CM)$$

$$\frac{\partial n}{\partial C} = \frac{\Delta n}{\Delta C} = -B_1(n)B_2(C)/(4C^2M)$$

$$\Delta n = -B_1(n)B_2(C)\Delta C/(4C^2M) = -\$\Delta C/(4C^2)$$

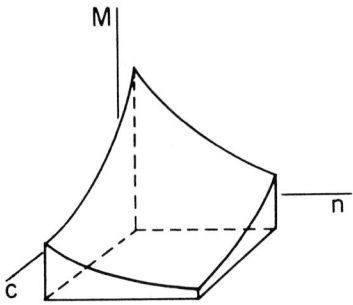

Fig. 3.26. The merit surface M(n, C). (From Mischke, *Computer-Aided Design*, p. 36. Used with permission.)

For an increase in capacity C, the equivalent change in the design factor (the change of equal merit) is given above. The Boolean functions may be replaced by unity or left in, as suits the engineer. The meaning of a zero Boolean function is that you are outside the marketplace and trades are meaningless.

The above function is called a *trade-off* function by some engineers. Since it is deducible from the merit function, in a way the merit function itself is a trade-off function.

Cost (as a negative or reciprocal quantity) makes a natural figure of merit; and the contributions of all other elements may be assessed by thinking in terms of dollars, one variable at a time. Situations exist in which the principal attention is focused on another parameter and cost is subordinate, although not completely out of the picture--it may be considered to be hovering in the background holding a veto power. In aircraft design, a minimum weight wing structure for a specified aerodynamic wing body is a desirable commodity but not at infinite cost to discover the optimal parameters. Cost may hover in the background as a Boolean variable (cost may not exceed available resources) and the figure of merit allowed to increase until the Boolean "chop" is exercised.

Engineering costs are overhead in the production of products, and the optimal framing configuration of structural steel for a building (minimizing the weight of steel present) may be so costly that an experienced designer's judgment in getting within 15 percent of optimal weight may be cheaper than an optimal solution and its cost of discovery.

Let us take the case of a motor-driven pump (cost $1000) with a reliability of 0.9 (which can be improved to a reliability of 0.99 by a single redundancy) and associated valves, piping, and controls for a total of $3000. A designer knows, in considering the development of a single motor-driven pump with a reliability of 0.99, that the market value of this pump cannot exceed $3000. Here is a direct market reading on the dollar worth of reliability: the improvement in reliability from 0.9 to 0.99 is worth only $2000 in the marketplace.

An approach to the creation of a figure of merit function should be clear at this point; however, the responsibility for generating a meaningful figure of merit lies with the designer, using all the inputs available and incorporating the line decisions of management.

3.7 QUALITATIVE PARAMETERS IN FIGURES OF MERIT

How does one construct a figure of merit for the taste of tea, for the beauty of an automobile, for the lines of a ship, for the convenience of use, for consumer eye appeal? The answer is that we have not yet discovered how.

Suppose a qualitative factor--novelty of approach--is to be a factor in a figure of merit. One way to meet the problem is to construct a figure of merit including all quantitative elements of merit and order the solutions according to decreasing merit with the cost of each alternative tabulated alongside. Now the designer, introducing the factor of novelty of approach, must take alternative solutions one by one and decide (with other competent persons) if the difference in novelty between the first and second alternative is worth the difference in cost. Can the product command a price differential in the marketplace to defray the increased cost of fabrication?

The answers to these questions are rooted in the cloudy area of understanding consumer tastes and in the impact of advertising in creating and extending markets. The answers will be the best judgment of those who have a history of creditable assessment of these factors.

If perfect judgment of novelty worth were available, would the final ordering be optimal? Consider a figure of merit with only two factors, X and Y. The merit surface is depicted in Fig. 3.27. The extremum appears as point p

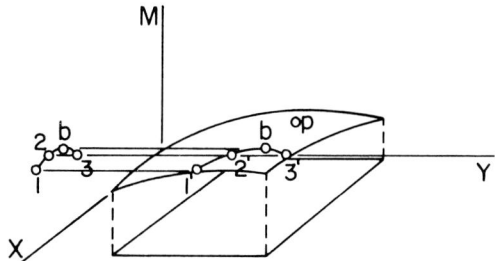

Fig. 3.27. A merit surface, an exploratory path on it, and the projection of the path onto the MX-plane. (From Mischke, *Computer-Aided Design*, p. 38. Used with permission.)

in the figure. Suppose Y eludes us (we feel it qualitative), and the merit function made with X (and an uncontrolled but not random Y) traces the locus 1'2'3' in merit space with the maximum exhibited at b. The designer's merit curve is the projection of 1'2'3' on the XM-plane, i.e., curve 1-2-3, which is not uniquely related to the merit surface. To discuss the Y-merit of alternatives 1, 2, 3 will not help discover the extremum point p or the optimal XY combination. Depending on the geometry of the merit surface, the X corresponding to b may be close to optimal or a long way from it. However, this is the best we can do short of shots in the dark. The quantification of parameter Y may or may not be worth a serious effort. It depends on the problem. The pressure to quantify a qualitative factor is a wholesome one. The rewards of success in the area may be very great.

Our best understanding arises in cases when complete quantification is possible and is implemented with accuracy. Anything less is less satisfactory, but it still may be the best we know how to do.

Ans: see sec 9.4

PROBLEMS

3.1. A right circular cylindrical container is to be fabricated from sheet steel by crimping and soldering seams. Management specifies that it wants the cheapest quart can that can be made from 0.020 in. tin-plated sheet steel stock. Specify a tactical function of merit by identifying all the parameters of which it is a function. How many of these are under the designer's control? Identify the regional and functional constraints.

3.2. A speed limit is to be established for a straight, limited access highway heavily used by commuters. Suggest a candidate for a tactical figure of merit function. If more than one candidate is suggested, debate the different candidates.

3.3. The best straight line is to be passed through some data points. Suggest a tactical figure of merit function. Is there disagreement within the class? Can it be resolved?

3.4. A steel skeleton is to be designed for a large building. What could the tactical figure of merit function be? Is there disagreement among the members of the class? Can it be resolved?

3.5. Most parachutists would agree that there is an appropriate time to pull the rip cord after bailing out of an airplane. Suggest a tactical figure of merit function.

3.6. The ability of a simply supported rectangular cross-section beam to carry a transverse load is proportional to the width of the cross section and to the square of the depth of the cross section. A circular log is to be sawed into a rectangular cross-section timber beam. If the merit of the cross section is taken as proportional to the largest transverse load the beam can carry, express a tactical figure of merit. How many decisions does the designer have? Using your figure of merit, ascertain the proportions of the cross section of the strongest timber that can be sawed from a right circular cylindrical log.

3.7. The ability of a rectangular cross-section, simply supported beam to resist transverse deflection (sagging) is proportional to the width of the cross section and proportional to the cube of the depth of the cross section. A circular log is to be sawed into a rectangular cross-section timber beam. If the merit of the cross section is taken as proportional to the ability to resist transverse deflection, express a tactical figure of merit. Express regional and functional constraints, if any. How many design decisions does the designer have? Use your figure of merit to ascertain the proportions of the stiffest timber that can be sawed from a right circular cylindrical log.

3.8. For the hoist problem of Sec. 3.2, prepare a figure similar to Fig. 3.5 that includes contours for different numbers of wires. Determine if the size associated with the largest figure of merit changes, and if so, the direction of change.

3.9. In the widget problem of Sec. 3.3, the cost of alternative (b) falls to $0.08 per relay inspected. At what probability of encountering a bad relay would an inspection plan be considered?

3.10. What is the expected value of a roll of a die? What is the expected value of the roll of a pair of dice?

3.11. Construct a histogram for the outcomes of a roll of a pair of dice. Use a class interval of unity and center the class intervals on the outcomes.

3.12. The material removed in a machining process in a sample of four was 0.120, 0.115, 0.093, 0.108 lb. What is the unbiased estimator of the mean of all the pieces machined and the unbiased estimator of the standard deviation of all the pieces machined (which numbered 100,000).

3.13. If, in Prob. 3.12, only ten pieces were in the production run and the sample 0.120, 0.115, 0.093, 0.108 was chosen at random without replacement, determine the unbiased estimators of the population mean and population standard deviation.

Figures of Merit

3.14. A continuous variable is distributed uniformly in the interval. The probability density function is given by

$$f(x) = 1 \quad (0 \leq x \leq 1)$$
$$ = 0 \quad \text{(elsewhere)}$$

What is the expected value of x? What is the expected value of x^2, i.e., a value of x chosen at random, then squared? Can you generalize to a formula for the expected value of x^n? What is the variance in x? What is the standard deviation?

3.2. $M_1 = $ greater flow $\left(\dfrac{cars}{mile}\right)\left(\dfrac{cars}{hour}\right)$

$M_2 = $ miles/hr

$M_3 = $ max miles/gallon ; max. passenger mile/gallon.

Garbage in equals garbage out.

CHAPTER 4

ENGINEERING AND THE DIGITAL COMPUTER

4.1 WHAT IS AN ALGORITHM?

An algorithm is a mathematical procedure, a step-by-step process for accomplishing a task. Broadly speaking, an apple pie recipe is an algorithm. In mathematical and computational work the step-by-step process or procedure may be quite technical or complex, but there are many simple ones too. Consider the area of a rectangle and its algebraic formulation, $A = xy$. Figure 4.1 depicts the locus of y as a function of x for a given area A. To find the

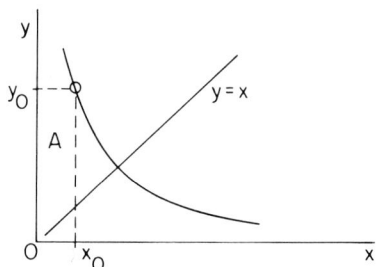

Fig. 4.1. The locus of side lengths x and y of a rectangle with area A.

size of the sides of a square with the same area, we can stand at an arbitrary point with y-coordinate y_0 on the locus of Fig. 4.1 and note that as we move toward the single point where $x = y$, the long side of the rectangle gets shorter and the short side of the rectangle gets longer. If we can invent a scheme that assures progress along the locus and terminates at $x = y$, we may be able to construct an algorithm. Consider an area of 13 square units. Let us place our arbitrary point on the locus at $x_0 = 1$ and $y_0 = 13$. The short side must be made longer and the long side, $y_0 = 13$, must be made shorter. An averaging process involving x_0 and y_0 can be used to accomplish this. Let us form a new side length x_1 by averaging x_0 and y_0:

$$x_1 = 0.5(x_0 + y_0) = 0.5(1 + 13) = 7$$

If one new side is $x_1 = 7$, the other side of necessity must be $y_1 = 13/7 = 1.8671$. Of these new sides, the longer is shorter than the original 13 and the shorter is longer than the original 1, so we are moving toward equality of the sides. A new x_2 would be

$$x_2 = 0.5(x_1 + y_1) = 0.5(7 + 1.8671) = 4.4286$$

Then $y_2 = 13/4.4286 = 2.9355$ and

$x_3 = 0.5(x_2 + y_2) = 0.5(4.4286 + 2.9355) = 3.6821 \quad y_3 = 13/3.6821 = 3.5306$

$x_4 = 0.5(x_3 + y_3) = 0.5(3.6821 + 3.6306) = 3.6064 \quad y_4 = 13/3.6064 = 3.6047$

$x_5 = 0.5(x_4 + y_4) = 0.5(3.6064 + 3.6047) = 3.6056 \quad y_5 = 13/3.6056 = 3.6055$

The length of the side of the square to four significant digits is 3.6056, the final digit 6 being in doubt. While we only rarely need to square rectangles, we often need to extract the square roots of numbers, and we have extracted the square root of 13.

A useful description of the procedure we have followed is depicted in Fig. 4.2. Such a picture is called a *logic flowchart* in computer jargon. It

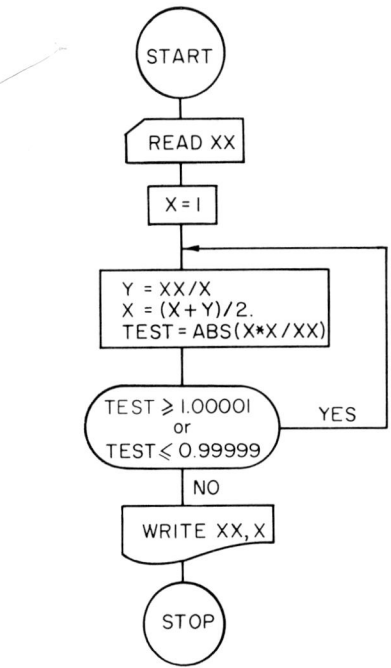

Fig. 4.2. A logic flowchart for an algorithm for finding the length of the side of a square of a specified area.

instructs us in (or recalls for us) the sequence of operations necessary to extract the square root of a positive number in an iterative and simpleminded fashion. The flow sheet contains a test to see if we are done. The longhand method above used the intuitive test of completion, which is the point at which $x_i = y_i$.

We can program the steps to carry out this ritual, using a common pocket scientific calculator, or a current model of a card programmable calculator to actually program a sequence of steps. The preparatory steps are storage.

The Digital Computer   75

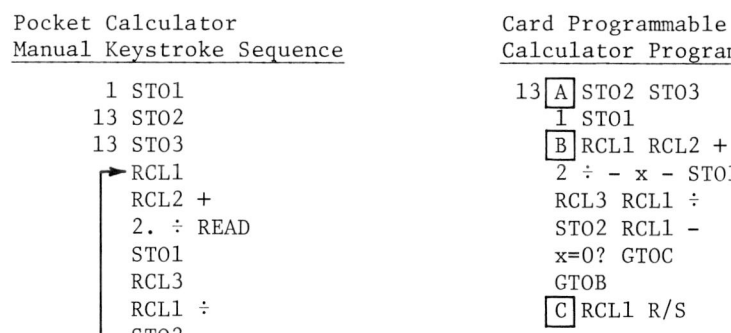

```
        Pocket Calculator                 Card Programmable
    Manual Keystroke Sequence             Calculator Program
              1  STO1                    13 [A] STO2  STO3
             13  STO2                     1     STO1
             13  STO3                    [B] RCL1  RCL2 +
           ┌── RCL1                       2  ÷  -  x  -  STO1
           │   RCL2 +                     RCL3  RCL1 ÷
           │   2.  ÷  READ                STO2  RCL1 -
           │   STO1                       x=0?  GTOC
           │   RCL3                       GTOB
           │   RCL1 ÷                    [C] RCL1  R/S
           └── STO2
```

As you proceed down the sequence from RCL1, remember that your current estimate of the answer ($\sqrt{13}$) is displayed after keystrokes 2.÷. When the illuminated display does not change from cycle to cycle as you depress the STO1 buttons, you have the estimate to the number of significant digits in the display. For example, finding $\sqrt{13}$, using a pocket calculator, the following sequence of estimates was obtained:

    7.00000000
    4.42857143
    3.68202765
    3.60634549
    3.60555136
    3.60555128
    3.60555128

In six iterations, eight digit accuracy was obtained, the last digit having no confirmation.

A FORTRAN program can be written to carry out the steps in this algorithmic iteration process. For those who have had programming it is direct reading. For those who have not had programming in FORTRAN, it is nearly direct reading.

```
         10 READ(5,2,END=99) XX
          2 FORMAT (E15.7)
            X=1
          1 Y=XX/X
            X=(X+Y)/2.
            TEST=ABS(X*X/XX)
            IF(TEST.GE.1.00001.OR.TEST.LE.0.99999)GO TO 1
            WRITE(6,3) XX,X
          3 FORMAT('THE SQUARE ROOT OF ',G15.7' IS EQUAL TO ',G15.7)
            GO TO 10
         99 STOP
            END
    $ENTRY
      1.3E 01
```

The program begins by reading the number whose square root is sought, and assigning it the variable name XX. Then it initializes X with the value of one.

The variable Y is calculated by dividing the number XX by the current best estimate of the square root, namely X. The new estimate of the square root is formed by averaging the values of X and Y and updating X to this value. The variable TEST is a number in the neighborhood of unity when the estimate is close to the square root. The program will go to statement 1 if the test is not within 0.00001 of unity and will continue around this loop until the test is satisfied, whereupon it will write out the number and its square root and go to statement labeled 10 to read another data card to do another problem. If the card reader sees a control card from another job, the program branches to statement labeled 99 and stops.

Thus we see that an algorithm may be presented as an essay, as an algebraic description, as a set of sequential manual steps on a calculator keyboard, or as a FORTRAN program. It is the stratagem that is the algorithm, although many people consider the steps to be the algorithm. Of interest to numerical analysis is convergence. Will this square root ritual converge on the correct answer? Always? If it does, another important question is, How rapidly does it converge? The answer to this question affects the cost of extracting a square root. What is the error when you stop?

Algorithms may be used by those with an intimate knowledge of the anatomy and detail of the method. Algorithms may be automated by machinery (or by the drill and teamwork of a football squad). Algorithms may be built on other algorithms and the opportunity to stand on the shoulders of our ancestors is available. The modern digital computer has the fundamental capability to add; and by various algorithmic processes this capability has been parlayed into the illusion of multiplication, exponentiation, and all that these imply. FORTRAN users stand on an algorithmic mountaintop and are in a position to implement these algorithms as well as their own.

4.2 ENGINEERING DESIGN AND THE DIGITAL COMPUTER

The advent of the computer was a development with far-reaching consequences for engineering design. Computers are used, and have been used since their inception, for analysis, since they calculate with speed and accuracy and without fatigue. The impact of the computer on engineering design has so far been modest, but exploitation of its potential in this field is under way. The development of the time-sharing mode of operation and remote access to large computers will eventually lead to the growth of a computer utility and will provide even small engineering offices access to the largest computers available. The use of these computers by some to improve their competitive position will result in their use by many. Large computers simply attract large problems.

The engineering graduate of today will function in a professional world in which the computer is an integral part of the design process. It is important for new engineers to be aware of what is known to be possible, for they will be called on to use their knowledge, imagination, and creativity to make even more possible.

The employment of the computer in engineering design is a blend of art and intellectual skill. If an experienced designer can look at a required wing-airfoil body and, on the basis of experience, make decisions concerning structural patterns and design a wing that is nearly optimal (against some criterion such as weight), what is the economic worth of knowing precisely the optimal configuration? When that question is answered we know what amount of additional resources can be committed to discovering the optimal parameters. If the original design effort is measured in years of a person's time and a redesign effort (to vary parameters in search of optimality) is measured in months, how many redesigns may be attempted? A half dozen might double design costs without finding the optimum. Can a computer do redesigns quickly and at a lesser cost than manual redesigns? If it can, then whatever the allowable expense in the pursuit of optimality, the computer will win an economic position in the design process.

The history of digital computation devices probably started with the abacus in 600 B.C., with the most progress occurring since 1945. Detailed histories are available in textbooks on computing to which the reader is referred for detail. An important point for the present discussion is the fact that a contemporary computer can add 256,000 eight digit numbers in less than a second and at a cost approximating a dime. The computer in its tasks can outperform a human with regard to speed, accuracy, and endurance. The machine, once taught, can remember how to perform an operation tomorrow or a year from now; and contrary to humans, later performances occur faster and more economically (due to equipment improvements in the interim). The computer can:

--take direction (the unconditional GO TO statement)
--make decision and branch (IF and computed GO TO statements)
--iterate (the DO statement)
--read and write alphanumeric information
--remember (store and recall on command)
--appear to remember specific drills (subprogram capability)
--draw and communicate graphically
--start, stop, or pause

It can do these things with speed, accuracy, and economy. These attributes of an apprentice are useful, and it is convenient for the engineering student to view the digital computer as just that--an apprentice, to be instructed and employed for tasks that can be delegated to it.

This chapter will not instruct in the grammar of a computer language but rather will speak to the literary aspects of a problem-oriented language. Students without an introductory knowledge of a scientific programming language such as Basic, FORTRAN or PL1 can acquire this knowledge by studying any of the textbooks designed for this purpose. For this chapter it is assumed that the reader can program in one of these languages. This text will use FORTRAN IV to illustrate the points being made.

Languages allow the user to make declarations useful in solving problems. In FORTRAN these include DIMENSION, FORMAT, COMMON, REAL, INTEGER, FUNCTION, SUBROUTINE, EXTERNAL, DOUBLE PRECISION, RETURN, END, statement number (label). Languages allow control statements. In FORTRAN these include unconditional GO TO, conditional GO TO, arithmetic IF, logical IF, iterative DO, CONTINUE, and CALL. Languages allow input/output statements such as READ, WRITE, PRINT, and the use of a literal. Languages allow arithmetic assignment statements such as

```
Y = 3.*A*A - 16.*B**0.3
```

Such statements may utilize internally provided functions such as SIN(X), COS(X), ATAN(X), EXP(X), SQRT(X), TANH(X), and ABS(X), to name a few. Their display in an arithmetic assignment statement causes the function to be replaced by its arithmetic equivalent. For instance,

```
Y = 4.*SIN(X)
```

will have SIN(X) replaced by its numerical equivalent, multiplied by 4, and the result stored in a memory location assigned to variable Y.

Use of the digital computer in engineering design is more than desirable; it is essential. The scope of engineering problems demands more than a programming knowledge for a successful resolution. The ability to organize the programming effort to permit both group and individual contribution is required. The ability to stand on the shoulders of those who have gone before is essential to the economic use of the computer.

All design problems are properly optimization problems. Many times we choose the first satisfactory (suitable, feasible, and acceptable) solution because the cost of generating other alternatives is too high. The computer has the potential of reducing the costs of generating alternatives.

The engineer who says, "Design problems are optimization problems, but we are usually forced to run with the first satisfactory alternative," may be talking one game and playing another. Engineering design problems demanding optimization can catch engineers short on two accounts: they aren't sufficiently familiar (1) with optimization methodology to apply it effectively and (2) with organizing a computer effort on a large scale with multiple individual input and contribution.

4.3 IOWA CADET: A COMPUTER AID TO DESIGN

Let us examine how engineers can involve the computer without changing their *modi operandi*. If a designer

1. defines the problem in quantitative terms
2. decides how to recognize a satisfactory solution
3. decides how to recognize solution merit so as to be able to distinguish between two satisfactory alternatives and choose the better
4. generates alternatives
5. through analysis establishes alternative merit
6. discovers the solution with highest merit using some search stratagem
7. then makes design decisions and implements them

The designer is naturally proceeding in a manner completely compatible with the capabilities of the digital computer.

A Computer-Augmented Design Engineering Technique (IOWA CADET) embodies these essential ideas:

--Meaningful competitive design is best accomplished when timely feedback is available from the marketplace. Many situations exist in which the time available or the enormity of resources committed will not allow feedback from the marketplace to influence design decisions properly. Under these circumstances the influence of the marketplace on design must be simulated; the vehicle of this simulation is the figure of merit.
--A *figure of merit* is simply a number whose magnitude constitutes an index to the merit or desirability of an alternative solution to a problem. Alternate names are criterion function, utility function, objective function.
--The designer seeks to maximize the figure of merit subject to the many constraints placed on the solution by nature; by law; by social, economic, and political factors; by geometric and material considerations.
--A designer who proceeds as outlined in 1-7 above is progressing in a manner completely compatible with the capabilities of the digital computer and the IOWA CADET algorithm.
--Computer time and engineering and programming time associated with a problem must be charged to design costs. If programming is carried out in the

usual fashion of solving only the specific problem at hand, the number of
projects in which computer assistance can be economically justified will be
few. However, if programming is carried out in such a way as to be useful
in subsequent tasks, the unit costs decline and the number of applications
of computer assistance will increase.
--At the project engineering level it will be generally fruitless to wait for
proprietary programs to be developed by talented organizations (e.g., computer manufacturers) because the number of general problems that can be
solved by canned proprietary programs is, at best, limited to a small fraction of existing problems. It will be necessary for the individual engineer
or engineering groups to develop, in terms of their own knowledge of common
or recurring problems, their own computer capabilities (programs).
--Documentation of any capability of substantial size to be used by many different persons must be carefully structured so as *to invite use*. Only in
this way will the capability grow by local contribution toward greater utility, eventually becoming custom-made to the local needs (and no outsider is
likely to do better).

The figure of merit, the means whereby the designer compares two alternatives and with high precision chooses the better, can be parlayed into an optimization stratagem. Mathematical programming optimization algorithms do
precisely this. Simple attributes of systems are often candidates for figures
of merit. Examples include design attributes such as design factor, cost, reliability, time interval, magnitude of displacement, and mass or weight. When
multiple factors contribute to the decision, the figure of merit is a function
of the multiple factors with the appropriate trade-offs quantitatively expressed. If you cannot arrive at a figure of merit, you cannot optimize, with
or without the computer.

The basic thrust in computer programming for engineering problems should
be to keep programming *general* by separating the strategy from the tactics.
If one desires to use a stratagem such as variable secant, Newton's method,
interval-halving, etc., to find the zero place (root) of a function, the technique is not to program the problem in one piece but to separate the operator
from the operand. Specifically, one can write a SUBROUTINE subprogram that
will implement the method with *any function of the proper class* and write a
FUNCTION subprogram to inform the subroutine exactly what mathematical function is to be operated on at this particular time. The programming of the
strategy is the more difficult task and this should be done only *once* and
thereafter used by the engineer and others to find the zero place of other
mathematical functions. The programming of the operand is often as simple as
depicted in Fig. 4.3, where FUNCTION F(X) is coded as

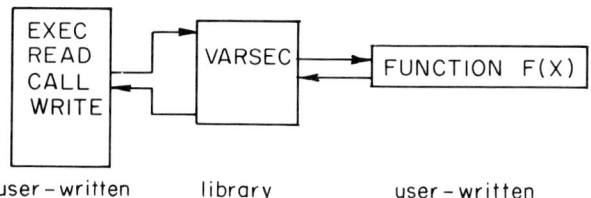

Fig. 4.3. Programming by separating the algorithm from the function to
be operated on. The subroutine VARSEC finds the root of a function of
one independent variable; the subprogram FUNCTION F(X) declares the mathematical function whose root is sought.

```
FUNCTION F(X)
F = (expression in x)
RETURN
END
```

This separation of the stratagem from details of a specific case is applicable to most mathematical operations and, indeed, to the optimization process itself. Figure 4.4 shows an optimization stratagem program, OPT: The engineer writes the executive program that calls OPT, which in turn calls and manipulates user-written subroutine MERIT and reports back to the executive program the optimal parameters of the mathematical function whose extremum is sought.

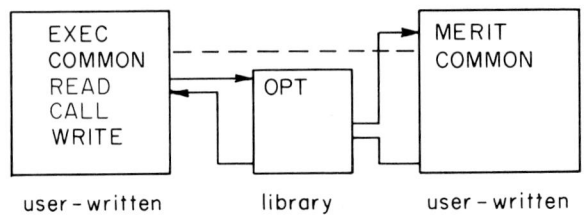

Fig. 4.4. Programming by separating the algorithm from the function to be operated on. The subroutine OPT finds the largest ordinate of a function; the subprogram MERIT declares the mathematical function whose extremum is sought.

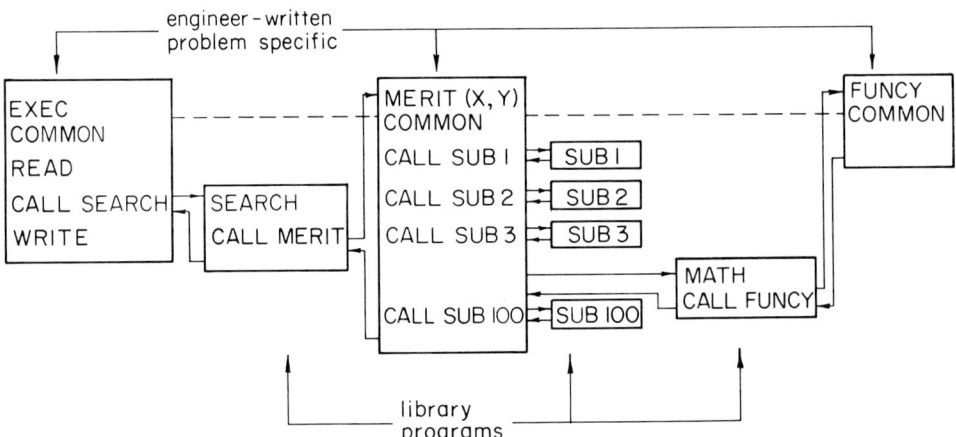

Fig. 4.5. General program organization of the IOWA CADET algorithm for computer-augmented design.

We can see a structure emerging as the result of the posture of separating the strategic programming from the tactical detail. Figure 4.5 depicts an executive program whose purpose is to READ in problem-specific information, call SEARCH, and WRITE the results of the search that happen to be problem-specific. SEARCH calls and manipulates another user-written, problem-specific routine, MERIT. This routine is really an executive program at a lower level whose purpose is to generate the figure of merit. For this purpose it calls engineering analysis routines that carry out the engineering analysis rituals, utilizing the specific information for the case at hand. MERIT also may have mathematical chores to accomplish and consequently it calls MATH, a mathematical operation strategy (performed in this case on user-written EQUAT). The structural form is that of a multilayer laminate in which engineer-written, problem-specific programs call and manipulate library programs the engineer has probably never seen. These in turn call and manipulate programs the engineer has written (which are problem-specific). The depth of the lamination stack is directly related to the complexity of the problem. To use this structure expeditiously, it is necessary to adopt an effective documentation scheme.

4.4 DOCUMENTATION

The following questions are foremost in the mind of the potential user of library routines and must be answered to user satisfaction in any effective documentation scheme.

--What is the number of arguments in the call list?
--What is the order of those arguments?
--What is the mode of those arguments?   *real or integer*
--What arguments are defined prior to the call; which are defined as a result of the call? *what do you have to identify.*
--Are additional subprograms required?
--What are the definitions of the arguments in the call list?
--What declarations are required in the calling program?
--Where can the user find the analysis on which the subroutine or function subprogram is based, including diagrams and additional information?
--What are the units of measurement of the call list arguments?
--Who wrote the subprogram and has test results?
--Have any subprogram names or COMMON labels been preempted?
--How much memory is utilized by the subprogram in core?

The conventions adopted for documentation in the IOWA CADET algorithm are a response to the user-need as expressed by the preceding 12 questions.

All subprogram arguments are presented in a form similar to

ME0000(A1,A2,A3,A4,B1,B2,B3)

wherein the arguments defined in the calling program are denoted A1, A2, A3, A4, and the call list arguments defined as a result of the call (and consequently returned to the calling program) are denoted B1, B2, B3. If arguments in the list are of integer mode, the call list arguments displayed in the documentation are

ME0000(A1,I2,A3,A4,B1,B2,J3)

wherein the second argument is a variable in integer mode, defined in the calling program prior to the call; and the last argument in the call list is a variable in integer mode, defined in the calling program as a result of the call. Thus the symbol I2 is used for an integer variable that must be defined in the calling program prior to the call, and J3 is used for the third variable defined as a result of the call. If the subprogram is considered to be a function in the mathematical sense, independent variables are documented as A's (real) and I's (integer) and dependent variables are documented as B's (real) and J's (integer).

```
                                                            Page 0000

  Descriptive short-title of subprogram
  ME0000(A1,A2,I3, A4, B1, J2)
                                                            Author

       ⎡Sentence or paragraph describing what the subprogram ⎤
       ⎢does.                                               ⎥
       ⎣                                                    ⎦

       ⎡Paragraph, which is more specific than that above,  ⎤
       ⎢identifying the independent and dependent information.⎥
       ⎣                                                    ⎦

       ⎡Paragraph reference source of the theory.           ⎤
       ⎢                                                    ⎥
       ⎣                                                    ⎦

       ⎡Paragraph noting special features or warnings.      ⎤
       ⎢                                                    ⎥
       ⎣                                                    ⎦

  CALLING PROGRAM REQUIREMENTS

       ⎡Listing of declarations necessary in the calling program.⎤
       ⎢                                                    ⎥
       ⎣                                                    ⎦

       ⎡Listing of names and arguments of any subprograms that must⎤
       ⎢be user-written and are necessary for the use of the subprogram.⎥
       ⎢Names of other documented subprograms called in turn by this⎥
       ⎣subprogram.                                         ⎦

  CALL LIST ARGUMENTS

       List of all dummy variables, meaning, units of measurement.

  PREEMPTED NAMES

       Listing of SUBROUTINE and FUNCTION names or COMMON labels used by
       the subprogram and consequently not available to the user.

  SIZE

       Memory utilization citing compiler(s).
```

Fig. 4.6. Usual organization of documentation page of a CADET program.

In general, subroutines will print out information (1) in the case of an erroneous call, (2) if a convergence monitor is requested by the user, and (3) if the objective of the subprogram is to create tabular information. An erroneous call will trigger error messages of the following form, indicating a *logic error* on the part of the user.

The Digital Computer 83

```
*****ERROR MESSAGE SUBROUTINE ME0000*****
     VALUE OF I2,   51, OUTSIDE ALLOWABLE RANGE 1 THRU 50
```

The message identifies the offended subprogram, the documentation name of the offending variable, the numerical value of the offending variable, and some indication of the nature of the offense so that the user need not necessarily refer to the documentation. Some subroutines will return an error variable, commonly at the end of the call list.

Subroutine documentation is purposely in the form that encourages users to select their own meaningful names for the variables.

The physical units of measurement of all call list arguments are expressed in the documentation in terms of fundamental dimensions force F, length L, time T, temperature θ, and charge Q if any consistent set of units may be used. If there are internal constants that limit the units to a particular set, then explicit units will be expressed (such as in., ft, lbf, s).

The format of the documentation page is the subject of Fig. 4.6. Examples of documentation pages are found in Appendix A.

4.5 ERROR MESSAGING AND TESTS

The FORTRAN compilers (particularly WATFOR and WATFIV) have excellent diagnostics of a grammatical and syntactical nature. They cannot detect logical errors because very little problem structure is imposed on the FORTRAN user. However, the laminar structure imposed on the engineer faced with a design (optimization) problem allows the detection of symptoms in many cases of logic error on the part of the user, and sufficient information can be given in such cases to point the user toward the error (and away from the library subprogram).

Subprograms should WRITE in only three circumstances, since the user loses control over what is written and where it is written and displayed. These circumstances are:

1. usage error by user (written on next available line of output)
2. convergence or display monitor (written on separate page so user can detect and discard at will)
3. objective of subprogram is to create tabular information display

The error message warning to the user of logic error performs most of the following duties:

--declares the offended subprogram by name
--declares which argument is offending by giving documentation name
--declares the numerical value of the offending argument
--declares why the value is offensive
--declares circumstances leading to subsequent abort or suspension of subprogram operation
--declares circumstances of subprogram frustration

Most important, the error message points *away* from the subprogram and *toward* the user's logical error. As an example,

```
*****ERROR SUBROUTINE ME0001*****
     VALUE OF A1, 0.130241E02, IS GREATER THAN UNITY
```

There are many processes in which the engineer desires to see evidence of the fulfillment of the evolution, e.g., attainment of a maximum or a minimum, the vanishing of a residual, an approach to a signal matrix, etc. This purpose is served by a monitor (or convergence monitor) displayed on a separate page (or not) at the user's discretion. A monitor from a successful run can easily be discarded. It builds confidence in the solution integrity but has no necessary place in the formal output or in its interpretation.

In statistical work, when a maximum-likelihood-parameter estimate for a parent distribution is determined by a maximization process, the user may be interested in inspecting (or plotting) the hazard function, the cumulative density function, the density function, or the reliability. A library routine that accepts the likelihood estimates and return points on all these curves can use a monitor with a tabular display so the graph from the plotter can be compared with data the user would otherwise never see.

When a program must be written by the engineer to fill a gap in the library, it is important that a test be carried out by a program that

--exercises every path in the logic flowchart with correct answers
--exercises every error message singly and in multiple
--verifies the return arguments now defined in the calling program as a result of the call

This test program is filed in the library capability. If the program just written is to be lightly used, it may be left as is, awkward programming and all, for subsequent use. If the program has a high usage potential it may be

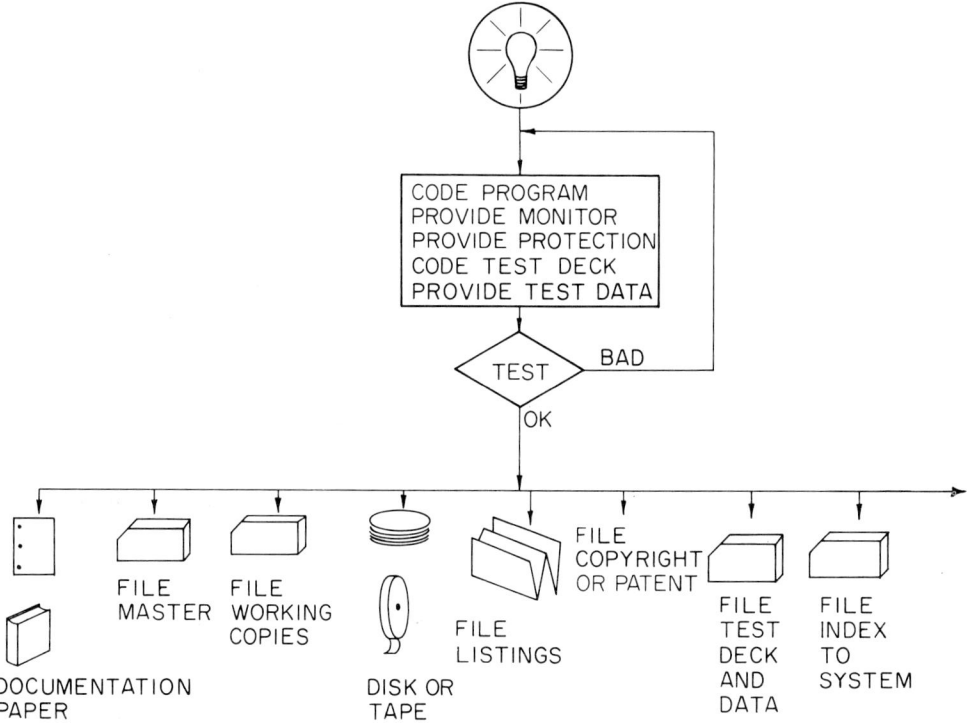

Fig. 4.7. Tasks associated with adding a program to a multiuser library.

given, with the test deck and a valid output to a professional technical programmer. The programmer may be permitted to change programming to conserve storage, expedite execution, improve numerical accuracy, improve rates of convergence, etc., as long as the improvements do not alter the test deck results. Engineers write programs for their own (and others') use, and their confidence is largely in themselves, their knowledge, and the confirmations resulting from the test deck runs. After surrendering their programs to technical programmers, they may see their methodology changed beyond confident comprehension; they now rely on their test programs to confirm their confidence in the modified programs. If the answers are the same (or numerically more precise), they can be reasonably confident that the technical programmers have made only tactical changes and left the strategy intact. They can now use the modified programs with the same confidence as they used the originals.

If a suggestion is made later that a subprogram could be made more useful with some modification, a technical programmer can add data cards that exercise the new, additional capability to the test deck, and from output triggered by the old data cards the programmer can see that none of the former capability has been disturbed.

This procedure allows a useful and effective division of labor between engineers and programmers. They each do what they know best, and the responsible engineer has the test deck as a final check against any esoteric methodology practiced by the programmer in the interests of economy of execution time and storage and in the pursuit of numerical accuracy. Figure 4.7 depicts the steps necessary to add a program to an existing design library.

After a program is successfully tested, the program cards should be sequenced and working copies made and filed. The master deck is so labeled and carefully filed for preservation as the primary source of code. The listing of the successful run is carefully read, dated, declared valid for the compiler used, and filed in its folder. The test deck is labeled and filed with the data. A documentation paper is prepared and inserted in the documentation volume. The working copies of the deck are lent to users and are source decks

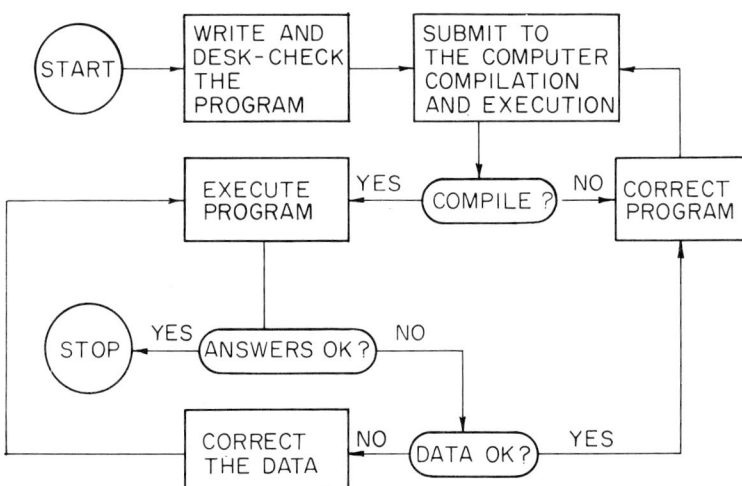

Fig. 4.8. Universal flowchart for creating a successful computer program.

for the computation center, which places the programs on disk or tape. Copyright or patent possibilities should not be overlooked as the legal status of computer software seems volatile. The minimum requirement is that you file the deck and listing of your first run and your successful run. Your bound notebook records the chronology of the concept and its embodiment in code. Figure 4.8 depicts a universal flowchart for creating a successful program.

4.6 SOME USEFUL NUMERICAL METHODS

A recurring problem is that of finding the zero place (root) of a function. This problem is important in itself and also because other problems can be transformed into root-finding problems. One basic approach was offered by Sir Isaac Newton. Consider a function such as that depicted in Fig. 4.9 in the neighborhood of a root. Consider also an ordinate located at x_0 with a

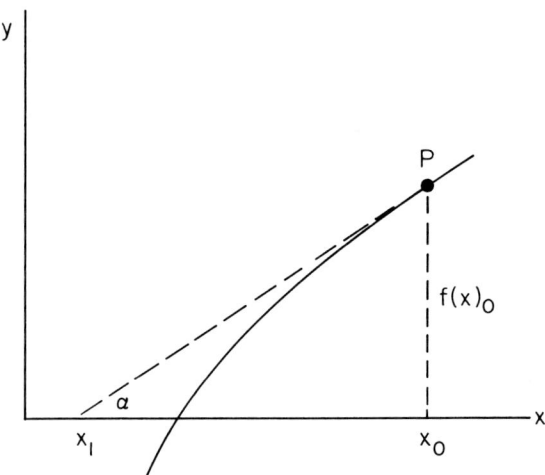

Fig. 4.9. A function and its tangent at point P.

magnitude of $f(x_0)$. A tangent to the curve is constructed at point P and a right triangle is formed by the tangent, ordinate, and x-axis. The point where the tangent crosses the x-axis is a better estimate of the root than x_0 and is denoted as x_1. The tangent of the angle α between the x-axis and the tangent is numerically equal to the mathematical slope of the tangent line, which is numerically the first derivative of the function evaluated at x_0, namely $f'(x_0)$:

$$\tan \alpha = f(x_0)/(x_0 - x_1) = f'(x_0)$$

Solving for the better estimate of the root x_1, we obtain

$$x_1 = x_0 - f(x_0)/f'(x_0)$$

An iterative process can be inaugurated, using the preceding relation on a repetitive basis.

$$x_{i+1} = [x - f(x)/f'(x)]_i \tag{4.1}$$

As an example of its use, let us determine the zero place of $\ln x$. Now

The Digital Computer

$$f(x) = \ln x \qquad f'(x) = 1/x$$

and the iteration formula becomes

$$x_{i+1} = \left[x - \frac{\ln x}{1/x} \right]_i = [x(1 - \ln x)]_i$$

Choosing $x_0 = 0.5$, the successive estimates of the roots are 0.500000, 0.846574, 0.987577, 0.999923, and 1.000000. The pocket calculator programs establishing these estimates are:

Pocket Calculator Manual Keystroke Sequence	Pocket Caculator Program
1 STO2	0.5 [A] STO1 LN CHS
0.5 STO1	STO2 -x-
►LN CHS	RCL1 - x=0?
RCL2 +	GTO3 RCL2
RCL1 ×	GTOA
└STO1 (read)	[3] RCL2 R/S

In this form the method is denoted the Newton-Raphson method. In numerical analysis studies you will investigate the conditions for convergence. The method does not perform well when the slope of the functional locus between point P and the root is very small; the method fails if at any place along the interval between P and the root, the slope vanishes. Questions concerning rate of convergence and error are also addressed. The method requires the ability to differentiate the function; the process may be easy, difficult, or impossible depending on the function and your ability.

As an example of a problem that involves some difficulty in differentiation, determine the root of $x^x - 5 = 0$. We set

$$f(x) = x^x - 5$$

$$f'(x) = x^x(1 + \ln x)$$

The iteration formula becomes

$$x_{i+1} = \left[x - \frac{x^x - 5}{x^x(1 + \ln x)} \right]_i$$

If $x_0 = 3$, the successive estimates of the root are 3.000000, 2.611736, 2.309411, 2.158945, 2.130242, 2.129373, and 2.129372. The programs establishing these estimates are:

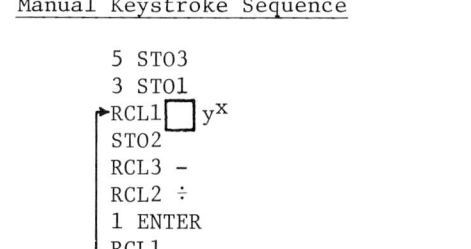

Pocket Calculator Manual Keystroke Sequence	Pocket Calculator Program
5 STO3	3 [A] STO1 RCL1
3 STO1	y^x 5 - RCL1
►RCL1 ☐ y^x	RCL1 y^x ÷
STO2	RCL1 LN 1 +
RCL3 -	÷ CHS RCL1 +
RCL2 ÷	-x- STO2
1 ENTER	RCL1 - x=0?
RCL1	GTO3 RCL2 GTOA

Pocket Calculator Manual
Keystroke Sequence (cont.)

Pocket Calculator
Program (cont.)

```
⎡ LN + ÷
⎢ CHS
⎢ RCL1 +
⎣ STO1 (read)
```

$\boxed{3}$ RCL2 R/S

As another example we will find the zero place of $x^2 - C$.

$$f(x) = x^2 - C$$
$$f'(x) = 2x$$

and the iteration formula becomes

$$x_{i+1} = \left[x - \frac{x^2 - C}{2x} \right]_i = \frac{1}{2}\left[x + \frac{C}{x} \right]_i$$

This iteration formula is the same as that evolved in our heuristic approach to the square root in Sec. 4.1.

The next estimate of a root is the arithmetic average of the length of one side of a rectangle, x_i, and the length of the other side, C/x_i, where C is the area of the rectangle. This algebraic-calculus approach of Newton-Raphson has considerably more power.

A heuristic method for extracting the fifth root is not at all clear, but the Newton-Raphson method responds routinely. Let us find the zero place of $x^n - C$.

$$f(x) = x^n - C$$
$$f'(x) = nx^{n-1}$$

and the iteration formula becomes

$$x_{i+1} = \left[x - \frac{x^n - C}{nx^{n-1}} \right]_i = \frac{1}{n}\left[(n-1)x + \frac{C}{x^{n-1}} \right]_i$$

If $n = 5$, $C = 10$, we are seeking the fifth root of 10 and the iteration formula becomes

$$x_{i+1} = \frac{1}{5}\left[4x + \frac{10}{x^4} \right]_i$$

Taking an initial estimate of $x_0 = 2$, the successive estimates of the root are 2.000000, 1.725000, 1.605878, 1.585435, 1.584894, and 1.584893. The calculator programs establishing these estimates are:

Pocket Calculator
Manual Keystroke Sequence

```
    2 STO2
   10 STO1
 ┌▶ RCL1
 │  RCL2
 │  x²x²÷
```

Pocket Calculator
Program

```
2 [A] STO1 4 ×
10 ENTER RCL1
x²x² ÷ + 5 ÷
-x- STO2 RCL1 -
x=0? GTO3 RCL2
```

The Digital Computer

```
Pocket Calculator Manual        Pocket Calculator
Keystroke Sequence (cont.)      Program (cont.)

   ┌ RCL2                         GTOA
   │ 4 × +                        ③ RCL2 R/S
   │ 5 ÷
   └ STO2 (read)
```

To repeat, the method of Newton-Raphson is simple but it does require the ability to differentiate the function. The CADET subroutine VARSEC approximates the tangent (slope) with a secant and thereby avoids the necessity of the user having to explicitly differentiate the function.

The Newton-Raphson method just described for the successive approximation of roots is called a *first order* method, since it uses a single attribute of the curve at point P (the tangent) to extrapolate to the root. Is there a way to "bend" the tangent in the same sense as the curve and obtain an even better approximation to the root for each iteration? If a function is easily twice differentiable, the *second order* Newton-Raphson method may be employed. Figure 4.10 shows a functional locus in the neighborhood of a root, the estimate of a tangent, the true root, and the intermediate estimate of a parabolic extrapolation from point P. Expand a Taylor series about $(x + \delta)$.

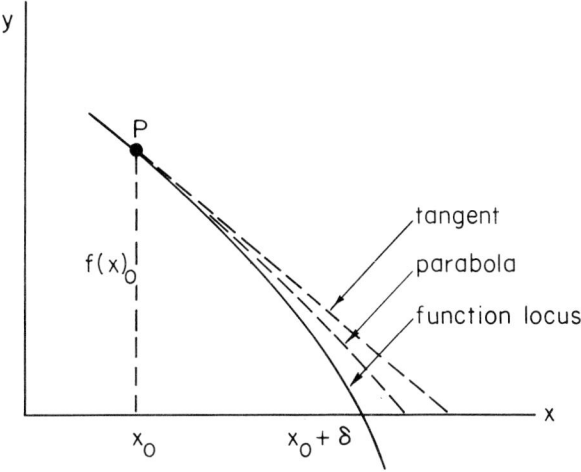

Fig. 4.10. A function, its tangent projection to a root estimate, and a parabolic projection to a root estimate.

$$f(x + \delta) = f(x) + \delta f'(x) + (1/2)\delta^2 f''(x) + \ldots$$

Let $(x + \delta)$ be the exact root; then $f(x + \delta) = 0$

$$0 = f(x) + \delta f'(x) + (1/2)\delta^2 f''(x) + \ldots$$

Neglecting higher order terms and rewriting, we obtain

$$0 = f(x) + \delta[f'(x) + (\delta/2)f''(x)]$$

The first order method approximates δ with $-f(x)/f'(x)$; to avoid the quadratic in δ (with the attendant two roots and the problem of which one is significant), we will approximate the δ within the brackets to $-f(x)/f'(x)$ and write

$$0 = f(x) + \delta\left[f'(x) - \frac{1}{2}\frac{f(x)f''(x)}{f'(x)}\right]$$

The solution for δ results in

$$\delta \approx \frac{-f(x)}{f'(x) - \frac{1}{2}\frac{f(x)f''(x)}{f'(x)}}$$

Whence the iteration formula for the Newton-Raphson second order method is

$$x_{i+1} = x_i + \delta_i = \left[x - \frac{f(x)}{f'(x) - \frac{1}{2}\frac{f(x)f''(x)}{f'(x)}}\right]_i \quad (4.2)$$

Let us apply this second order method to the problem of determining the fifth root of 10.

$f(x) = x^5 - 10$
$f'(x) = 5x^4$
$f''(x) = 20x^3$

General
Card Programmable
Calculator Program
_____

x_0 [A] STO1 GSBC RCL1 GSBE ×
RCL1 GSBD ÷ 2 ÷ CHS
RCL1 GSBD + 1/x RCL1
GSBC × CHS RCL1 + -x-
STO2 RCL1 - x=0? GTO3
RCL2 GTO A [3] RCL2 R/S
[C] f(x) RTN ⎱
[D] f'(x) RTN ⎬ Problem specific
[E] f''(x) RTN ⎱ user programmed

The iteration formula becomes

$$x_{i+1} = \left[x - \frac{x^5 - 10}{5x^4 - \frac{2(x^5 - 10)}{x}}\right]_i$$

Using $x_0 = 2$ as before, the successive estimates of the fifth root of 10 are 2.000000, 1.620690, 1.584928, and 1.584893. The pocket calculator programs establishing these estimates are:

Pocket Calculator
Manual Keystroke Sequence

    5 STO1
    10 STO2
    2 STO3
 ┌─ RCL1 □ y^x
 │ RCL2 -
 │ STO4
 │ 2 ×
 │ RCL3 ÷
 │ CHS

Pocket Calculator
Program

2. [A] STO1 5 y^x 10 -
    STO3 2 × RCL1 ÷
    CHS RCL1 $x^2 x^2$ 5 ×
    + 1/x RCL3 × CHS
    RCL1 + -x- STO2
    RCL1 - x=0?
    GTO3 RCL2 GTOA
    [3] RCL2 R/S

The Digital Computer

                Pocket Calculator Manual   Pocket Calculator
                Keystroke Sequence (cont.)   Program (cont.)

                    ⎡ RCL3 $x^2 x^2$
                    ⎢ 5 × + 1/x
                    ⎢ RCL4 ×
                    ⎢ CHS
                    ⎢ RCL3 +
                    ⎣ STO3 (read)

The method is more rapidly convergent than the first order method; however the complication of the iteration formula increases, and for manual execution, the first order method may well be preferable.

The zero place of $\ln x$ can likewise be found, using the second order method.

$f(x) = \ln x$

$f'(x) = 1/x$

$f''(x) = -1/x^2$

The iteration formula becomes

$$x_{i+1} = \left[x \left(1 - \frac{\ln x}{1 + 0.5 \ln x} \right) \right]_i$$

Successive estimates of the root are, beginning with $x_0 = 0.5$, 0.500000, 1.030394, 0.999998, and 1.000000. The programs establishing these estimates are:

                Pocket Calculator   Pocket Calculator
                Manual Keystroke Sequence   Program

                    0.5 STO1   0.5 [A] STO1 LN
                   ⎡ LN   RCL1 LN 0.5 ×
                   ⎢ RCL1 LN   1 + ÷ CHS 1 +
                   ⎢ 0.5 ×   RCL1 × -x-
                   ⎢ 1 + ÷   STO2 RCL1 -
                   ⎢ CHS   x=0? GTO3
                   ⎢ 1 +   RCL2 GTOA
                   ⎢ RCL1 ×   [3] RCL2 R/S
                   ⎣ STO1 (read)

Some class problems comparing the time necessary to determine a root to (say) six significant figures will be instructive on preferences for manual root finding.

Integration is a recurring engineering problem and numerical methods are used when necessary, because of

--digital computer usage
--inability to integrate the function any other way
--having data only and no knowledge of the function

A method suggested by Thomas Simpson and bearing his name consists of placing *equally* spaced ordinates in the integration interval and passing parabolic

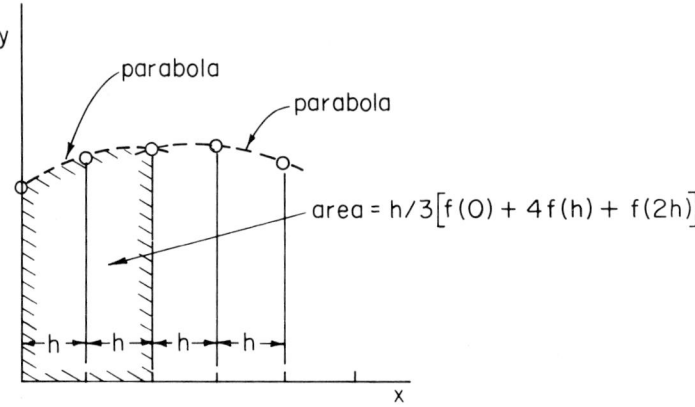

Fig. 4.11. Equally spaced ordinates of and parabolic approximations to the function.

curves through each contiguous set of three ordinates. Figure 4.11 shows the parabolic approximation, the true curve, and the ordinate spacing h. Modules of this sort are repeated as often as desired within the integration interval. The area under the parabolic curve is

$$A = \int_0^{2h} (a + bx + cx^2)\, dx = \left[ax + \frac{bx^2}{2} + \frac{cx^3}{3} \right]_0^{2h}$$

$$= 2ah + 2bh^2 + (8/3)ch^3 \tag{4.3}$$

The constants a, b, and c can be evaluated by Cramer's rule from

$$f(0) = a + 0 + 0$$

$$f(h) = a + bh + ch^2$$

$$f(2h) = a + 2bh + 4bh^2$$

Explicit expressions for a, b, and c are

$$a = f(0)$$

$$b = \frac{4f(h) - f(2h) - 3f(0)}{2h}$$

$$c = \frac{f(2h) - 2f(h) + f(0)}{2h^2}$$

Substituting a, b, and c into Eq. (4.3) yields

$$\text{area} = (h/3)[f(0) + 4f(h) + f(2h)]$$

For successive sets of three points in the integration interval we add

$$\begin{aligned}
\text{area} = &(h/3)[f(0) + 4f(h) + f(2h) \qquad\qquad\qquad\qquad\qquad] + \\
&(h/3)[\qquad\qquad\quad + f(2h) + 4f(3h) + f(4h) \qquad\qquad\qquad] + \\
&(h/3)[\qquad\qquad\qquad\qquad\qquad\qquad + f(4h) + 4f(5h) + f(6h)] +
\end{aligned}$$

$$\text{area} = (h/3)[f(0) + 4f(h) + 2f(2h) + 4f(3h) + 2f(4h) + \ldots]$$

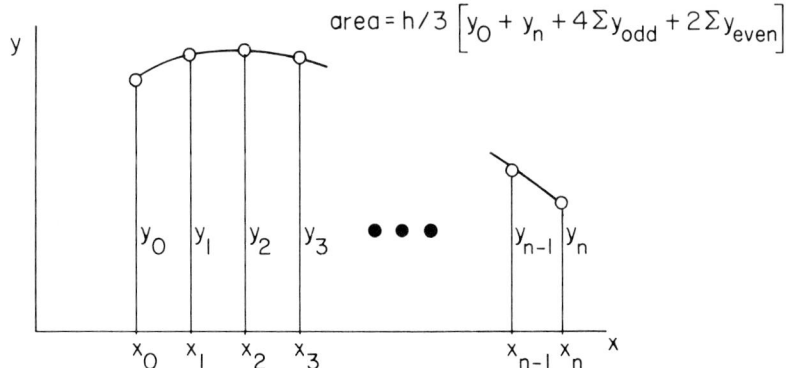

Fig. 4.12. Ordinates evaluated in Simpson's rule integration.

More simply expressed in terms of ordinates, we have

$$\text{area} = (h/3)(y_0 + 4y_1 + 2y_2 + 4y_3 + 2y_4 + \ldots + 4y_{n-1} + y_n)$$

This equation may be simplified by assembling the odd and even subscripted ordinates:

$$\text{area} = (h/3)(y_0 + y_n + 4\Sigma y_{odd} + 2\Sigma y_{even}) \tag{4.4}$$

where n is an even number. Simpson's rule requires an even number of intervals and an odd number of ordinates of the curve, uniformly spaced (see Fig. 4.12). As an example of the application of Simpson's rule in a domain in which we can check our answer, consider evaluating the integral

$$\int_0^\pi \sin x \, dx$$

choosing seven ordinates, we write

x	sin x	Extension
0	$0.000000 \times 1 =$	0.000000
30	$0.500000 \times 4 =$	2.000000
60	$0.866025 \times 2 =$	1.732051
90	$1.000000 \times 4 =$	4.000000
120	$0.866025 \times 2 =$	1.732051
150	$0.500000 \times 4 =$	2.000000
180	$0.000000 \times 1 =$	0.000000
		11.464102

The Simpson rule area is

$$\text{area} = \frac{h}{3}(11.464102) = \frac{\pi/6}{3}(11.464102) = 2.000863$$

The actual area has the magnitude

$$\int_0^\pi \sin x \, dx = -[\cos x]_0^\pi = -1(-1 - 1) = 2$$

As impressive in precision as is Simpson's rule with just seven ordinates, good accuracy must not be assumed for it is not assured. Consider the integral

$$\int_0^\pi \sin 9x\, dx$$

Again choosing seven ordinates, we write

x	sin 9x			Extension
0	0	×	1 =	0
30	-1	×	4 =	-4
60	0	×	2 =	0
90	1	×	4 =	4
120	0	×	2 =	0
150	-1	×	4 =	-4
180	0	×	1 =	0
				-4

The value of the integral by Simpson's rule is

$$\int_0^\pi \sin 9x\, dx = \frac{h}{3}(-4) = \frac{\pi/6}{3}(-4) = -0.698132$$

Formal quadrature gives

$$\int_0^\pi \sin 9x\, dx = -\frac{1}{9}[\cos 9x]_0^\pi = -\frac{1}{9}(-1 - 1) = \frac{2}{9} = 0.222222$$

The lesson here is that the curve must be "smooth enough" so that the parabolic approximations are "excellent." A smooth curve is accomplished by increasing the number of ordinates in the interval until the real curve and the parabolic segments are practically congruent. It cannot be done without upper bound, since the number of calculations increases. Each calculation is approximate due to computer truncation error. As a practical matter, the accuracy will first increase with increasing numbers of ordinates, then decrease. When working on problems that cannot be checked, knowledge of the properties of numerical analysis is vital. Such a course is part of your mathematical preparation in engineering.

As an example of an integral to be evaluated with no knowledge of an alternative route to the answer in closed form, consider

$$\int_0^1 x^x\, dx$$

Choosing nine ordinates in the integration interval, we write

x	x^x		Extension
0	1.000000 × 1 =		1.000000
0.125	0.771105 × 4 =		3.084422
0.250	0.707107 × 2 =		1.414214
0.375	0.692248 × 4 =		2.768992
0.500	0.707107 × 2 =		1.414214
0.625	0.745461 × 4 =		2.981845
0.750	0.805927 × 2 =		1.611855
0.875	0.889728 × 4 =		3.558910
1.000	1.000000 × 1 =		1.000000
			18.834452

Thus

$$\int_0^1 x^x\, dx = \frac{h}{3}(18.834452) = \frac{0.125}{3}(18.834452) = 0.784769$$

The Digital Computer

Ability to put a tolerance or largest error qualification on a number such as this is also part of numerical analysis. In this case, redoing the quadrature with five ordinates (nested in the above table) and comparing the result (0.788863) suggests that the error is not larger than 0.0003.

As a final example of quadrature carried out numerically with no knowledge of the function, consider the radial loads experienced by a ball bearing measured every 60°. They are 600, 860, 860, 600, 340, 340, 600 lbf. In antifriction bearing theory you will discover that the equivalent steady load on the bearing inflicting the same amount of damage as the varying load is the 3.3 power mean load. In algebraic terms,

$$F_{eq} = \left\{ \int_0^{2\pi} [P(\theta)]^{3.3} \, d\theta \right\}^{1/3.3}$$

We do not know the function $P(\theta)$ and we have measures of it only at seven equally spaced places. We use Simpson's rule as follows

θ	P	$P^{3.3}$		Extension
0	600	$1.471971(10^9) \times 1 =$		$1.471971(10^9)$
60	860	$4.828863(10^9) \times 4 =$		$19.315452(10^9)$
120	860	$4.828863(10^9) \times 2 =$		$9.657726(10^9)$
180	600	$1.471971(10^9) \times 4 =$		$5.887884(10^9)$
240	340	$0.225881(10^9) \times 2 =$		$0.451762(10^9)$
300	340	$0.225881(10^9) \times 4 =$		$0.903524(10^9)$
360	600	$1.471971(10^9) \times 1 =$		$1.471971(10^9)$
				$39.160290(10^9)$

By Simpson's rule,

$$\int_0^{2\pi} P^{3.3} \, d\theta = \frac{h}{3}(39.160290)(10^9) = \frac{\pi/3}{3}(39.160290)(10^9) = 13.669520(10^9)$$

$$F_{eq} = [13.669520(10^9)]^{1/3.3} = 1178.8 \text{ lbf}$$

4.7 WRITING COMPUTER PROGRAMS

As as exercise, let us write a FORTRAN computer program that will add to the CADET capability and implement the Newton-Raphson first order root-finding method. Fundamentally we are creating a SUBROUTINE subprogram that will implement the iteration formula,

$$x_{i+1} = \left[x - \frac{f(x)}{f'(x)} \right]_i$$

We wish to accomplish programming that will find the zero place for any function we can explicitly differentiate. Our plan of attack is shown in Fig. 4.13. To carry out this plan the following FORTRAN program is written.

```
      SUBROUTINE NEWRA1(I,GUESS,XROOT,YROOT)
      GO TO 100
    1 IF(I.EQ.0) GO TO 3
      WRITE(6,2)
```

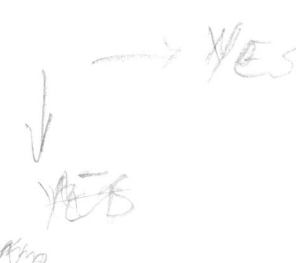

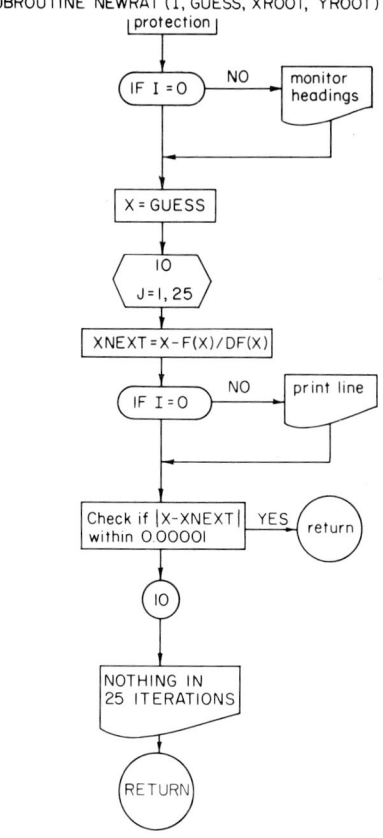

Fig. 4.13. Logic flowchart for Newton-Raphson root-finding algorithm.

```
2 FORMAT('1CONVERGENCE MONITOR IOWA CADET SUBROUTINE NEWRA1',/,/,
  1'      X             F(X)           DF(X)              XNEXT',/)
3 X=GUESS
  DO 10 J=1,25
  DENOM=DF(X)
  XNUM=F(X)
  IF(DENOM.NE.0.) GO TO 5
  WRITE(6,4)X
4 FORMAT(' *****ERROR MESSAGE SUBROUTINE NEWRA1*****',/,
  1'       FUNCTION SUBPROGRAM DF(X) SUPPLIED A VALUE OF ZERO',/,
  2'       AT X=',G15.7,'THEREBY DEFEATING NEWTON RAPHSON STRATEGY')
  RETURN
5 XNEXT=X-XNUM/DENOM
  IF(I.EQ.0) GO TO 7
  WRITE(6,6)X,XNUM,DENOM,XNEXT
6 FORMAT(4G15.7)
7 XROOT=XNEXT
  YROOT=XNUM
  IF(ABS(X-XNEXT).LE.0.00001) RETURN
  X=XNEXT
```

The Digital Computer

```
   10 CONTINUE
      WRITE(6,8)
    8 FORMAT(' *****ERROR MESSAGE SUBROUTINE NEWRA1*****',/,
     1'          UNABLE TO DISCOVER ROOT IN 25 ITERATIONS, INVESTIGATE')
      RETURN
  100 IERR=0
      IF(I.EQ.0.OR.I.EQ.1) GO TO 102
      IERR=1
      WRITE(6,101)I
  101 FORMAT(' *****ERROR MESSAGE SUBROUTINE NEWRA1*****',/,
     1'          ARGUMENT I1,',I15,' IS NOT 0 OR 1')
  102 IF (IERR.NE.0) RETURN
      GO TO 1
      END
```

Note the elements of protection designed to help the user who forgets the conditions of use or makes logic errors that abuse the routine. To test the routine, an executive test deck is written to see how the programming performs.

We choose a problem that we have solved before with the pocket calculator: find the zero place of $\ln x$ (Sec. 4.6). We program FUNCTION F(X) and FUNCTION DF(X) as follows to particularize the problem:

```
      FUNCTION F(X)
      F=ALOG(X)
      RETURN
      END

      FUNCTION DF(X)
      DF=1./X
      RETURN
      END
```

The test program is coded as

```
    C     TEST DECK FOR SUBROUTINE NEWRA1
      1 READ(5,2,END=99)I,GUESS
      2 FORMAT(I2,F10.5)
        WRITE(6,3)I,GUESS
      3 FORMAT(' NEW TEST',/,/,
     1'  I1=',I5,'   A2=',G15.7)
        CALL NEWRA1(I,GUESS,XROOT,YROOT)
        WRITE(6,4)XROOT,YROOT
      4 FORMAT(' B1=',G15.7,10X,'B2=',G15.7,/,/)
        GO TO 1
     99 STOP
        END
```

For the test the following data is supplied:

```
      $ENTRY
         1    0.5
         0    0.5
         1    1.2
         0    1.2
         3    1.2
```

The last data card exercises the error message for incorrect I1.

```
NEW TEST

I1=    1   A2=  0.5000000

CONVERGENCE MONITOR IOWA CADET SUBROUTINE NEWRA1

        X              F(X)             DF(X)           XNEXT

    0.5000000      -0.6931471         2.000000        0.8465735
    0.8465735      -0.1665582         1.181231        0.9875774
    0.9875774      -0.1250042E-01     1.012578        0.9999225
    0.9999225      -0.7748904E-04     1.000077        0.9999999
    0.9999999      -0.5960464E-07     1.000000        1.000000
B1=    1.000000                   B2= -0.5960464E-07

NEW TEST

I1=    0   A2=  0.5000000
B1=    1.000000                   B2= -0.5960464E-07

NEW TEST

I1=    1   A2=    1.200000

CONVERGENCE MONITOR IOWA CADET SUBROUTINE NEWRA1

        X              F(X)             DF(X)           XNEXT

    1.200000       0.1823214         0.8333334        0.9812142
    0.9812142     -0.1896447E-01     1.019145         0.9998224
    0.9998224     -0.1776376E-03     1.000177         0.9999999
    0.9999999     -0.5960464E-07     1.000000         1.000000
B1=    1.000000                   B2= -0.5960464E-07

NEW TEST

I1=    0   A2=    1.200000
B1=    1.000000                   B2= -0.5960464E-07

NEW TEST

I1=    3   A2=    1.200000
*****ERROR MESSAGE SUBROUTINE NEWRA1******
       ARGUMENT I1,           3 IS NOT 0 OR 1
B1=    1.000000                   B2= -0.5960464E-07
```

In the first output above note that the convergence monitor duplicates previous results (in the XNEXT column) from our pocket calculator program for finding the zero place of $\ln x$. Appendix A includes an example of a suitable documentation sheet for subroutine NEWRA1.

As an example of the use of programs such as NEWRA1, let us identify the interval in which the design factor of the hoist problem of Sec. 3.2 is >1, i.e., the domain of choice for the designer; and in the process obtain points

The Digital Computer

on the figure of merit function within the interval of plotting. The merit function was

$$M(d, m) = \frac{7200d - 4467d^3}{2124/m + 816d^2}$$

The function for which the roots are desired is

$$1 = \frac{7200d - 4467d^3}{2124/m + 816d^2} \quad \text{or} \quad y = \frac{7200d - 4467^3}{2124/m + 816d^2} - 1$$

and the value of d making y = 0 is the root(s) sought. The first derivative of y with respect to d is

$$\frac{dy}{dd} = \frac{(2124/m + 816d^2)[7200 - 3(4467)d^2] - (7200 - 4467d^3)(2)(816d)}{(2124/m + 816d^2)^2}$$

Rather than fuss with simplifying this expression, code it. Our program is

```
C      EXECUTIVE PROGRAM FOR HOIST PROBLEM
       COMMON WIRES
   1   READ(5,2,END=99)WIRES,GUESS1,GUESS2
   2   FORMAT(3F10.5)
       CALL NEWRA1(1,GUESS1,XROOT1,YROOT1)
       CALL NEWRA1(1,GUESS2,XROOT2,YROOT2)
       DELTA=(XROOT2-XROOT1)/10.
       X=XROOT1
       DO 100 J=1,11
       Y=F(X)+1.
       WRITE(6,3)X,Y
   3   FORMAT(' AT DIAMETER',G15.7,' DESIGN FACTOR IS',G15.7)
       X=X+ DELTA
 100   CONTINUE
       GO TO 1
  99   STOP
       END

       FUNCTION F(D)
       COMMON WIRES
       F=(7200.*D-4467.*D*D*D)/(2124./WIRES+816.*D*D)-1.
       RETURN
       END

       FUNCTION DF(D)
       COMMON WIRES
       X1=2124./WIRES+816.*D*D
       X2=7200.-3.*4467.*D*D
       X3=7200.*D-4467.*D*D*D
       X4=2.*816.*D
       DF=(X1*X2-X3*X4)/X1/X1
       RETURN
       END
```

The data supplied to the program are

$ENTRY
2.   0.25   1.0

The results are

CONVERGENCE MONITOR IOWA CADET SUBROUTINE NEWRA1

X	F(X)	DF(X)	XNEXT
0.2500000	0.5545397	5.146616	0.1422516
0.1422516	-0.6227005E-01	6.222570	0.1522587
0.1522587	-0.3905296E-03	6.143799	0.1523222
0.1523222	-0.2384186E-06	6.143279	0.1523222

CONVERGENCE MONITOR IOWA CADET SUBROUTINE NEWRA1

X	F(X)	DF(X)	XNEXT
1.000000	0.4552708	-4.566560	1.099696
1.099696	-0.3497165E-01	-5.241167	1.093023
1.093023	-0.1304150E-03	-5.200975	1.092998
1.092998	0.2861023E-05	-5.200821	1.092998

AT DIAMETER 0.1523222   DESIGN FACTOR IS 0.9999998
AT DIAMETER 0.2463897   DESIGN FACTOR IS 1.535880
AT DIAMETER 0.3404573   DESIGN FACTOR IS 1.967010
AT DIAMETER 0.4345248   DESIGN FACTOR IS 2.271324
AT DIAMETER 0.5285923   DESIGN FACTOR IS 2.438852
AT DIAMETER 0.6226598   DESIGN FACTOR IS 2.470152
AT DIAMETER 0.7167273   DESIGN FACTOR IS 2.373630
AT DIAMETER 0.8107948   DESIGN FACTOR IS 2.162612
AT DIAMETER 0.9048623   DESIGN FACTOR IS 1.852760
AT DIAMETER 0.9989299   DESIGN FACTOR IS 1.460156
AT DIAMETER 1.092997   DESIGN FACTOR IS 1.000008

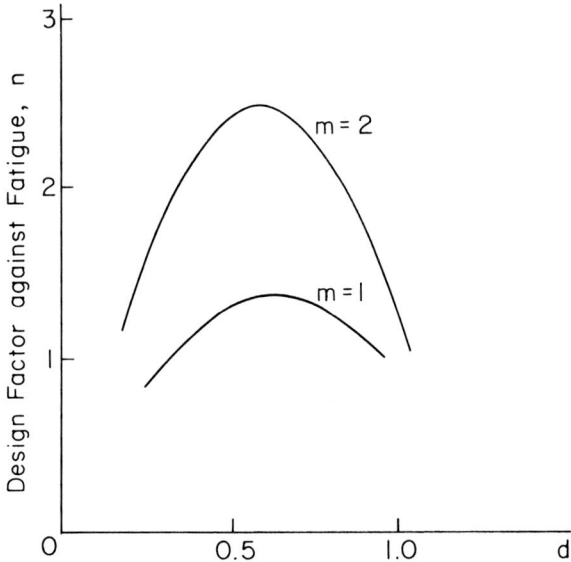

Fig. 4.14. Figure of merit function for mine hoist problem using one and two supporting wires.

Figure 4.14 shows the graphical solution for comparison with the previous result for a single hoist wire.

The next item of interest, after the feasible domain is established by two calls of NEWRA1 and a plot resulting from a DO-loop interrogation of F(X), is "what is the maximum merit in the feasible interval." It, too, can be ascertained by a "one-liner" delegation of the maximization chore to a library subroutine. In this case the golden section stratagem of Sec. 8.7 would be useful. It is also possible to have the computer plot the curves in Fig. 4.14.

Note that the executive program consists of a series of one-liners, simply calls to discharge (delegate) routine chores. This procedure is important for the programming engineer, whose work should center on the next logical tactical step that can be discharged in a brief declarative programming step. The computer support capability to which the programmer contributes is growing properly if it continues to increase the number of simple delegations of chores that are available.

PROBLEMS

4.1. Use Simpson's rule to evaluate the following integrals:

$$\int_0^\pi \sin^2 x \, dx \qquad \int_0^1 \frac{e^u - e^{-u}}{u} \, du$$

$$\int_0^{\pi/2} \sqrt{1 - 0.1 \sin^2 x} \, dx \qquad \int_0^1 x^2 (1 - x)^3 \, dx$$

$$\int_0^{\pi/2} \frac{dx}{2 + \cos x} \qquad \int_0^{\pi/2} \frac{dx}{\sqrt{1 - 0.1 \sin^2 x}}$$

$$\int_0^1 \frac{1 + x^2}{1 + x^4} \, dx \qquad \int_0^\pi \frac{dx}{5 - 4 \cos x}$$

$$\int_0^\pi \sin^{1.4} x \, dx \qquad \int_0^{\pi/2} \frac{dx}{3 \cos^2 x + 2 \sin^2 x}$$

4.2. In the analysis of critical speeds of shafts, the following frequency equation occurs: $\cos x (\cosh x) = 1$. Using the Newton-Raphson method, determine the first six angles x that satisfy the frequency equation.

4.3. A cantilevered shaft with fixed bearing at one end gives rise to the frequency equation, $\cos x (\cosh x) = -1$. Using the Newton-Raphson method, determine the first six angles x that satisfy the frequency equation.

4.4. Compare the solutions to Probs. 4.2 and 4.3 and examine the behavior of the differences between adjacent roots.

4.5. Prepare a logic flowchart for an algorithm to determine whether three given line segments can be assembled into a triangle.

4.6. Prepare a logic flowchart for an algorithm to determine whether four given line segments that represent links of a four-bar planer mechanism can be assembled into a four-bar linkage in which the smallest link can rotate 360°.

4.7. Prepare a logic flowchart for an algorithm for determining the quadrant of an angle in the range $-2\pi \leq \theta \leq 2\pi$.

4.8. Convert the FORTRAN program associated with Fig. 4.2 into a function subprogram SSQRT(X) and test its performance in comparison with your computer FORTRAN internal function SQRT(X).

4.9. Write a FORTRAN program to implement the Newton-Raphson second order method of root-finding and compare execution times with a first order program.
4.10. Write a FORTRAN program to implement Simpson's rule integration.
4.11. Write a FORTRAN program to implement Durand's rule integration.
4.12. Compare the performances of the programs of Probs. 4.9, 4.10, and 4.11.

> For example is not proof.
>
> **Hebrew Proverb**

CHAPTER 5

EMPIRICISM: THE BASIS OF MATHEMATICAL MODELING

5.1 INTRODUCTION

The engineering designer must be able to predict the consequences of an anticipated act of specification. When the act is as simple as specifying the diameter, length, emergent length, and material of an electric cattle-fence post, the prediction can be made on the basis of experience. The geometry of a 50-T crane hook, jet-engine pod struts on an airliner, or railroad-locomotive driving axle is another matter. Engineers use mathematical models of reality to predict. These models do not come about in the manner of cooking recipes; rather they are built using all available knowledge and are often tailored specifically to the case at hand. New knowledge of the real world is gained from experience and experiment. Experience may include planned and unplanned events. Experiments are usually planned events, organized in such a way as to reveal a great deal of information for the effort involved.

The first step in building a mathematical model, gathering experimental data, may have been done previously by someone other than the designer. The second step is to give the data graphical presentation so that communication of ideas is efficient. Because of the need for later mathematical manipulation, it is advantageous to use mathematical curve-fits that represent the data well. From these curve-fits, interrelationships are found that, when coupled with first principle, lead to the formulation of a mathematical predictive model of use to the designer.

The gathering of experimental data is a very expensive process, and economy of effort is realized by the designing of experiments. Such design involves establishing model and analog; establishing the most representative value of a measurable attribute; establishing measures of dispersion; determining errors in measurement and how they propagate to dependent (calculated) quantities; and answering questions such as "How much data do I take," "What do I do with it," and "With what confidence may I make a conclusion?"

If we have no prior experience with a phenomenon, the first step after formulating a design concept to meet a need is to gather data (see Fig. 5.1). To gather pertinent data we have to be familiar with standards, design of experiments, instrumentation and its calibration, and the established procedures for creating a measurement. Our success depends also on understanding the nature of definition, the concept and use of established units, and variability in the universe, and we might have to exploit some notions of similitude.

The design concept is united with many of the foregoing ideas to create data--quantitative descriptions of phenomena. The data, our laboratory experience, and the properties of line graphs permit a graphical display. Such displays impress the conceptual senses of observers, and suggest functional (mathematical) relationships.

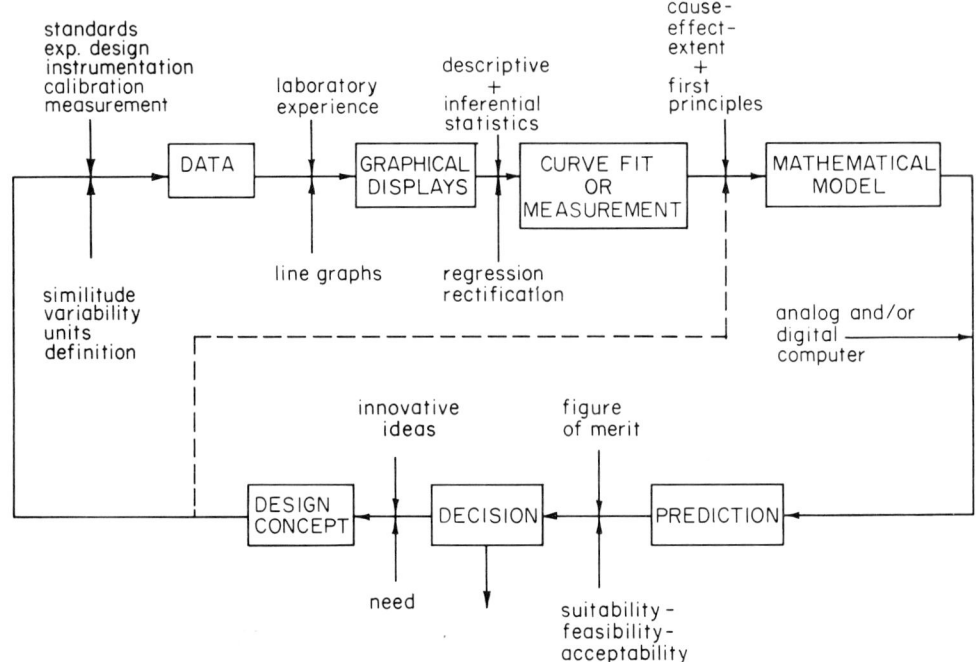

Fig. 5.1. Delineation of preliminary design, showing the intimate relationship of empiricism, prior knowledge, and concept in the creation of a mathematical model.

The graphical display of data suggests functional relationships between measured variables; and with ideas such as rectification, regression, and descriptive and inferential statistics, we can construct a quantitative approximation to reality that is convenient in mathematical manipulation. The result might also be a quantitative measurement complete with its qualifications. The result is a mathematical approximation to a cause-effect-extent relationship in nature. This result, previously discovered effect models, and first principles allow the desired mathematical model to be formulated.

If all the effects involved in a design concept have been previously quantified by empirical investigation, data-taking can be avoided and the dashed line path from concept to the mathematical model depicted in Fig. 5.1 can be traversed with an accompanying reduction in time and expense. Much of the effort in teaching and learning is directed at enabling the engineer to traverse the broken line more often than not.

The newly created mathematical model has to be exercised to force it to give the desired predictive information. Often this requires the assistance of a digital or an analog computer. The prediction enters the suitability-feasibility-acceptability considerations along with other inputs, and the original concept may become a retained satisfactory alternative. Whether another concept needs to be generated and explored is another decision to be made.

In the general process of thinking and carrying out the actions depicted in Fig. 5.1, ten achievements of empiricism permit the process and ensure the validity of the result. They are

--recognition of the importance and structure of definition
--graphical communication of the results of experiments
--mathematical curve fitting
--reduction in the number of variables
--the use of partial derivatives
--physical models and the notion of similitude
--international agreement on a system of units
--recognition of variability
--mathematical models of cause-effect-extent
--the notion of accountability and the subsequent enunciation of first principle

These are the foundations on which useful mathematical models are built. We will discuss them in turn. Subsequent courses will illuminate many nooks and crannys, often in such detail that the larger picture can get lost. Here, as we examine the methodology, it is possible to grasp the global view.

Before we begin experimental work we should be familiar with two views of matter. Matter has particulate attributes that are sometimes important. At other times matter can be viewed as continuous (nonparticulate). The notion of a continuum of matter is useful and important.

A quantity of a gas confined in a container exerts a time-invariant force intensity, commonly denoted as *gas pressure*, on the container wall. Whether the container is large or small, the attribute called pressure is time-invariant. Large or reasonably small pressures are also time-invariant. Matter in the gaseous form is well established to be a swarm of independent discrete particles called molecules, and their collisions with the container wall in aggregate give rise to the manifestation called pressure.

A discrete particle approach to the quantitative formulation of a mathematical prediction for pressure was carried out by James Clerk Maxwell. By relating the impulse delivered to the container wall to the change in momentum in the elastic collision between the wall and a single molecule and by the process of integration (adding the impulses of all molecules of all speeds and in all directions), he was able to derive an equation of the form

$$PV = (1/3)nmu^2$$

where n is the number of molecules, m is the molecular mass, and u is the root-mean-square average molecular velocity. In the light of Jacques Charles's experiments, $(1/3)nmu^2 = RT$, where R is the gas constant and T is the absolute thermodynamic temperature. A nonparticulate approach leading to the same equation originates with a combination of Robert Boyle's equation with Charles's equation:

$$PV = RT$$

When the amount of gas in a container is small (high vacuum) both equations break down in the pressure prediction, since the observed pressure is a time variable and both equations predict a time-invariant pressure. This breakdown occurs when the mean free path of a molecule (the distance between collisions on the average) is of length comparable to the smallest significant length in the problem. Superficially, one might remark that this is strange, since Maxwell's approach does indeed presume a continuum by demanding that all molecular speeds are possible and form a continuous distribution between zero and infinity. Another continuum of molecular direction is also presumed (any direction is just as likely as any other direction). The process of continuous integration rather than summation is used--another dependence on the idea of a continuum.

The continuum hypothesis is that *matter is continuously distributed* and it may be forever subdivided with the resulting parts retaining the identity and properties of the original piece. How does this compare with the overwhelming evidence of the molecular nature of matter? For many phenomena of interest to the engineer, *matter behaves as if it were continuous*. The presumption of a continuum of matter leads to an understanding of engines and turbines that is confirmed by experiments in nature. This confirmation does not mean that matter really is continuous, but deductions presuming a continuum of matter are precise, consistent, and useful models of natural phenomena that are effective in prediction and, consequently, valuable in engineering.

Success with the incorporation of continuum ideas into mathematical models of phenomena inevitably led to concepts that are dependent on the continuum idea. If we define the mass density of matter as the limit,

$$\rho = \lim_{\Delta x, \Delta y, \Delta z \to 0} \Delta m / (\Delta x \Delta y \Delta z)$$

then, in a continuum, the limit is approached asymptotically and it is real. When matter appears to be continuous in the large but is really discontinuous in the small, the limit does not exist. If the limit does not exist, the concept of density is devoid of meaning, i.e., without significance. A gas kineticist often views temperature as an index or measure of the mean kinetic energy of the molecules. If the container is a room, the sample taken in one place (of say 10^{10} molecules) will agree in mean kinetic energy with a sample of 10^{10} molecules taken in another part of the room, within our ability to measure. However, a sample of 10 molecules taken just slightly removed from another sample of 10 molecules will not agree in mean kinetic energy. Physically, therefore, the concept of temperature as a gross property, descriptive of the gas in any part of the room, is no longer unique (hence no longer a property) in the neighborhood of a point. It is no longer useful in describing, explaining, or understanding the behavior of matter in the small.

Another concept is pressure, which is the limit of the quotient of force divided by area as the area vanishes,

$$p = \lim_{\Delta A \to 0} \Delta F / \Delta A$$

This concept is no longer useful when the limit is approached. The same may be said for normal tensile stress, normal compressive stress, shearing stress, and other concepts of the mechanics of materials formulated with the implicit presumption that matter is continuous. High vacuum cannot be described with any precision by reporting the pressure, the density, or the temperature; these ideas, so useful in macroscopic phenomenological description, become useless in a high vacuum.

The concept of shearing stress in a fluid, defined by Newton for macroscopic fluid flow phenomena, breaks down when the presumed continuum of matter is violated. In a high vacuum, gas viscosity is a useless term. This is not to say that a very low pressure fluid flowing in a duct does not exert a fluid drag force on that containing duct, but it does mean that it cannot be predicted by Newtonian models of fluid flow.

The notion of tensile stress in a metal rod in tension is useful in the large. If the ultimate tensile stress of a rod is determined by experiment to be 60,000 psi (divide the ultimate load by the original cross-sectional area), one can find that similar rods will exhibit similar strengths. If the rod appears homogeneous in the large and is isotropic (has consistent and identical properties in every direction), the presumption of a continuum of matter does not hinder the model in any way; the tabulations of ultimate strengths made for engineering use are consistent with this idea. But the property of ultimate strength of the crystals and grains of which the rod are constituted is very different and direction dependent. When the crystals are tested, their ultimate strengths are found to be significantly higher and their properties differ in different directions (their isotropy disappears) even though we are nowhere near molecular size. The rod is likely to be a mixture of crystals and grains of varying size and orientation and different chemical identity. The rod is exceedingly complex in the small and much less complex in the large. When its gross properties are significant (in the large), continuum bases of understanding lead to decisions that are valid.

--The character of a mathematical model (based on a continuum or a discrete hypothesis) is not necessarily related to its usefulness in predicting states of matter.

5.2 IMPORTANCE OF DEFINITION

Two kinds of definitions are important for the quantitative understanding of phenomena. The first kind of definition is of the dictionary sort that delineates an entity, such as a beam, a proton, a liquid, etc. The classical definition of an entity (1) places it in a general class of similar things and (2) distinguishes it from all other members of its class. (3) Also, the converse must be true (consistent). For example, "A *cube* is a solid bounded by six equal squares." The cube is made a member of that class of things called solids, and it is distinguished from all other members of the class of solids by being bounded by six equal squares. The converse is true, i.e., a solid bounded by six equal squares is a cube.

"A *cam* is a mechanism for converting regular rotary motion into irregular rotary or reciprocating motion" does not satisfactorily define an entity because a slider-crank mechanism (piston, connecting rod, and crank) also converts regular rotary motion into reciprocating motion; the converse is not true because the definition does not sufficiently distinguish the cam from all other members of its class (mechanism). We can define matter by the statement, "*Matter* is anything that occupies space." Then, to be a satisfactory definition of an entity it is important that the converse, "Anything that occupies space is matter" be accepted as consistent. Consider a definition of the entity, liquid: "A *liquid* is matter that takes the shape of its container." It is expected that anything that takes the shape of its container is a liquid, such as compressed air in a tank or granulated sugar in a hopper. Since air and sugar are not generally accepted among empiricists to be liquids, we may conclude that the definition for liquid above is unsatisfactory.

The second kind of definition of interest is that of a measurable attribute of an entity, such as the *length* of a beam, the *speed* of a proton, the *viscosity* of a liquid. Observers consider the entities involved in a phenomenon, investigate important measurable attributes, and seek quantitative understanding. Observers in different frames of reference may obtain different numerical values for a measurable attribute. If generalities are to be inferred from experimental work, the generalities must be the same for all observers in all reference frames. For instance, an observer in a moving train traveling uniformly in a straight line observes the speed of a ball rolling in the aisle and reports a speed of 2 ft/s. An observer on the ground may attribute a speed of 90 ft/s to the ball. The two measurements of the same attribute (speed) of the ball differ. Is this because the observations are wrong or because the observers are measuring different quantities while convinced that they are measuring the same attribute, namely speed of the ball?

To make any headway in this process of collectively taking data and seeking understanding it is vital that we measure the same attribute of the same entity. All observers must agree on what speed is and how it is to be measured. To do this, all definition of measurable attributes must be *operational*, i.e., contain within them an idealized statement of how the attribute is to be measured. Operational definitions are the connection between the physical world and the verbal-symbolic world. Concepts such as force, length, time, temperature, and charge are not successfully defined as entities; they are understood in terms of human experience *and* operational definition. Failure to use operational definitions for measurable attributes of entities introduces confusion into an already complex undertaking.

> "The *velocity* of a particle in a frame of reference is the first time derivative of its position vector in that frame of reference."

As long as the experimenter attempting to measure velocity can establish that the method being used is equivalent (ideally) to the assessment of the first time derivative of the position vector of the particle, then the velocity of the particle is being measured. In the example of the ball on the train, the speed of the ball (with respect to the train) will be reported by the observer on the train and the observer on the ground as 2 ft/s, differing somewhat only due to experimental error. The definition, "The *viscosity* of a liquid is its resistance to flow," is nonoperational because it does not contain an idealized statement of how resistance to flow is to be measured. The definition, "*Homogeneity* is the property of uniformity of matter," is likewise nonoperational because it does not contain an idealized statement of how uniformity of matter is to be measured.

We can, at this point, list two rules concerning definitions:

1. The definition of an entity and its converse shall be consistent. The definition of entity places it in a class of similar things, then distinguishes it from all other members of its class.
2. Definitions of measurable attributes of entities shall be operational, i.e., contain within them an idealized statement of how the attribute is to be measured.

Events are the hallmark of change. If nothing changed, it is doubtful if the concept of time would have been formulated, and *future* would be a word devoid of meaning. Observers can determine whether two events are simultaneous or successive. Time has been conceived as a measure of the departure from simultaneity. The time of an event is the precise instant of the event as indicated by a clock. An interval of time is the duration between events as measured by the difference in clock indications simultaneous with the events. Any phenomenon that repeats itself can be used as a measure of time, for the measurement itself consists of counting the repetitions. The basis for a clock is a phenomenon such as an oscillating pendulum, a resonating electrical circuit, the rotation of the earth on its axis, or the revolution of the earth in its orbit.

The concept of time coupled with suitable clocks helps the experimenter grapple with change. The concept of time is considered fundamental, and a unit of time is a fundamental unit of measurement. The measurement of time is a counting process.

Change manifests itself in changes in geometry. The concept of length is important to assessing geometric changes. With a length standard, the length of a body can be measured by counting the number of times the standard can be laid end to end along the body. The metre originally was a standard of length intended to be 1 ten-millionth of the arc of the meridian through Paris from pole to equator. This intended length was the interval between scratches placed on a standard bar called the standard metre bar. Thereafter the metre *was* the distance between the scratches. Still later it was agreed to use 1,650,763.73 wavelengths of the orange-red line of spectrum of krypton 86 in vacuum as the metre. The fact that this standard distance does not go 10^7 times into the intended meridian mark represents a small discrepancy between intent and accomplishment; nevertheless, an arbitrary agreed-upon universal standard exists. The measurement of length is a counting process based on a comparison.

From length and time, derived concepts such as velocity, acceleration, area, and volume are possible. Observation of gravitational phenomena led to the concepts of force and mass, and Newton's laws of motion show that force may be expressed in terms of mass or vice versa, leading to the addition of force or mass to the list of fundamental concepts. Study of electrical phenomena led to the concept of charge and its addition to the list of fundamental concepts. Study of thermal phenomena led to the concept of temperature, considered fundamental. Each of the fundamental concepts has associated with it a fundamental unit; measurement of these fundamental quantities consists of a comparison with the fundamental unit and a counting process.

These "measuring sticks" are fundamental tools of empiricists. In their investigation of phenomena they seek relations between the numbers they obtain in the measuring process. Of all possible independent variables $\{x_j\}$, only a few have a significant (measurable) influence on the dependent variable of interest y. These few are denoted as $(x_1, x_2,...,x_n)$, characterized as independent abscissas, and envisioned as mutually perpendicular in space. At the point defined by $\{x_n\}$ the dependent variable has a value, visualized as an ordinate erected at $(x_1, x_2,...,x_n)$. To have a unique measurement of y, it is necessary that

1. measurements of y and $\{x_n\}$ are reproducible
2. y is single-valued in the absence of measurement error

If these requirements are met, there exists in the experimenter's eye a unique hypersurface, a function defined by the ordinate y erected at $\{x_n\}$.[1] If the hypersurface is unique, it can be described within the region of measurement and used to predict values of the dependent variable, given the independent variable set. The experimentally defined hypersurface can be described in many ways, some of which are subjects of ensuing sections.

The salient attribute of a measurement is that it is a statement of the result of a human observation. "I have weighed the sample and obtained a result of 1.32 lbf." Since all measurements are conveyed to someone else, it is necessary to convey the matter of confidence. Bluntly, is someone else going to believe you? It is better to say, "I have weighed the sample and found the weight to be 1.32 lbf with a 99 percent confidence that it lies between 1.31 and 1.33 lbf." This is still a matter of opinion and the competent experi-

---

1. Let X and Y be nonempty sets. Let f be a collection of ordered pairs (x, y) with x ε X, y ε Y. Then f is a function from X to Y if for every x ε X there is assigned a unique y ε Y.

menter will provide the listener with sufficient statistical information so that the reader can conclude whether the reporter's confidence is to be shared.

Experience and observation of the physical world through human senses and their extensions give rise to concepts that are understood by humans but are incapable of being verbalized to anyone's satisfaction. One such nonverbalized concept is that of length. A dictionary would say that length is the linear magnitude of anything measured from end to end. This reminds us of the concept, but the understanding that you have was gained from experience. The operational definition of length is expressed in terms of how many times the standard metre can be laid off contiguously in some (geometrically) significant direction. By this process a length is a measurable attribute of a body. Similarly, the concept of force has its operational definition in terms of acceleration of a kilogram body. A balance is an instrument that allows a standard kilogram body to be compared through secondary standards to the sample whose mass is to be assessed. The concept of temperature has its operational definition in terms of a particular kind of thermometer, the concept of time has its operational definition in terms of a cesium clock, etc.

Figure 5.2 is an attempt to indicate the levels of abstraction involved in one area of physics called thermodynamics. The zero level is the nonverbalized concepts that are shared by a majority of all mature persons. Thermodynamics (not exclusively) takes us to the first level of abstraction by means of operational definitions that are closely related to measurement standards. Quantities such as work, elevation potential energy, and heat (at the second level of abstraction) are understood in terms of operational definitions using first level of abstraction terms. Thus work is understood in terms of operational definitions of force and length.

The third level of abstraction involves the general concept of energy, which is understood in terms of second level concepts of work and heat. Energy is significant because $\S(dQ - dW)$ can be shown experimentally to vanish, leading to the recognition of the thermodynamic property, *energy*, and permitting the formulation of the first law of thermodynamics, one of the more sweeping statements that can be made about the physical world.

The fourth level of abstraction is represented by *entropy*, which is understood in terms of an operational definition involving energy and temperature. Entropy is significant because $\S dQ/T$ or $\S d(E + W)/T$ has important properties (experimentally determined) that led to the recognition of the thermodynamic property, *entropy*, and permitted the formulation of the second law of thermodynamics, perhaps the most sweeping statement that has been made concerning the physical world.

The point of this examination of levels of abstraction is to show the dependence on operational definitions for understanding. As a secondary benefit the reader can begin to understand why upperclass students gripe about difficulties in grasping the concept of entropy. It is four levels of abstraction removed from their zero level experience of some 20 years.

Operational definitions are the connection between the physical world and the verbal-symbolic world. Symbols are abstractions for ideas and are used for communication and thinking purposes. Symbols come in several species. They can stand for object or class, such as B for brake, Δ for triangle. They can stand for a measurable attribute of some entity, such as ℓ for length of a shaft, γ for weight density of substance. They can stand for process, such as square root, $\sqrt{}$, or integration, $\int$. They can stand for a state of nature, such as potential V or force F. They can stand for a pure number, such as e, π, or n.

Mathematical models often relate (or utilize) symbols of several kinds.

Empiricism 111

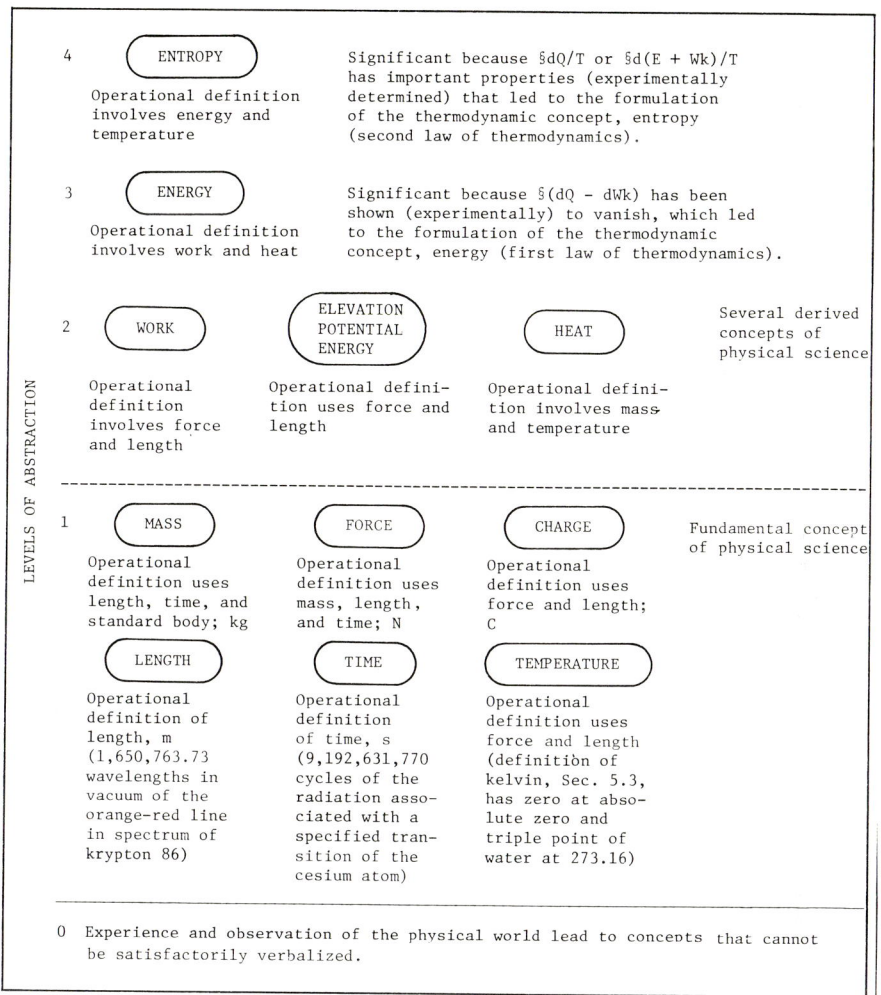

Fig. 5.2. Levels of abstraction associated with concepts in thermodynamics.

For example, the familiar equation from physics of electricity

$$v = n(d\phi/dt)(10^{-8})$$

includes state of nature v, process d/dt, measurable attribute ϕ, and number (n and 10^{-8}).

Remember that the definition of an entity and its converse must be consistent (true), since this is the only way that empiricists recognize entities. The definitions of measurable attributes of entities must be operational; i.e., they must contain a statement of an idealized method of measurement. A measurement is fundamentally a counting process involving fundamental units (standards). Measurements are characterized by the most representative values as well as confidence limits. To be useful, any defined dependent variable must be single-valued.

Precision refers to the difference between a measured quantity (measurement) and the most representative value of that quantity. *Accuracy* refers to the difference between a measured quantity (measurement) and the true value of the quantity. The most representative values of measurements are the basis for *deterministic* statements concerning the measurements of the entity involved and statements concerning cause, effect, and extent of phenomena. Measurements themselves are "almost reproducible" and therefore require *statistical* statements as companions to most representative values.

--The definition of an entity and its converse shall be consistent. The definition of an entity places the subject in a class of similar things and then distinguishes it from all other members of its class.
--The definition of a measurable attribute of an entity shall be operational, i.e., contain within it a statement of an idealized method of performing the measurement.
--Experience and observation give rise to concepts that cannot be satisfactorily verbalized, e.g., size, push or pull, departure from simultaneity, hotness. Fundamental concepts of physical science involve measurable attributes of entities, e.g., volume, area, length, position, velocity, acceleration, mass, temperature, etc. Operational definitions are the connection between the physical world and the verbal-symbolic world.

5.3 GRAPHICAL COMMUNICATION

A number may be represented by a unique point on a line graph. The position of such a point on a line depends on the scale of the line graph. For example, the number 3 is depicted on several line graphs of different scales in Fig. 5.3. The graph of 3 is A and the coordinate of A is $\cdot 3$; B is the graph of 3 and 3 is the coordinate of B; C is the graph of 3 and 3 is the coordinate of C. In Fig. 5.4, P is the plotting origin, ξ is the coordinate of P and x is the coordinate of Q. The *scale factor* $s_{f(x)}$ is defined as

$$s_{f(x)} = \frac{\overline{PQ}}{f(x) - f(\xi)} \tag{5.1}$$

where $\overline{PQ}$ is the directed distance from P to Q.

A line graph is said to be *linear* if $f(x) = x$, in which case the scale factor is

$$s_x = \frac{\overline{PQ}}{x - \xi}$$

Fig. 5.3. Three line graphs of the coordinate 3.

Fig. 5.4. A linear line graph.

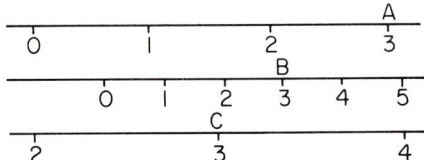

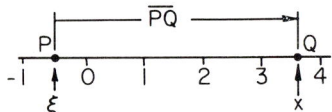

Empiricism

It is convenient to choose a preferred plotting origin such that $f(\xi) = 0$; in the linear case, if the coordinate of P is 0, then

$$s_x = \frac{X}{x} = \frac{\text{(directed distance from plotting origin to some graph Q)}}{\text{(coordinate of Q)}}$$

Figure 5.5 depicts this situation, with the preferred plotting origin denoted with an open dot.

Another useful line graph is the *logarithmic* line graph, commonly known as a logarithmic scale, depicted in Fig. 5.6. A line graph is logarithmic if $f(x) = \ln x$ and the scale factor is

$$s_{\ln x} = \frac{\overline{PQ}}{\ln x - \ln \xi} = \frac{\overline{PQ}}{\ln x/\xi}$$

If the plotting origin is chosen so that $f(\xi) = 0$, the coordinate of the plotting origin is unity and the scale factor becomes

$$s_{\ln x} = \frac{X}{\ln x} = \frac{\text{(directed distance from plotting origin to some graph Q)}}{\ln(\text{coordinate of Q})}$$

Figure 5.7 depicts this situation. The base 10 logarithm can be used, i.e., $f(x) = \log x$, and the scale factor is

$$s_{\log x} = \frac{\overline{PQ}}{\log x - \log \xi} = \frac{X}{\log x}$$

Another useful line graph is the squared scale (Fig. 5.8). The scale function is $f(x) = x^2$ and the scale factor becomes

$$s_{x^2} = \frac{\overline{PQ}}{x^2 - \xi^2}$$

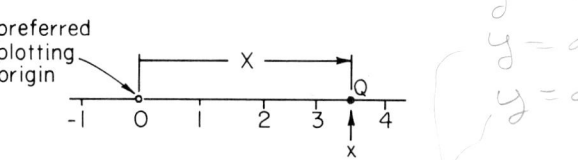

Fig. 5.5. A linear line graph showing the location of the preferred plotting origin and the plotting distance X.

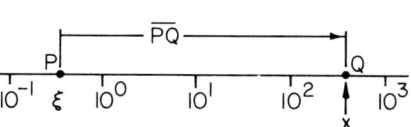

Fig. 5.6. A logarithmic line graph.

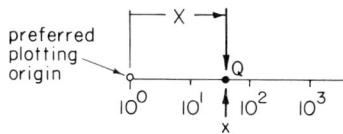

Fig. 5.7. A logarithmic line graph showing the location of the preferred plotting origin and the plotting distance X.

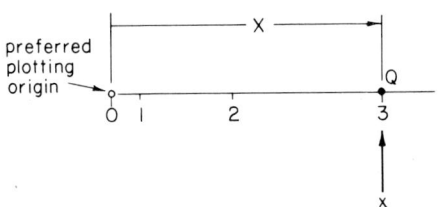

Fig. 5.8. A squared scale line graph showing the location of the preferred plotting origin and the plotting distance X.

Taking the plotting origin at $f(\xi) = 0$ means the coordinate of the plotting origin is zero and

$$s_x^2 = \frac{X}{x^2} = \frac{\text{(directed distance from plotting origin to some graph Q)}}{\text{(coordinate of Q)}^2}$$

Two scales placed perpendicularly may be used to graph a point with two coordinates. In Fig. 5.9(a) the point (1, 5) is indicated, among others. The distance from the plotting origin of the ordinate (y = 0) to y = 5 is denoted Y, and the distance from the plotting origin of the abscissa (x = 0) to x = 1 is denoted X. If X = 0.5 in. and Y = 1 in., then

$$s_x = X/x = 0.5/1 = 0.500 \text{ in.} \qquad s_y = Y/y = 1/5 = 0.200 \text{ in.}$$

If the plotted points on the graph are columnated along a straight line, the mathematical function $y = f(x)$ is said to be linear and the function $Y = g(X)$ is also linear. The slope $\Delta Y/\Delta X$ is called the physical slope and has meaning only in terms of a particular plot. The slope $\Delta y/\Delta x$ is called the mathematical slope and is the same for all plots.

Figure 5.9(b) depicts the situation with a linear scale for the abscissa x and a logarithmic scale for the ordinate y. The point (2, 100) is indicated, among others. The distance from the ordinate plotting origin (y = 1) to y = 100 is denoted Y, and the distance from the abscissa plotting origin (x = 0) to x = 2 is denoted as X. If X = 1/2 in. and Y = 1 in., then

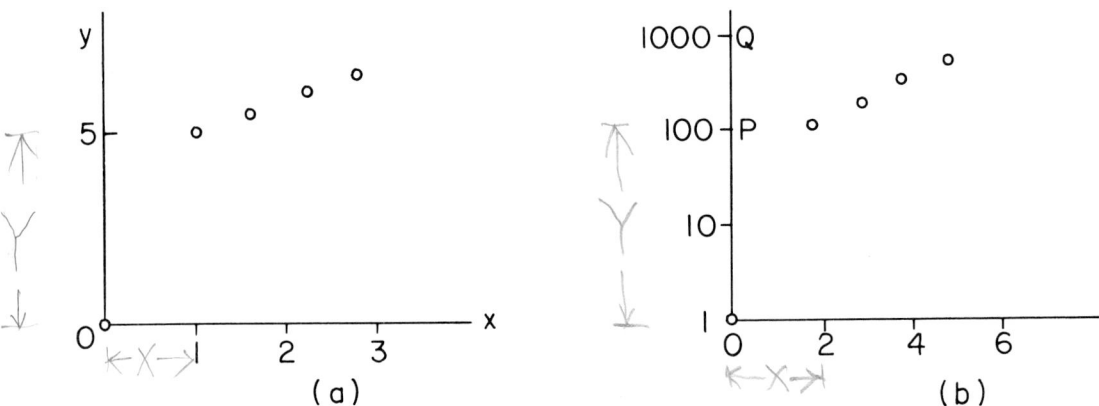

Fig. 5.9. The use of two line graphs to plot data points having two coordinates.

Empiricism

$$s_x = X/x = (1/2)/2 = 0.250 \text{ in.}$$

$$s_{\ln y} = Y/\ln y = 1/\ln 100 = 0.217 \text{ in.}$$

In logarithmic scales the length of a decade (tenfold increase) is useful. Points P and Q on the ordinate of Fig. 5.9(b) bound a decade on a logarithmic scale and are 0.5 in. apart.

$$s_{\ln y} = \overline{PQ}_{10}/(\ln y/\xi) = 0.5/\ln 10 = 0.217 \text{ in.}$$

or, alternatively, using base 10 logarithms,

$$s_{\log y} = \overline{PQ}_{10}/(\log y/\xi) = 0.5/\log 10 = 0.5 \text{ in.}$$

length of decade $\overline{PQ}_{10} = s_{\ln y}(\ln 10) = 0.217(2.30) = 0.500 \text{ in.}$

$$= s_{\log y}(\log 10) = 0.5(1) = 0.500 \text{ in.}$$

and we note that the base 10 scale factor is numerically the length of a decade, a useful interpretation. If the data string of Fig. 5.9(b) is columnated along a straight line, the mathematical function y = f(x) is nonlinear.

The discovery of appropriate coordinate transformations to linearize the plotted data string is called *rectification*, which is a step in the process of mathematical curve fitting.

Example 1. Construct a line graph with a Cartesian (linear) scale, using a scale factor of 0.5 in. and displaying coordinates in the range $-1 \le x \le 5$.

The first step is to establish the measurements from the plotting origin to the cardinal scale ticks. From Eq. (5.1) simplified for a linear scale,

$$X = s_x x = 0.5x \text{ in.}$$

The linear distance to the cardinal scale ticks are

$S_x = 0.5$
$S_x = \frac{X}{x} \text{ or } X = S_x x$

$X_{-1} = 0.5(-1) = -0.5 \text{ in.}$ $X_3 = 0.5(3) = 1.5 \text{ in.}$

$X_0 = 0.5(0) = 0 \text{ in.}$ $X_4 = 0.5(4) = 2.0 \text{ in.}$

$X_1 = 0.5(1) = 0.5 \text{ in.}$ $X_5 = 0.5(5) = 2.5 \text{ in.}$

$X_2 = 0.5(2) = 1.0 \text{ in.}$

The line graph depicted in Fig. 5.10(a) has been constructed from these measurements.

Example 2. Take quantitative information from the line graph depicted in Fig. 5.10(b) and ascertain the coordinate of R.

A reading of the coordinate of R from the line graph results in an interpolation estimate of $x_R = 4.5$. The estimate can be done with greater precision by measuring with an engineer's scale. Using points P and Q, the distance between them of 1 in., and Eq. (5.1), the scale factor can be determined.

$$s_x = \overline{PQ}/(x - \xi) = 1.00/(3 - 1) = 0.500 \text{ in.}$$

The distance PR is 1.77 in.; using Eq. (5.1) for points P and R and solving for x we obtain

$$S_x = \frac{\overline{PR}}{(x_R - \xi)} \quad \text{or} \quad x_R - \xi = \frac{\overline{PR}}{S_x}$$

$$x_R = \frac{\overline{PR}}{S_x} + \xi$$

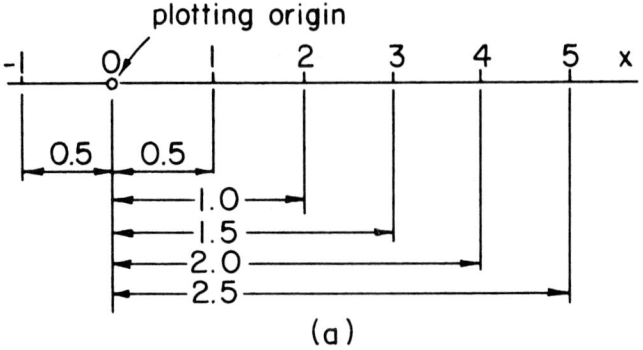

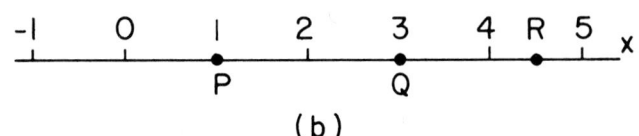

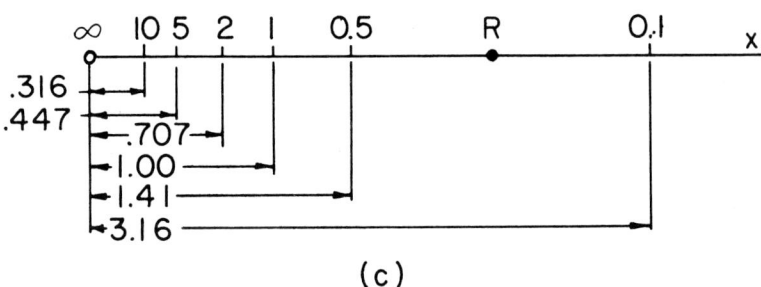

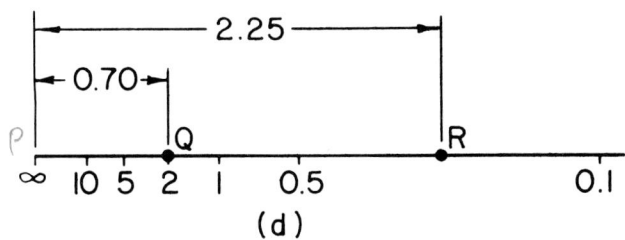

Fig. 5.10. (a) Line graph from Example 1. (b) Line graph from which quantitative information is taken in Example 2 by estimating a coordinate and by measurement. (c) Nonlinear line graph from Example 3. (d) Nonlinear line graph from which quantitative information is taken in Example 4 by estimating the coordinate of R and by measurement.

$$x_R = \xi + \overline{PR}/s_x = 1 + 1.77/0.500 = 4.54$$

Example 3. Construct a nonlinear scale wherein $g(x) = 1/\sqrt{x}$, using a scale factor of 1 in. in the coordinate range $0.1 \le x < \infty$.

The distances from the plotting origin to any coordinate tick of interest is given by $X = s_{g(x)} g(x)$; if $g(x) = 1/\sqrt{x}$, then

$$X = s_{1/\sqrt{x}}(1/\sqrt{x}) = 1(1/\sqrt{x}) \text{ in.}$$

The distances from the plotting origin to the cardinal scale ticks can be calculated:

$X_{0.1} = 1(1/\sqrt{0.1}) = 3.16$ in. $X_5 = 1/\sqrt{5} = 0.447$ in.

$X_{0.5} = 1/\sqrt{0.5} = 1.41$ in. $X_{10} = 1/\sqrt{10} = 0.316$ in.

$X_1 = 1/\sqrt{1} = 1.00$ in. $X_\infty = 0$ in.

$X_2 = 1/\sqrt{2} = 0.707$ in.

Figure 5.10(c) shows the $1/\sqrt{x}$ scale constructed with the above measurements.

Example 4. Take information from the nonlinear line graph depicted in Fig. 5.10(d) and ascertain the coordinate of R.

A reading of the coordinate of R from the line graph results in an interpolation estimate of $X_R = 0.22$. Better precision is obtained by measuring with an engineer's scale. The plotting origin is at coordinate ∞, since Eq. (5.1) particularized for this case is

$$s_{1/\sqrt{x}} = \frac{\overline{PQ}}{(1/\sqrt{x}) - (1/\sqrt{\xi})}$$

If $1/\sqrt{\xi}$ is set at zero, then ξ is the coordinate of the plotting origin and is infinite. The distance from the plotting origin to point Q by measurement is 0.70 in. The coordinate of Q is 2; therefore the scale factor is

$$s_{1/\sqrt{x}} = X/(1/\sqrt{x}) = 0.70/(1/\sqrt{2}) = 0.99 \text{ in.}$$

The distance from the plotting origin to R is 2.25 in. by measurement. The coordinate of R is

$$x_R = \left(\frac{s_{1/\sqrt{x}}}{X}\right)^2 = \left(\frac{0.99}{2.25}\right)^2 = 0.194$$

Measurements of a dependent variable taken under conditions of reproducibility are considered to lie in a unique surface or hypersurface. Indeed, the hypersurface is defined by both the specific measurements taken and those that could have been taken. We will never know where those that could have been taken would fall. We are confronted with a finite number of points and the problem of delineating the geometry of the hypersurface sufficiently so that it is useful in *prediction*.

Consider a constant volume gas thermometer depicted symbolically in Fig. 5.11. A definite amount of oxygen is confined in the sensing bulb, which is connected to a Bourdon tube pressure gauge by a capillary tube. The sensing element is immersed in air-saturated water with ice at atmospheric pressure

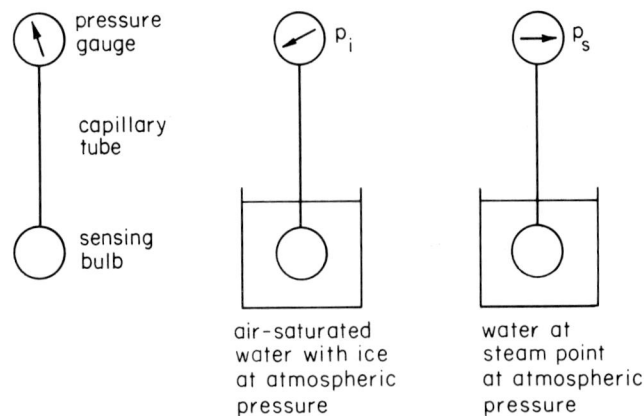

Fig. 5.11. A conceptual model of a constant volume gas thermometer.

(the state of which is unique to water and called the ice point). The oxygen pressure p_i is read from the pressure gauge. The sensing element is then immersed in boiling water at atmospheric pressure (the state of which is called the steam point) and the pressure p_s of the oxygen is recorded.

Let us plot the ratio p_s/p_i on a linear scale as ordinate and the oxygen pressure at the ice point, p_i, on a linear scale as abscissa. Some data points are shown in Fig. 5.12. We have discovered that the dependent variable p_s/p_i is reproducible and single-valued when displayed together with the oxygen pressure at the ice point. Since a unique p_s/p_i exists for every p_i, it follows that the definition of a mathematical function has been satisfied and that

$$f(p_i, p_s/p_i) = 0$$

Since we have satisfied the requirements of a function only at a finite set of points, the data collected cannot uniquely determine $f(p_i, p_s/p_i)$. Should the

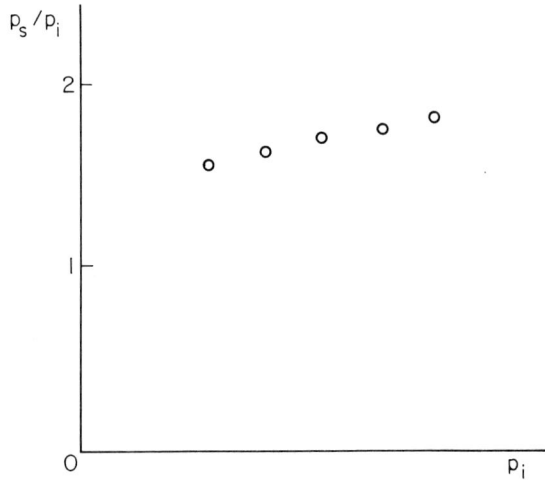

Fig. 5.12. Data obtained with several charging pressures p_i.

Empiricism

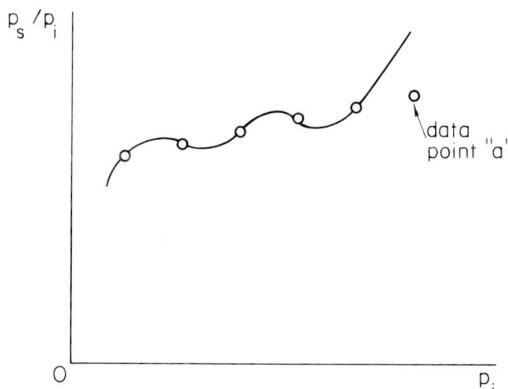

Fig. 5.13. The folly of passing a curve through a finite set of data points is revealed when the next data point a is found and plotted.

individual data points in Fig. 5.12 be connected to indicate that there is a functional relationship between p_s/p_i and p_i for oxygen at different charging pressures? Conceptually, in the $(p_i, p_s/p_i)$-plane is a curve that is defined by these observations and those that could have been taken. The idea that the points fall on a continuous curve is reasonable and appealing. What kind of curve should be passed through the data? Figure 5.13 shows one of an infinite number of possibilities. Clearly the faired curve contains every data point. If we add one more data point a, we can conclude that this curve is one of an infinite set defined by the data but (1) not one of the infinite set defined by the data augmented by point a and (2) not the unique curve defined by all the possible data that could be taken.

If we must "change our minds" about the geometry of the functional relationship every time we augment data, we are refining a previously less complete picture. This does not save our wounded ego, but that is the way it is. Every time we add to the store of knowledge, we change the status quo, either by reinforcing previously advanced ideas or by undermining them and suggesting a change of viewpoint.

Suppose we connect the data points with straight-line segments (1) to indicate that there is a functional relationship as yet not completely understood and (2) to avoid misleading the reader into concluding that understanding is better than it really is. If data points are close enough together and are the result of the superposition of the measurements of different observers using different techniques, we may be willing to assume the risk of fairing in a smooth curve with the expectation that new data will fall on or near the drawn curve.

Note that the existence of the curve in the first place is a *human* conviction based on the experience of the apparent reproducibility of the data, which in the last analysis is the result of an inspection of a finite set and therefore constitutes generalizing from the particular. Individual observers are spared some of the embarrassment of this unscientific behavior because they rarely give explicit mathematical form to their faired-in curves but have merely introduced something pleasing to the eye, reinforcing a conviction of a functional relationship without expounding a unique relationship.

Figure 5.14 shows the pressure-volume-temperature (pvT) surface for a substance such as water that expands on freezing. The determination of this surface required exhaustive data taking. Note that the dependent variable (pressure) is not single-valued everywhere and discontinuities in slope exist

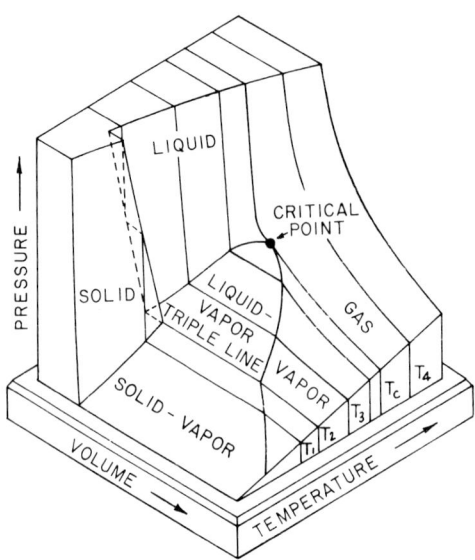

Fig. 5.14. The pvT surface for a substance that expands on freezing. (From *Thermodynamics* by Lee and Sears. Copyright 1963, Addison-Wesley, Reading, Mass., p. 90. Used with permission of the publisher.)

in the surface. In spite of the long and exhaustive experimental effort, the surface is still not uniquely established, but its geometry is well described. Figure 5.15 shows the pvT surface for a substance that contracts on freezing.

Returning now to the gas thermometer, we have an opportunity and therefore a problem. As the oxygen charging pressure gets lower and lower, the da-

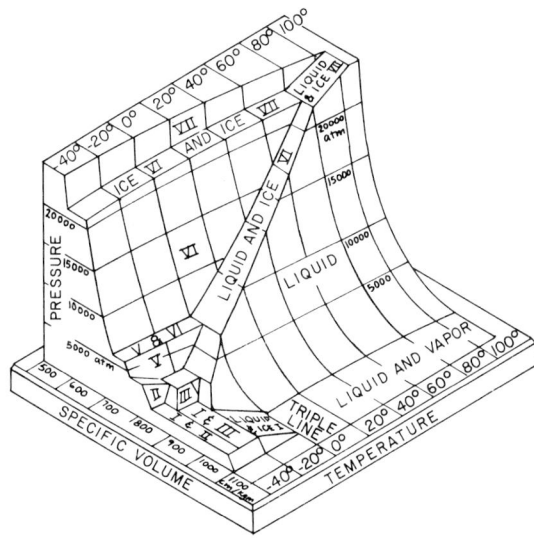

Fig. 5.15. The pvT surface for a substance that contracts on freezing. (From *Thermodynamics* by Lee and Sears. Copyright 1963, Addison-Wesley, Reading, Mass., p. 89. Used with permission of the publisher.)

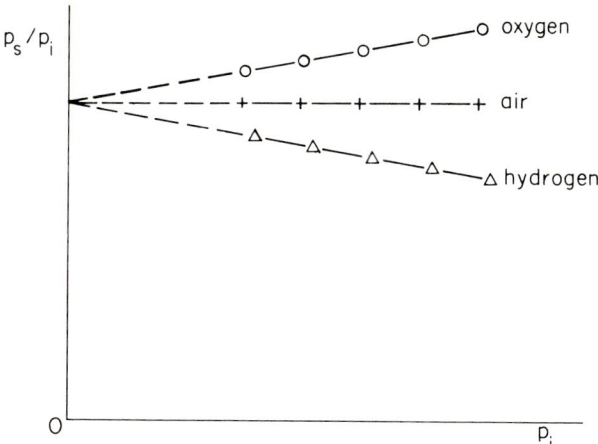

Fig. 5.16. Different data point strings associated with thermometric substances of differing chemical composition.

ta points approach an ordinate intercept. When the oxygen charging pressure becomes small and approaches zero it becomes experimentally difficult to obtain good measurements. The ordinate intercept cannot be determined experimentally because our apparatus will not work in a vacuum. If we had used another kind of gas (say hydrogen) or a mixture of gases (say air), we would obtain different data.

A different string of data points occurs for each kind of thermometric substance, as depicted in Fig. 5.16. As observers experimented with the gas thermometer, they saw that the value of p_s/p_i was less of a function of the kind of gas as the charging pressure p_i was reduced. Was it possible that all gases were approaching the same behavior as their pressures were reduced? It appeared that these strings of data were all aimed at the same ordinate intercept. If we cannot measure p_s/p_i at zero pressure, how do we determine the ordinate intercept for each gas and compare them? The answer lies in extrapolation; i.e., projecting a curve-fit beyond the region in which data is available. Mechanically, straight lines and circular arcs extrapolate easily. How does one extrapolate? Is each data string here a straight line? The experimental evidence is that the data strings are linear on Cartesian coordinates. What straight line best fits the data? Does best fit mean least error? The strings of data points in Fig. 5.16 seem to establish straight lines. What is the best estimate of a straight line? The usual approach is to select the straight line for which the data points have a least sum of squared deviations. Such a line is called a *least square line*. The line so determined is not the "line established by nature" but the best line that our data (which includes experimental error) justifies. When statistical considerations are included it is possible to draw *confidence bands* at a specified confidence level (say 0.95). See Fig. 5.17. We can be assured that there are 95 chances in 100 that the true line lies within the confidence bands. The confidence bands extended to the ordinate intercept give confidence limits on the extrapolation to the ordinate intercept. The results of the best experimental work on this problem gives the intercept as

$$\lim_{p_i \to 0} p_s/p_i = 1.36609 \pm 0.00004$$

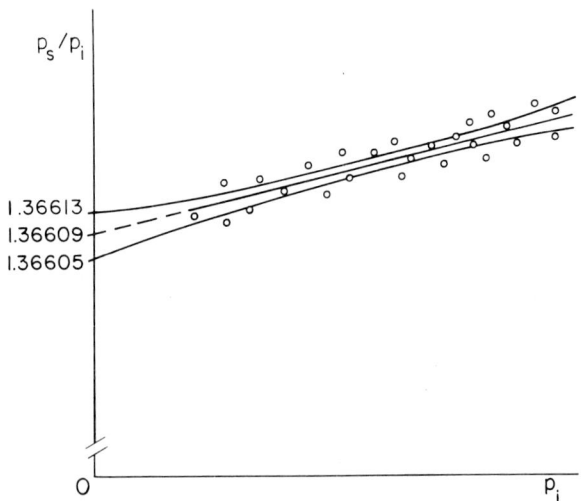

Fig. 5.17. The extrapolation of a data string to zero pressure and the associated uncertainty in the intercept.

If this limit is valid for all gases, it is ideally possible to construct a thermometer that gives temperature measurements *independent of the nature of the thermometric substance*. Such a lack of operational definition of temperature played havoc with progress in the field of thermodynamics until the problem was recognized. Now temperature is defined operationally as

$$T/T_i = \lim_{p_i \to 0} (p/p_i)_v$$

Prior to 1954, the Kelvin temperature scale was defined as having 100 divisions between the steam point and the ice point (0.0098 units below the triple point). Thus

$$T_s - T_i = 100 \qquad T_s/T_i = 1.36609$$

The two equations can be solved simultaneously to yield

$$T_s = 373.16 \pm 0.03 \qquad T_i = 273.16 \pm 0.03$$

After 1954 the Kelvin scale of absolute temperature had its definition changed from the historical development just presented and the Celsius temperature scale was defined. Because of reproducibility problems with different laboratory determinations of the ice point and the steam point, one fixed point, the triple point of water, was established as exactly 1/100 of a degree Celsius (0.01°C). The best gas thermometer determination of the Kelvin temperature of the triple point of water at that time was 273.16 K. By definition a change in temperature of 1 K is a change of 1°C; consequently the Kelvin temperature of the ice point is 273.16 - 0.01 or 273.15 to five significant digits. The Celsius temperature of the ice point is 273.15 degrees less or 0.00°C, not exactly but to two significant decimal digits.

The Rankine absolute temperature scale is defined as having graduations 5/9 the size of the Kelvin scale, so the Rankine temperature of the ice point is 273.15(9/5) or 491.67 R to five significant figures. The Fahrenheit ice point is 32°, so $T_R - t_F = \phi$. At the ice point, 491.67 - 32 = ϕ = 459.67, and the conversion equation from Fahrenheit to Rankine is

$$T_R = t_F + 459.67$$

to five significant digits. The corresponding equation for converting Celsius temperature to Kelvin is

$$T_K = t_C + 273.15$$

to five significant digits.

The nonoperational definition of any attribute of caloric (phlogiston) confused investigators (hence their inductions, thus the science of thermodynamics) for a century. A nonoperational definition of temperature (the property of a thermodynamic system indicating whether one system is in thermodynamic equilibrium with another before they are brought into communication through a diathermic wall) hampered scientists for nearly two centuries. ("What experience and history teach us is this . . . that people and governments never have learned anything from history or acted upon principles deduced from it," said George William Hegel, 1770-1831. We are also indebted to Percy Williams Bridgman for the concept of an operational definition.)

Some empiricists are not seeking to define in detail the geometry of a surface or hypersurface. This objective is denied them by the enormity of certain problems and the hopelessness of trying to gather systematic data for use in describing the surface. One problem of this sort confronts investigators interested in fatigue failure of metal hardware, more specifically, steels. Consider a bar that is axially loaded (line of action through centroid of prism cross section), the load varying with time. Let the load variation be repetitive. It is common in these cases to resolve the load (or stress) into a steady (or mean) component F_m (or s_m) and an amplitude (or variable) component F_a (or s_a), as depicted in Figs. 5.18 and 5.19. The endurance limit in tension for a steel, S_e, is defined as the largest completely reversed normal stress that the material can endure and exhibit infinite life. The ultimate strength is the nominal normal stress based on original cross-sectional area at which tensile rupture occurs. As you can see from Fig. 5.18, an infinite number of combinations of mean and variable stress can be imposed on a prism in axial loading. If, by test, we cause failure to occur, we can plot the failure as a point on the $s_a s_m$-plane, as indicated in Fig. 5.20. The task of the empiricist is to draw a locus that separates all the

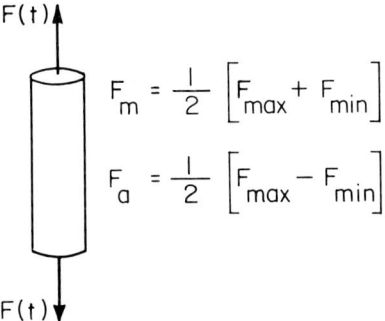

Fig. 5.18. A right circular cylindrical bar with time-varying sinusoidal load applied.

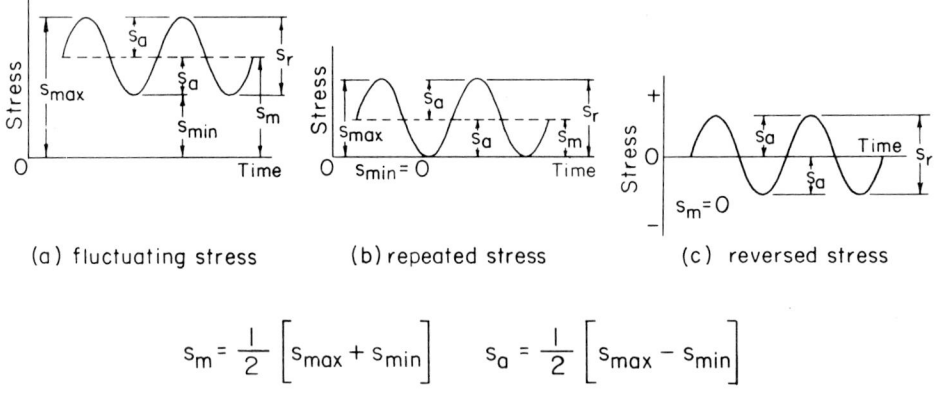

$$s_m = \frac{1}{2}\left[s_{max} + s_{min}\right] \qquad s_a = \frac{1}{2}\left[s_{max} - s_{min}\right]$$

Fig. 5.19. Tensile stresses associated with the rod of Fig. 5.18. (From *Mechanical Engineering Design* by J. E. Shigley, Fig. 5-23. Copyright 1977, McGraw-Hill, New York. Used with permission of the publisher.)

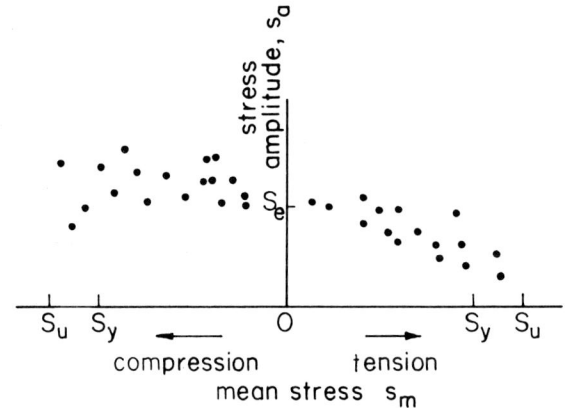

Fig. 5.20. Combinations of mean stress s_m and stress amplitude s_a, resulting in failure of the rod of Fig. 5.18. (Shigley, *Mechanical Engineering Design*, adapted from Fig. 5-24b.)

failure combinations of s_a and s_m from the combinations that do not result in failure. One such line is due to C. R. Soderberg and is shown in Fig. 5.21. The first requirement imposed on the designer is that the allowed stress combinations will be *safe*. For this the designer requires (in this particular case) that the half plane be divided into two domains: one containing all the known failures and the other containing all the known safe combinations. To make the interpretation simple we will allow the unsafe domain to include some safe combinations, but no failures are permitted in the safe domain. The second requirement is that the designer know roughly where this dividing line is and thus assess the margins against failure for unexpected (rare, unusual) circumstance and nonconservative discrepancies in the modeling procedure.

In the general case of material failure, many complications occur of which you are unaware at this point. The designer simplifies the model so that there are fewer independent variables and pulls in the boundary locus so

Empiricism

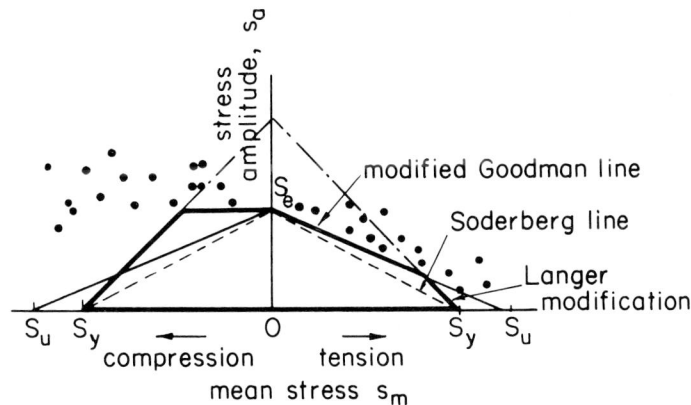

Fig. 5.21. Attempts to partition the safe domain from the failure domain by Goodman, Soderberg, and Langer. (Shigley, *Mechanical Engineering Design*, adapted from Fig. 5-24b.)

that a design may be safe, even if a real parameter of the failure phenomenon is not measurable. Your sequence of courses in material science, metallurgy, and introductory design will illuminate these points in great detail.

Thus the empiricist seeks to describe a surface or hypersurface in a hyperspace in sufficient detail from finite available data so that the geometry is sufficiently well described (1) to allow interpolation for predictive purposes with confidence and tolerable error, or (2) to divide an infeasible region from a feasible region sufficiently to establish the probability that a point in the infeasible domain lies within the surface.

--A line graph is the basis of graphical communication with quantitative content.
--A distinction can be observed between the mathematical slope of a line and the physical slope of a line.
--All curve-fits to data have associated confidence intervals that should be properly indicated.
--An empirically defined surface may seek to well predict or well represent a functional relation that gives evidence of being present, or it may seek to partition feasible combinations (decision sets) from infeasible ones.

5.4 MATHEMATICAL CURVE FITTING

The empiricist may be successful in posing questions to nature and recording the quantitative responses. Let us presume that the questions and answers are meaningful and the problems of experimental error have been resolved. However valid the data and however neatly recorded, we still face a problem of stating the interrelationship between the significant parameters for ourselves and communicating it to others. A sheet of tabulated information does not come alive and quickly communicate the recorded dependencies. Graphical communication and representation of data affords some condensation of space. Representing the data by an equation, empirically developed, leads to even greater saving in space; a great convenience to memory; and most important, convenience and flexibility in logical manipulation.

Data points can be represented as dots, hollow dots, intersecting line segments, rectangles, triangles, etc. Many symbolisms are available to tag

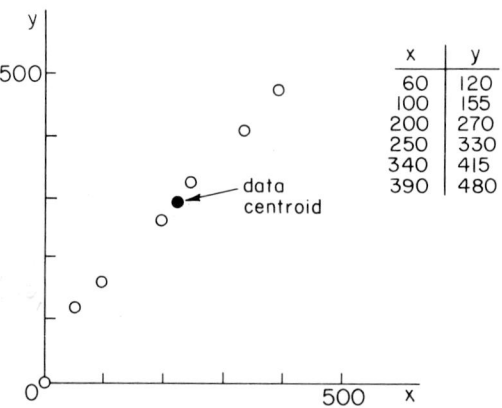
Fig. 5.22. Data rectified when plotted on linear (Cartesian) coordinates.

data points at a glance.

Suppose that data on Fig. 5.22 are plotted on Cartesian coordinates with x as the abscissa and y as the ordinate. The trend of the data is fairly straight with point-to-point line segments nearly constituting a straight line. If the observer is satisfied with a straight-line representation of the data (if such an approximation is satisfactory to the purpose), the problem is to draw a straight line that "best" represents the data. Of all possible straight lines, which is the best? One measure of goodness-of-fit is the sum of the squares of the deviations from the fitted straight line. Let us presume that the data relationship is linear and the deviations from linearity are entirely due to errors in the y-measurement. That is to say, the x-measurement is either error-free or the error in x is very small compared with the error in y. In this case, Fig. 5.23 indicates in an exaggerated form the errors associated with each data point in connection with a candidate straight line. Denoting the i-th abscissa as x_i, the i-th ordinate as y_i, and the error associated with the i-th measurement as e_i, we can write

$$e_i = y_i - a - bx_i = y_i - \hat{y}$$

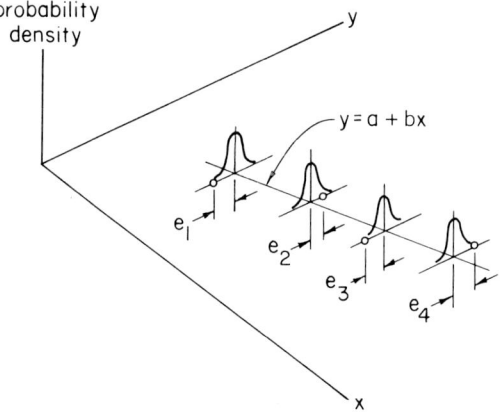
Fig. 5.23. The rationale in establishing a least squares straight line.

Empiricism

where b is the slope of the straight line y = a + bx and a is the ordinate intercept. Our measure of the goodness-of-fit can be the sum of the squares of all the e_i:

$$S = \sum_{i=1}^{n} e_i^2 = e_1^2 + e_2^2 + \ldots + e_n^2 = \sum_{i=1}^{n} (y_i - a - bx_i)^2$$

The best fit is the line that makes S a minimum. The parameters a and b can be adjusted to bring this about, and the constants so determined are called the least squares regression parameters. If we differentiate S with respect to a and equate to zero and differentiate S with respect to b and equate to zero, the optimal parameters a and b are established. The differentiation with respect to a is most interesting for it reveals a useful property of the best line.

$$\frac{\partial S}{\partial a} = \sum_{i=1}^{n} 2(y_i - a - bx_i)(-1) = 0$$

$$\sum_{i=1}^{n} y_i - \sum_{i=1}^{n} a - \sum_{i=1}^{n} bx_i = 0$$

If every term in the last equation is divided by n, the number of data points, $\sum y_i/n$ is recognized as the average ordinate or the y-component of the centroid of the pattern of data points. Likewise, $\sum x_i/n$ is recognized as the x-component of the centroid of the pattern of data points. Thus the preceding equation can be written

$$\bar{y} = a + b\bar{x}$$

The interesting result is that the centroid of the data points lies on the best least square line. This fact is useful in drawing a best straight line by eye. The *eyeball-best-line* can be drawn through the centroid with a transparent straightedge, Fig. 24(a), and oriented so that the fidelity of the line to the data is satisfying to the eye (one degree of freedom). For many purposes the best line drawn by eye (two degrees of freedom) is sufficiently accurate; see Fig. 24(b). The centroid property is not general to curves other than straight lines.

If we choose to draw an eyeball-best straight line, the next step is to establish the equation for the drawn line, i.e., determine a and b of y = a +

Fig. 5.24. Two common techniques for establishing a straight line that represents data.

bx. Locate two points on the drawn line that will be the basis for establishing the equation of the line and read the coordinates x_1, x_2, y_1, and y_2. The points selected should be near the extremes of the line segment so as to reduce the influence of the coordinate reading error. From the equation of a straight line (from analytical geometry),

$$y = a + bx \tag{5.2}$$

we form two equations with the coordinates (x_1, y_1) and (x_2, y_2) as follows

$$y_1 = a + bx_1$$
$$y_2 = a + bx_2 \tag{5.3}$$

for semi $y = \ln y$, $x = x$
for log-log $y = \log y$, $x = \ln x$

These two simultaneous equations with slope a and the ordinate intercept b may be solved for a and b, using Cramer's rule:

$$b = \frac{\begin{vmatrix} y_1 & 1 \\ y_2 & 1 \end{vmatrix}}{\begin{vmatrix} x_1 & 1 \\ x_2 & 1 \end{vmatrix}} = \frac{y_1 - y_2}{x_1 - x_2} \quad a = \frac{\begin{vmatrix} x_1 & y_1 \\ x_2 & y_2 \end{vmatrix}}{\begin{vmatrix} x_1 & 1 \\ x_2 & 1 \end{vmatrix}} = \frac{x_1 y_2 - x_2 y_1}{x_1 - x_2}$$

$b = \frac{\Delta y}{\Delta x}$

$a = \frac{y_1 x_1 - y_1 x_2}{x_2 - x_1}$

$y = a + bx$

Substitution of these values for a and b into Eq. (5.2) yields

$$y = a + bx = \frac{x_1 y_2 - x_2 y_1}{x_1 - x_2} + \frac{y_1 - y_2}{x_1 - x_2} x$$

A simpler relation can be obtained from subtracting Eq. (5.3) from Eq. (5.2) and substituting the value for b found from Cramer's rule,

$$y - y_1 = b(x - x_1) = \frac{y_1 - y_2}{x_1 - x_2}(x - x_1)$$

or more easily remembered,

$$\frac{y - y_1}{x - x_1} = \frac{y_1 - y_2}{x_1 - x_2}$$

If the coordinates read from the drawn best line are (80, 138) and (350, 435), then

$$\frac{y - 138}{x - 80} = \frac{138 - 435}{80 - 350} \qquad y = 1.11x + 49.5$$

or regression line.

Least square parameters can be established as follows. Display the sum of the squared discrepancies between the best line and the data points.

$$S = \Sigma(y - a - bx)^2$$

Differentiating S with respect to a and b yields

$$\frac{\partial S}{\partial a} = 2 \Sigma(y - a - bx)(-1)$$

$$\frac{\partial S}{\partial a} = 2 \Sigma(y - a - bx)(-x)$$

Empiricism

Equating the partial derivatives to zero results in

$$\Sigma y = na + b\Sigma x$$

$$\Sigma xy = a\Sigma x + b\Sigma x^2$$

Simultaneous solution for the parameters a and b yields

$$a = \frac{\Sigma x^2 \Sigma y - \Sigma x \Sigma xy}{n\Sigma x^2 - (\Sigma x)^2} \qquad (5.4a)$$

$$b = \frac{n\Sigma xy - \Sigma x \Sigma y}{n\Sigma x^2 - (\Sigma x)^2} \qquad (5.4b)$$

Note that it is possible to establish a and b from the numerical data without plotting, which is necessary with the eyeball-best-line method. (Careful engineers *always* plot the data to judge their satisfaction with the straightness of the data string and appropriateness of regression line.) If the transformation h(y) applied to the y-coordinates of data pairs and g(x) applied to x-coordinates of data pairs results in rectification of the data string of form

$$h(y) = a + bg(x)$$

then Eqs. (5.4a) and (5.4b) become

$$a = \frac{\Sigma g^2(x) \Sigma h(y) - \Sigma g(x) \Sigma g(x) h(y)}{n\Sigma g^2(x) - [\Sigma g(x)]^2} \qquad (5.4c)$$

$$b = \frac{n\Sigma g(x) h(y) - \Sigma g(x) \Sigma h(y)}{n\Sigma g^2(x) - [\Sigma g(x)]^2} \qquad (5.4d)$$

The method of least squares allows confidence bounds on the parameters a and b to be determined and expressed, as well as confidence bounds of the line as a whole to be displayed. Methods for doing this are part of a first course in statistics or measurements. The advent of the pocket calculator has made this method attractive.

Straight lines are easily drawn and easily extrapolated, and the algebraic representation is simple. For these reasons, rectification of curves is sought as a means to simplify the determination of nonlinear curve fitting equations. The proper transformation applied to ordinate and abscissa will rectify curves, but the transformation may not be known a priori.

Rectification is a powerful curve fitting tool. If you have a set of paired data $\{(x, y)_n\}$ that is nonlinear when plotted in the xy-plane, you may be able to discover transformations that can be applied to each set, i.e., the x's or the y's or both, that will result in a linear plot. The linear plot is the authority to equate the ordinal transformation h(y) to the abscissa transformation g(x) such that

$$h(y) = a + bg(x)$$

If you know the coordinates of two points on the rectified locus, then

$$h(y_1) = a + bg(x_1) \qquad h(y_2) = a + bg(x_2)$$

The constant b is the mathematical slope of the rectified line. By solving the above equations simultaneously for b, we obtain

$$b = \frac{h(y_1) - h(y_2)}{g(x_1) - g(x_2)}$$

In preparing the graphical plot, scale factors must have been chosen such that

$$Y = s_{h(y)} h(y) \qquad X = s_{g(x)} g(x)$$

[margin notes: SCALE FACTOR & SCALE VALUES; $S_g(x) = X_2 - X_0 / x_2 - x_0$; $S_h(y) = Y_2 - Y_0 / y_2 - y_0$; if Y_0 & X_0 are not at plotting origin]

Consequently

$$b = \frac{\left(\dfrac{Y_1}{s_{h(y)}} - \dfrac{Y_2}{s_{h(y)}}\right)}{\left(\dfrac{X_1}{s_{g(x)}} - \dfrac{X_2}{s_{g(x)}}\right)} = \frac{s_{g(x)}}{s_{h(y)}} \left(\frac{Y_1 - Y_2}{X_1 - X_2}\right) = \frac{s_{g(x)}}{s_{h(y)}} \left(\frac{\Delta Y}{\Delta X}\right) \qquad (5.5)$$

In words this means

$$\begin{pmatrix}\text{mathematical}\\ \text{slope of}\\ \text{rectified line}\end{pmatrix} = \begin{pmatrix}\text{abscissa-scale factor}\\ \text{ordinate-scale factor}\end{pmatrix} \begin{pmatrix}\text{physical or}\\ \text{measured slope}\\ \text{of rectified line}\end{pmatrix}$$

The expression for a is

$$a = \frac{1}{s_{h(y)}} \left(\frac{X_2 Y_1 - X_1 Y_2}{X_2 - X_1}\right) \qquad (5.6)$$

[margin note: for log-log $y = e x^b$ (or similar)]

These constants can also be ascertained from Eqs. (5.4a) and (5.4b).
Points are located on a graph by measurement or by coordinate. To plot points by *measurement*, the horizontal and vertical displacements of the point from the plotting origins are measured with an engineer's scale or other finely divided precision scale. Accuracy by this method is high. To plot points by *coordinate*, use the graphs of the coordinates (scale ticks) as the basis for locating the plotted points. Precision suffers because of interpolation but convenience is high, especially on graphs with nonlinear scales. Commercially prepared paper with common transformations, such as log-log or semilog, are convenient in plotting by coordinate. In establishing the equation of a locus from an existing graph, the prudent engineer measures rather than reads coordinates in the interest of precision.

Example 5. Plot by measurement; scale by measurement with linear scales.
Consider the data set

$$\{x_n\} = \{0, 1, 2, 3\} \qquad \{y_n\} = \{4, 1, 4/9, 1/4\}$$

which is "clean" (free of experimental error). The engineer decides to plot the data using two Cartesian scales that are laid out as depicted in Fig. 5.25(a). The scale factors are identified by point c on the abscissa and point d on the ordinate.

$$s_{g(x)} = [X/g(x)]_c = 3/3 = 1 \text{ in.} \qquad s_{h(y)} = [Y/h(y)]_d = 2/2 = 1 \text{ in.}$$

The data will not plot linearly on these coordinates; either from trial and error or physical insight the investigator suspects that the transformation $g(x) \equiv x$ and $h(y) \equiv 1/\sqrt{y}$ will rectify the data string. The displacement of the data from the respective plotting origins is $X = s_{g(x)} x$ in. and $Y = s_{h(y)} y$ in. for abscissa and ordinate, respectively, and the following chart is prepared.

Empiricism

			X	Y
x	y	$1/\sqrt{y}$	(in.)	(in.)
0	4	0.5	0	0.5
1	1	1.0	1	1.0
2	4/9	1.5	2	1.5
3	1/4	2.0	3	2.0

Using these measurements, X and Y, the circled points in Fig. 5.25(a) are located and plotting by measurement has been accomplished. The data string has been rectified to the satisfaction of the engineer and the eyeball-best-line has been drawn. Another engineer who sees the plot, finds it useful, and wants an equation representing the eyeball-best-line, selects two points on the line (not data points) widely spaced in the range of the data. These points are labeled 1 and 2 in Fig. 5.25(a). The distances of these points from the respective plotting origins are obtained by measuring with an engineer's scale.

$X_1 = 1/2$ in. $X_2 = 5/2$ in. $\Delta X = 2$ in.

$Y_1 = 3/4$ in. $Y_2 = 7/4$ in. $\Delta Y = 1$ in.

The slope of the line $h(y) = a + bg(x)$ is given by Eq. (5.5) as

$$b = [s_{g(x)}/s_{h(y)}](\Delta Y/\Delta X) = (1/1)(1/2) = 0.5$$

The ordinate intercept is given by Eq. (5.6) as

$$a = \frac{1}{s_{h(y)}}\left(\frac{X_2 Y_1 - X_1 Y_2}{X_2 - X_1}\right) = \frac{1}{1}\left[\frac{(5/2)(3/4) - (1/2)(7/4)}{5/2 - 1/2}\right] = 0.5$$

The equation relating x and y is recovered by writing

$$h(y) = a + bg(x) = 0.5 + 0.5g(x) = [1 + g(x)]/2$$

Substituting $1/\sqrt{y}$ for $h(y)$ and x for $g(x)$ yields

$$1/\sqrt{y} = (1 + x)/2 \text{ or } y = 4/(1 - x)^2$$

Had the engineer decided to use ordinates and abscissas with different scale factors, the plotted points and eyeball-straight-line would appear as depicted in Fig. 5.25(b). Establishing the equation of the eyeball-best-line, using points 1 and 2 proceeds as follows. The scale factors are determined using points c and d:

$s_{g(x)} = [X/g(x)]_c = 1.5/3 = 0.5$ in. $s_{h(y)} = [Y/h(y)]_d = 4/2 = 2$ in.

The distances of points 1 and 2 from the respective plotting origins are obtained by measuring with an engineer's scale.

$X_1 = 1/4$ in. $X_2 = 5/4$ in. $\Delta X = 1$ in.

$Y_1 = 3/2$ in. $Y_2 = 7/2$ in. $\Delta Y = 2$ in.

The slope of the line $h(y) = a + bg(x)$ is given by Eq. (5.5):

$$b = s_{g(x)}/s_{h(y)}(\Delta Y/\Delta X) = (0.5/2)(2/1) = 0.5$$

The ordinate intercept is given by Eq. (5.6):

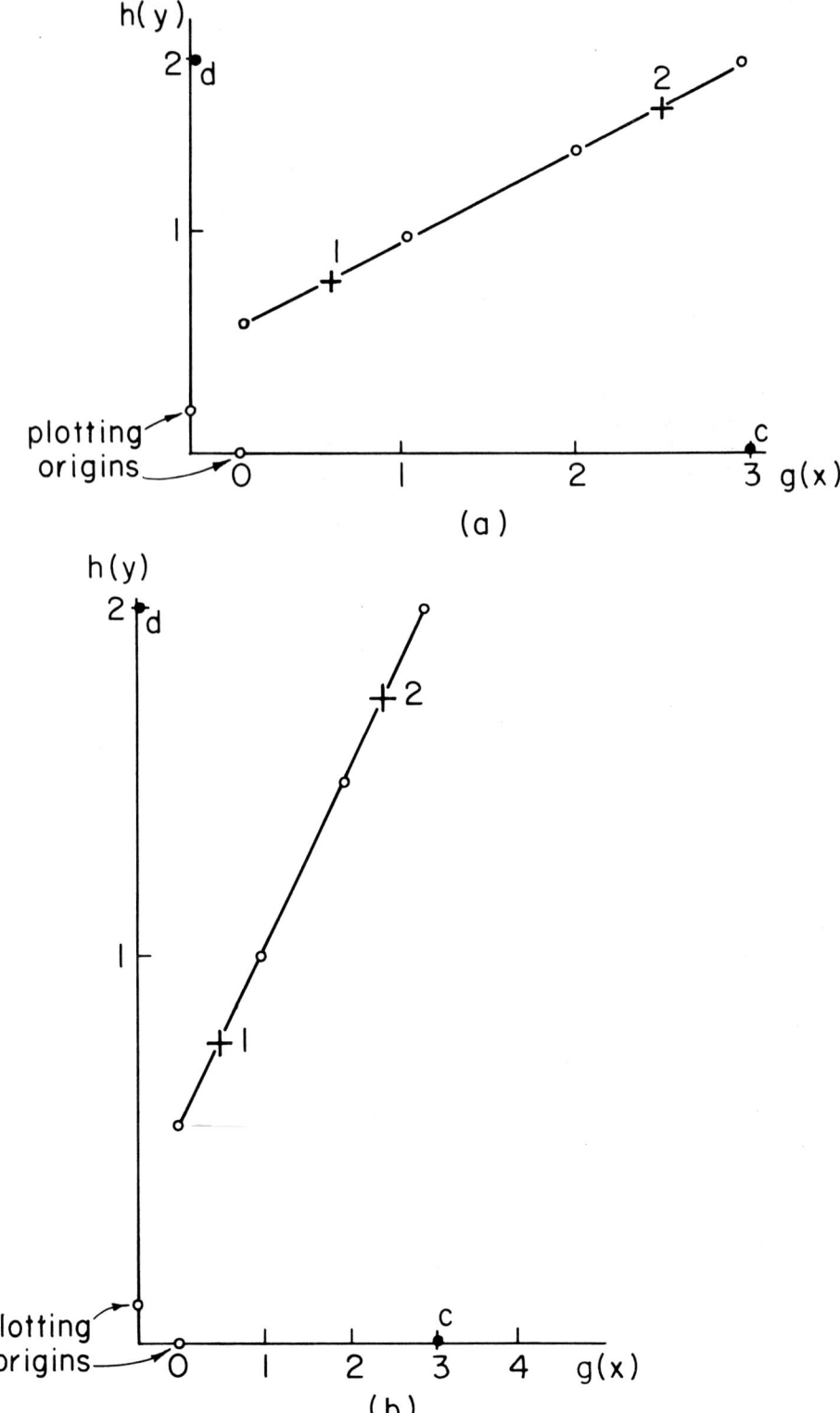

Fig. 5.25. (a) Data points are plotted by measurement in Example 5; equal scale factors are used for ordinate and abscissa. (b) Data points of Example 5 plotted using unequal scale factors.

Empiricism

$$a = \frac{1}{s_{h(y)}} \left(\frac{X_2 Y_1 - X_1 Y_2}{X_2 - X_1} \right) = \frac{1}{2} \left[\frac{(5/4)(3/2) - (1/4)(7/2)}{5/4 - 1/4} \right] = 0.5$$

and the equation $y = 4/(1 - x)^2$ is recovered as before.

Example 6. Plot by coordinate; scale by measurement with nonlinear scales. Consider again the data set

$$\{x_n\} = \{0, 1, 2, 3\} \quad \{y_n\} = \{4, 1, 4/9, 1/4\}$$

The engineer has determined again that $g(x) \equiv x$ and $h(y) \equiv 1/\sqrt{y}$ will rectify the data and wishes to plot by coordinate using a nonlinear scale for ordinate and a linear scale for abscissa that will rectify the data. For the abscissa the engineer decides to set $s_X = 0.5$ in. The scale ticks of the abscissa are established from

$X = s_X x = 0.5x$ in. $X_2 = 0.5(2) = 1.0$ in.

$X_0 = 0.5(0) = 0$ in. $X_3 = 0.5(3) = 1.5$ in.

$X_1 = 0.5(1) = 0.5$ in.

Scale ticks for the ordinates are established from the decision to make $s_{1/\sqrt{y}} = 2$ in. and it follows that

$Y = s_{1/\sqrt{y}} (1/\sqrt{y}) = 2/\sqrt{y}$ in.

$Y_{0.25} = 2/\sqrt{0.25} = 4.00$ in. $Y_3 = 2/\sqrt{3} = 1.155$ in.

$Y_{0.30} = 2/\sqrt{0.30} = 3.651$ in. $Y_4 = 2/\sqrt{4} = 1.000$ in.

$Y_{0.50} = 2/\sqrt{0.50} = 2.828$ in. $Y_5 = 2/\sqrt{5} = 0.894$ in.

$Y_1 = 2/\sqrt{1} = 2.000$ in. $Y_{10} = 2/\sqrt{10} = 0.632$ in.

$Y_2 = 2/\sqrt{2} = 1.414$ in. $Y_\infty = 0$ in.

and the coordinate scales can be created as shown in Fig. 5.26. The data, plotted by coordinates, are shown as circled points in the figure with the eyeball-best-line drawn. Another engineer, who sees the published data and needs the equation of the eyeball-best-line, recovers the abscissa scale factor from $s_{g(x)} = X/x$. The measurement from the plotting origin of the abscissa to point c, using an engineer's scale, is 1.5 in. The coordinate of c is 3; therefore

$$s_{g(x)} = (X/x)_c = 1.5/3 = 0.50 \text{ in.}$$

The measurement from the plotting origin of the ordinate to point d, using an engineer's scale, is 4 in.; since the coordinate of d is 0.25,

$$s_{h(y)} = \left(\frac{Y}{1/\sqrt{y}} \right)_d = (Y\sqrt{y})_d = 4\sqrt{0.25} = 2 \text{ in.}$$

Establishing the equation of the eyeball-best-line from points 1 and 2 on the line involves measuring their displacements from the plotting origins with an engineer's scale:

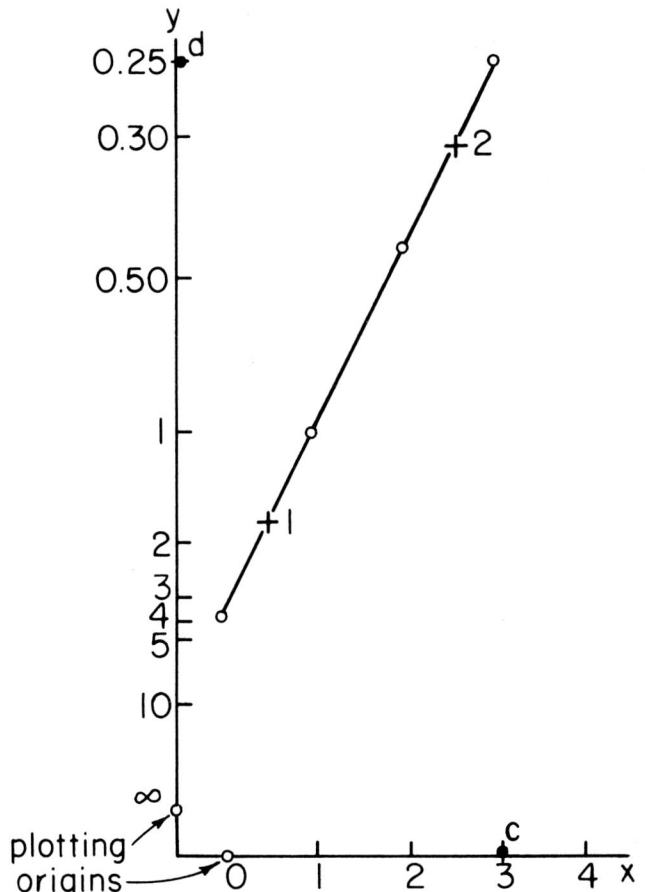

Fig. 5.26. Plotting by coordinate, then recovering equation of rectified line.

X_1 = 0.25 in. X_2 = 1.25 in. ΔX = 1 in.

Y_1 = 1.5 in. Y_2 = 3.5 in. ΔY = 2 in.

The slope of the line $h(y) = a + bg(x)$ is given by Eq. (5.5):

$b = [s_{g(x)}/s_{h(y)}](\Delta Y/\Delta X) = (0.5/2)(2/1) = 0.5$

The ordinate intercept is given by Eq. (5.6)

$$a = \frac{1}{s_{h(y)}}\left(\frac{X_2 Y_1 - X_1 Y_2}{X_2 - X_1}\right) = \frac{1}{2}\left[\frac{(5/4)(3/2) - (1/4)(7/2)}{5/4 - 1/4}\right] = 0.5$$

Now

$h(y) = a + bg(x) = 0.5 + 0.5g(x) = [1 + g(x)]/2$

Substituting $1/\sqrt{y}$ for $h(y)$ and x for $g(x)$ yields

$1/\sqrt{y} = (1 - x)/2$ or $y = 4/(1 - x)^2$

Empiricism

Since rectification problems begin with data and the object is the discovery of a satisfactory mathematical curve-fit, knowledge of the algebraic forms rectified by particular transformation schemes is not generally available. The discovery of the proper transforms to rectify data for approximation with algebraic expressions is heuristic. The knowledge of a few such transformations and the corresponding algebraic form that is rectified is useful. If the data set is transformed by applying the same transformation (the natural logarithm) to $\{x_n\}$ and $\{y_n\}$ and $\{\ln x_n\}$ and $\{\ln y_n\}$ is rectified when plotted on Cartesian coordinates, then the rectification is the authority to write

$$h(y) = a + bg(x)$$

where the constants a and b can be determined by an eyeball-best-line through the data using Eqs. (5.5) and (5.6) or with Eqs. (5.4c) and (5.4d) employing least squares technique. Substituting the transform specifics into the above equation yields for the natural logarithmic transformation,

$$\ln y = a + b \ln x = a + \ln x^b$$

Raising e to the $\ln y$ power gives

$$\exp(\ln y) = \exp(a + \ln x^b) = \exp(a)\exp(\ln x^b) = \exp(a)x^b$$

or $y = cx^b$. For the base 10 logarithmic transform applied to ordinate and abscissa,

$$\log y = a' + b' \log x$$

where

$$a' = \frac{1}{s_{\log y}}\left(\frac{X_2 Y_1 - X_1 Y_2}{X_2 - X_1}\right) = \frac{a}{\ln 10} \qquad b' = b$$

Raising 10 to the $\log y$ power results in

$$10^{\log y} = 10^{a'} \cdot 10^{\log x^b}$$

Then $y = 10^{a'} x^b$ or $y = cx^b$, as before, demonstrating the interchangeability of the natural and base 10 logarithmic transform for purposes of rectifying data.

Alternatively, the data pairs x_i, y_i can be plotted by coordinate on log-log paper and the data string will be rectified if the coordinate y is a power of x. One may suspect that the transformation will rectify the data string because of the physics of the problem and merely be seeking conformation, or one may inquire if y is a power of x and interpret the straightness or lack of it as an answer.

As an example, consider the data

x	y	$\ln x$	$\ln y$	$\ln^2 x$	$(\ln x)(\ln y)$
1.75	3.0	0.560	1.099	0.313	0.615
3.20	4.7	1.163	1.548	1.353	1.800
8.00	7.7	2.079	2.041	4.324	4.245
20.5	13.5	3.020	2.603	9.123	7.861
35.0	21.0	3.555	3.045	12.640	10.821
72.0	31.0	4.277	3.434	18.290	14.686
Σ		14.655	13.769	46.044	40.031

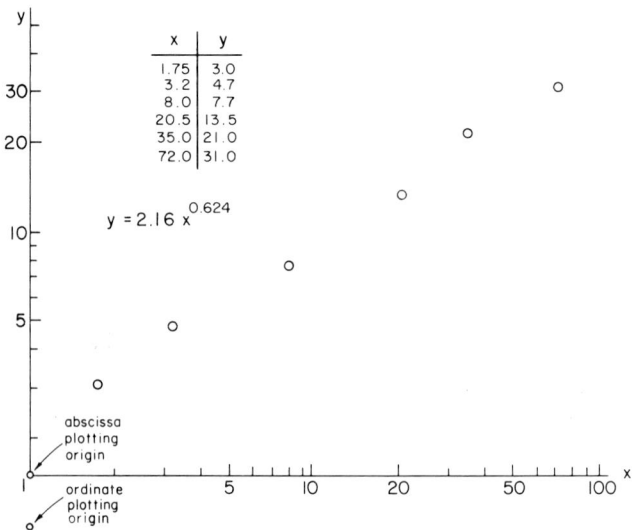

Fig. 5.27. Data rectified by applying a logarithmic transformation to both coordinates (figure is 0.408 actual size).

If x and y are plotted on commercial 2-cycle by 2-cycle paper (or a portion of a larger paper), they result in a string of data points in nearly linear array (see Fig. 5.27). The scale factors of the graph are

$$s_{\ln x} = X/\ln x = 3.75/\ln 10 = 1.63 \text{ in.}$$

$$s_{\ln y} = Y/\ln y = 3.75/\ln 10 = 1.63 \text{ in.}$$

The physical slope by measurement is 0.6246; consequently the mathematical slope is

$$b = (s_{\ln x}/s_{\ln y})(\Delta Y/\Delta X) = (1.63/1.63)(0.6246) = 0.6246$$

The value of the constant a reading directly from an eyeball-best-line on the graph is 2.14. It may be obtained using Eq. (5.4c) in the form

$$a = \frac{(\Sigma \ln^2 x)(\Sigma \ln y) - \Sigma \ln x \cdot \Sigma (\ln x)(\ln y)}{n \Sigma \ln^2 x - (\Sigma \ln x)^2} = \frac{46.044(13.789) - 14.655(40.031)}{6(46.044) - 14.655(14.655)}$$

$$= 0.770$$

Equation (5.4d) is particularized as

$$b = \frac{n(\Sigma \ln x \cdot \ln y) - (\Sigma \ln x)(\Sigma \ln y)}{n \Sigma \ln^2 x - (\Sigma \ln x)^2} = \frac{6(40.031) - (14.655)(13.769)}{6(46.044) - 14.655(14.655)} = 0.624$$

Now

$$h(y) = a + bg(x) = 0.770 + 0.624 g(x)$$

$$\ln y = 0.770 + 0.624 \ln x$$

$$\exp(\ln y) = \exp(0.770 + 0.624 \ln x) = \exp(0.770) x^{0.624}$$

The curve-fit is $y = 2.159 x^{0.624}$, using least square parameters.

Empiricism   137

Another transformation that rectifies data associated with growth and decay is the natural logarithm transform of, say, the y's with no transformation (the identity transform) of the x's. Rectification of plotted points on an x-$\ln$ y graph or commercial semilog paper is authority to say

$$h(y) = a + bg(x)$$

where

$$a = \frac{1}{s_{h(y)}}\left(\frac{X_2 Y_1 - X_1 Y_2}{X_2 - X_1}\right) \qquad b = \frac{s_{g(x)}}{s_{h(y)}}\left(\frac{\Delta Y}{\Delta X}\right)$$

If the y-transformation is the Naperian base logarithm, then $\ln y = a + bx$. Raising e to the $\ln y$ power yields

$$\exp(\ln y) = \exp(a + bx) = \exp(a)\exp(bx)$$

or $y = ce^{bx}$. If the y-transformation is base 10 logarithm, then $\log y = a' + b'x$, where

$$a' = \frac{1}{s_{\log y}}\left(\frac{X_2 Y_1 - X_1 Y_2}{X_2 - X_1}\right) = \frac{a}{\ln 10} \qquad b' = \frac{s_x}{s_{\log y}}\left(\frac{\Delta Y}{\Delta X}\right) = \frac{b}{\ln 10}$$

Replacing $\log y$ with $\ln y/\ln 10$, a' with $a/\ln 10$, and b' with $b/\ln 10$, we have

$$\frac{\ln y}{\ln 10} = \frac{a}{\ln 10} + \frac{b}{\ln 10} x$$

and we recover $y = ce^{bx}$. However, beginning again with $\log y = a' + b'x$ and raising 10 to the $\log y$ power yields

$$10^{\log y} = 10^{(a' + b'x)} = 10^{a'} \cdot 10^{b'x}$$

or $y = c' 10^{b'x}$, which may be useful. It is equivalent to $y = ce^{bx}$. We learn that exponential curves of the form $y = ce^{bx}$ or $y = ck^{b'x}$ are rectified by taking the logarithm of the dependent variable. For example, consider the case in which natural logarithmic transform has been applied to the y-coordinate only.

x	y	$\ln y$	x^2	$x \ln y$
0.6	3.4	1.22	0.360	0.73
1.4	5.6	1.72	1.96	2.41
2.0	8.8	2.17	4.00	4.35
3.1	18.0	2.89	9.61	8.96
3.8	33.0	3.50	14.44	13.29
4.8	60.6	4.10	23.04	19.70
Σ 15.70	129.4	15.61	53.41	49.44

We choose to plot the data on semilog paper with x on the Cartesian axis (abscissa), as depicted in Fig. 5.28. From the plot,

$$s_x = X/x = 2.5/5 = 0.500 \text{ in.}$$

$$s_{\ln y} = Y/\ln y = 2/\ln 10 = 0.869 \text{ in.}$$

The physical slope of the rectified line is $3.02/2.5 = 1.21$, so the mathematical slope is

$$b = \frac{s_x}{s_{\ln y}}\left(\frac{\Delta Y}{\Delta X}\right) = \frac{0.500}{0.869}(1.21) = 0.696$$

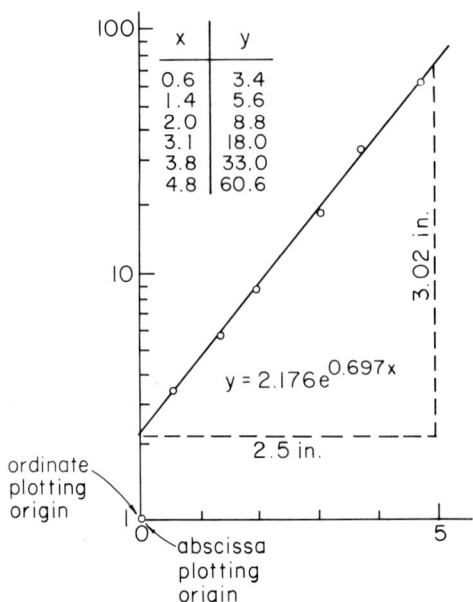

Fig. 5.28. Data rectified by applying a logarithmic transformation to one coordinate (the ordinate). (Figure is 0.638 actual size.)

or from Eq. (5.4d),

$$b = \frac{n(\Sigma x \ln y) - \Sigma x (\Sigma \ln y)}{n \Sigma x^2 - (\Sigma x)^2} = \frac{6(49.44) - 15.70(15.61)}{6(53.41) - 15.70(15.70)} = 0.697$$

If the constant a cannot be read from the graph, one can use Eq. (5.6) or Eq. (5.4c) in the following form:

$$a = \frac{\Sigma x^2 (\Sigma \ln y) - \Sigma x (\Sigma x \ln y)}{n \Sigma x^2 - (\Sigma x)^2} = \frac{53.41(15.61) - 15.70(49.44)}{6(53.41) - 15.70(15.70)} = 0.778$$

c = exp(0.778) = 2.176

Direct reading from the graph indicates c = 2.2, approximately.

As the result of reading and study of the material in this section the student should appreciate that

--Since straight lines are the easiest curves to handle, many schemes that rectify the data string are employed.
--Curve fitting techniques that are associated with data string rectification include eyeball-best-line, eyeball-best-line through data centroid, and least squares methods of parameter determination.
--The transformations used to rectify data strings need not be the same for both variables involved.
--The authority for the functional relationship developed depends on the goodness-of-fit of the rectified data string to a straight line and any physical insight associated with the transformation involved.
--When an ordinate transformation h(y) is applied to y-coordinates of data and an abscissa transform g(x) is applied to x-coordinates of data and the data string is rectified to the observer's satisfaction, the best straight line through the data is given by

Empiricism

$$h(y) = a + bg(x) \tag{5.7}$$

where

$$a = \frac{1}{s_{h(y)}}\left(\frac{X_2 Y_1 - X_1 Y_2}{X_2 - X_1}\right) = \frac{\Sigma g^2(x)\ \Sigma h(y) - \Sigma g(x)\ \Sigma g(x)h(y)}{n\ \Sigma g^2(x) - [\Sigma g(x)]^2} \tag{5.8}$$

$$b = \frac{s_{g(x)}}{s_{h(y)}}\left(\frac{Y_2 - Y_1}{X_2 - X_1}\right) = \frac{n\ \Sigma g(x)h(y) - \Sigma g(x)\ \Sigma h(y)}{n\ \Sigma g^2(x) - [\Sigma g(x)]^2} \tag{5.9}$$

5.5 REDUCTION OF NUMBER OF VARIABLES

The amount of experimental work needed to describe phenomena to the necessary precision for intended use can be measured by the number of experiments required to sufficiently describe the hypersurface

$$x_1 = f(x_2, \ldots, x_n)$$

or, alternatively, to define the function

$$g(x_1, x_2, \ldots, x_n) = 0$$

For illustration let us say that it takes p points to establish a curve to our satisfaction and that p contours on a graph specify the role of another variable to our satisfaction. What number of experiments are necessary to describe a phenomenon involving n parameters, i.e., $x_1, x_2, \ldots, x_n$? If we take a sheet of graph paper and plot x_1 as the ordinate and x_2 as the abscissa, as shown in Fig. 5.29(a) with all other parameters $x_3, x_4, \ldots, x_n$ held constant, it takes p points to establish a locus to our satisfaction. The role of parameter x_3 can be introduced by performing the p experiments with one value of x_3, say $(x_3)_1$. This is done p times resulting in p contours on the graph paper. To this point we have expended p^2 points. Note that the introduction of the third variable increased the experimental effort exponentially. In Fig. 5.29(b) we have p curves of p points each. The role of the fourth variable x_4 may be revealed by constructing Fig. 5.29(b) p times with p different values of x_4. In other words, the role of the fourth variable may be shown with a chapter of p pages containing p curves of p points each, as depicted in Fig. 5.29(c).

The next parameter x_5 requires a book of p chapters of p pages of p curves of p points each. The parameter x_6 requires a shelf of p books of p chapters of p pages of p curves of p points each. Each additional parameter that is significant to the description of the phenomenon requires bookcases, rooms, floors, buildings, etc., to delineate the phenomenon. If the requisite number of points per locus is p and n is the number of parameters describing the phenomenon, the number of experimental points necessary is

$$N = p^{n-1} \tag{5.10}$$

where N is the number of experiments. If $p = 5$ and $n = 7$, then $N = 5^6 = 14{,}625$ experiments. If an experimental determination costs $100 or $1000 per point, understanding the phenomenon can be an expensive proposition. Is there any way to reduce this investment of time and money?

If a phenomenon is described by $f(N, k, d, G, D) = 0$ and we know that $k = d^4 G/(8D^3 N)$, we could plot k versus $d^4 G/(D^3 N)$, observe a straight-line data string, and establish the constant 8 and the above equation with only p points rather than the p^4 points necessary to the parameter-by-parameter approach.

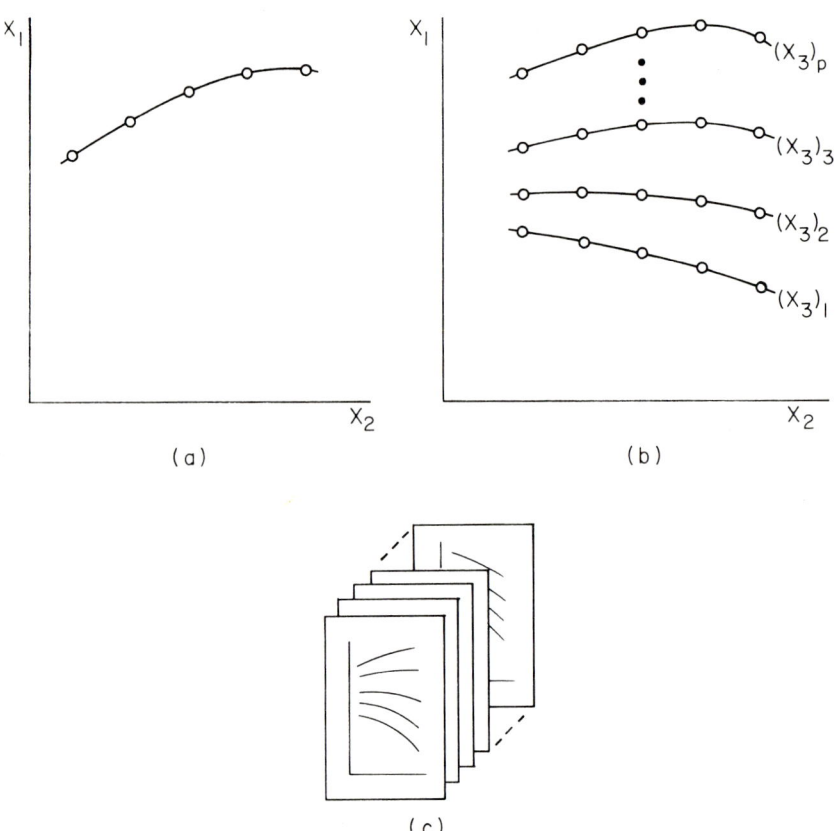

Fig. 5.29. (a) A plot relating X_1 to X_2 with all other variables held constant. (b) A family of loci resulting from X_3 being held at different levels. (c) A chapter of graphs resulting from X_4 being held at different levels.

Likewise, we could plot k versus d^4/D^3 for constant G/N and obtain the same result with p^2 points. If we plot Nkd versus GD we discover nothing but confusion. Knowing the answer, we can spot the groups of variables that are naturally clustered and as a result of working with them reduce the dimensionality of the problem. Alas, we do not know the solution in advance. What then would suggest, in the absence of the answer, a natural clustering of parameters that would *always* reduce the dimensionality of the problem and *always* be fruitful to use in pursuing the understanding that will lead to the building of a satisfactory mathematical model? We are indebted to E. Buckingham for the pregnant idea of clustering parameters for dimensional reasons.

The units of stress are lbf/in.^{2} and the units of modulus of elasticity are lbf/in.^{2}; consequently the strain, $\varepsilon = S/E$ (the quotient of stress and modulus of elasticity), is unitless. Or the dimensions of stress are force/length^2 and the dimensions of modulus of elasticity are force/length^2; hence ε, the strain, is dimensionless. To help in visualizing the strain, its units are reported as in./in. The idea of Buckingham's is even more basic than this example suggests. Some units, which may be regarded as fundamental, are not expressible in terms of the others. For instance, time may be expressed in seconds, hours, days, or centuries, but the fundamental idea of time and its

Empiricism

measure (counting on clocks) is common to all these units. *Time*, the concept for ordering the flow of events, is unique. The dimension time (symbolized as T) is fundamental; i.e., $\dim(t) = T$.

The concept used for quantitative description of size, formalized in Euclid's plane geometry, is *length* (symbolized as L); and the fundamental notion of length is common to units such as metres, inches, light-years, and angstroms.

The more subtle concept of a common property of all matter, *mass*, characterizes a body in gravitational and mechanical responses. This idea is not expressible in terms of length and time and thus is also fundamental. The concepts of *temperature* and electrical *charge* are likewise fundamental.

Ideas such as *velocity*, which has dimensions of length/time (or L/T), are expressible in terms of fundamental dimensions and they themselves are called *secondary* or *derived* quantities. The concept of force may seem fundamental but it is expressible in terms of mass, time, and acceleration as MLT^{-2}. If one desires to call force a fundamental dimension, then mass is derived and expressible as force over acceleration or $FL^{-1}T^2$. Because both these viewpoints are convenient in engineering mechanics, ==investigators speak of the MLT system or the FLT system to communicate the fundamental set used==.

Buckingham suggested that a group of dimensionless parameters be formed from the set of pertinent parameters describing a phenomenon. Such groups are to be formed by multiplication, probably explaining the use of the symbol π-term. A π-term is formed as follows:

$$\pi = x_1^{e_1} x_2^{e_2} x_3^{e_3} \ldots x_n^{e_n}$$

All the parameters from the descriptive set are multiplied together after being raised to the powers of $e_1, e_2, \ldots, e_n$. His argument is that the proper selection of the exponents would lead to a dimensionless term π. To choose the set $\{e_n\}$, the rule of dimensional homogeneity is adopted. The rule states that analytically derived equations modeling phenomena must be valid for all systems of units.

For a body falling from rest the relation

$$s = (1/2)gt^2 \qquad S = V_i t + \tfrac{1}{2} a t^2 \quad \text{where } V_i = 0 \ \& \ a = g$$

is dimensionally homogeneous. The body falls without regard to artificial concepts such as units, and if the phenomenon that is unique is to be modeled by a unique equation, the expression must be independent of units. The expression $s = 16.1 t^2$ is nonhomogeneous and violates the rule of dimensional homogeneity, since t *must* be in seconds and s *must* be in feet. The distance fallen in one second in vacuum near the earth's surface is

$$s = (1/2)gt^2 = (1/2)(32.2)(1)^2 = 16.1 \text{ ft}$$

for an "English" observer and

$$s = (1/2)gt^2 = (1/2)(980)(1)2 = 490 \text{ cm}$$

for a "metric" observer. The distances are identical when scaled off in the appropriate units. This example has another surprising attribute:

$$gt^2/s = 2$$

which is a pure number. A body falling freely in a vacuum near the surface of the earth falls so as to maintain $gt^2/s = 2$. If you wish to describe the phenomenon of a freely falling body, you might conclude from preliminary work that the phenomenon is describable in terms of the parameters s, t, and g, or that f(s, t, g) = 0. If we use the parameter-by-parameter approach, we conduct p experiments at constant g and repeat p times with different g (at grossly different elevations above the earth or on different planets, moons, or asteroids). With p^2 experiments (some at considerable cost), we can obtain the approximation s = $(1/2)gt^2$. If we make the dimensionless group $\pi = gt^2/s$, we are seeking the relationship $f(\pi) = 0$. An experiment with only p points can establish that

$$f(\pi) = gt^2/s - 2.00$$

or $gt^2/s = 2.00$.

Let us apply the rule of dimensional homogeneity to this falling body problem, f(s, t, g) = 0. The dimensionless group(s) will be found from

$$\pi = s^{e_1} t^{e_2} g^{e_3} \qquad (5.11)$$

Let us apply the rule of dimensional homogeneity. Present in the parameter set is the fundamental dimension length. The fundamental dimension L appears in π as L^0, in s as L^1, in t as L^0, and in g as L^1; thus Eq. (5.11) is expressible as

$$L^0 = (L^1)^{e_1}(L^0)^{e_2}(L^1)^{e_3}$$

resulting in the exponential equation,

$$0 = (1)e_1 + (0)e_2 + (1)e_3$$

Note that the coefficient of the exponents is the exponent of the fundamental dimension in each parameter. The fundamental dimension time appears in the parameter set, so Eq. (5.11) is expressible as

$$T^0 = (T^0)^{e_1}(T^1)^{e_2}(T^{-2})^{e_3}$$

and the exponential equation is

$$0 = (0)e_1 + (1)e_2 - (2)e_3$$

Since no other fundamental dimensions are resident in the parameter set, the exponential equations are

$$e_1 + e_3 = 0 \qquad e_2 - 2e_3 = 0$$

We have two equations and three unknowns. One may be chosen arbitrarily; choose one as a matter of convenience. Since we are interested in an expression for s, let us "force" s to appear in the dimensionless group to the first power. Setting $e_1 = 1$

$$1 + e_3 = 0 \qquad e_2 - 2e_3 = 0$$

from which $e_3 = -1$ and $e_2 = -2$ and the dimensionless group is determined as

$$\pi = s^{e_1} t^{e_2} g^{e_3} = s^1 t^{-2} g^{-1} = s/(gt^2)$$

What have we done? We have been given the set of parameters that describes the phenomenon. We have identified the number of fundamental dimensions present in the parametric set. We have written that many exponential equations. We made as many convenient arbitrary assignments of exponents as was mathemat-

Empiricism 143

ically possible and determined the remaining exponents. What follows is a formalization of this process that answers the question of exactly how many dimensionless groups are in the parameter set.

5.6 ESTABLISHING A COMPLETE SET OF DIMENSIONLESS VARIABLES

Again, consider the parameter set to be $\{x_n\}$ and the function sought to be $g(x_1, x_2, \ldots, x_n) = 0$. Let the number of fundamental dimensions in the parameter set be f and denoted as $(d_1, d_2, \ldots, d_f)$. Let the quantity k_{ij} be the exponent of dimension d_i in variable x_j. We can now form a *matrix of dimensions*

$$\begin{array}{c|ccccccc} & x_1 & x_2 & x_3 & x_4 & \cdots & x_n \\ \hline d_1 & k_{11} & k_{12} & k_{13} & k_{14} & \cdots & k_{1n} \\ d_2 & k_{21} & k_{22} & k_{23} & k_{24} & \cdots & k_{2n} \\ d_3 & k_{31} & k_{32} & k_{33} & k_{34} & \cdots & k_{3n} \\ \vdots & & & & & & \\ d_f & k_{f1} & k_{f2} & k_{f3} & k_{f4} & \cdots & k_{fn} \end{array}$$

The elements of the dimensional matrix above are seen to be the coefficients of the exponential equations. The exponential equations are f in number and appear as

$$k_{11}e_1 + k_{12}e_2 + k_{13}e_3 + \ldots + k_{1n}e_n = 0$$

$$k_{21}e_1 + k_{22}e_2 + k_{23}e_3 + \ldots + k_{2n}e_n = 0$$

·

$$k_{f1}e_1 + k_{f2}e_2 + k_{f3}e_3 + \ldots + k_{fn}e_n = 0$$

There are f exponential equations and n exponents. If the selection of the ordering $x_1, x_2, \ldots, x_n$ across the top of the dimensional matrix is made so that if x_1 were to appear only in π_1, it could be used to control the magnitude of π_1; if x_2 were to appear only in π_2, it could be used to control the magnitude of π_2; etc.; then the arbitrary designation of exponents should be made in the same preferential order, i.e., $e_1, e_2, \ldots$ and proceed as far as allowable. In fact the designation

$$e_1 = 1, \; e_2 = 0, \; e_3 = 0, \ldots$$

$$e_1 = 0, \; e_2 = 1, \; e_3 = 0, \ldots$$

$$e_1 = 0, \; e_2 = 0, \; e_3 = 1, \ldots$$

·

would place x_1 only in π_1 and no other π-term, x_2 only in π_2 and in no other π-term, x_3 only in π_3 and in no other π-term, etc.

Let us be more specific and say that we are examining a phenomenon with seven parameters and four fundamental dimensions. We have (superficially) three choices in the exponents, i.e., e_1, e_2, and e_3.

The exponential equations become

$$k_{14}e_4 + k_{15}e_5 + k_{16}e_6 + k_{17}e_7 = -k_{11}e_1 - k_{12}e_2 - k_{13}e_3$$
$$k_{24}e_4 + k_{25}e_5 + k_{26}e_6 + k_{27}e_7 = -k_{21}e_1 - k_{22}e_2 - k_{23}e_3$$
$$k_{34}e_4 + k_{35}e_5 + k_{36}e_6 + k_{37}e_7 = -k_{31}e_1 - k_{32}e_2 - k_{33}e_3$$
$$k_{44}e_4 + k_{45}e_5 + k_{46}e_6 + k_{47}e_7 = -k_{41}e_1 - k_{42}e_2 - k_{43}e_3$$

If, for a set e_1, e_2, and e_3, there is to be a unique set e_4, e_5, e_6, and e_7, it is essential that the determinant of the coefficients of the exponents e_4, e_5, e_6, and e_7 does not vanish; i.e.,

$$\begin{vmatrix} k_{14} & k_{15} & k_{16} & k_{17} \\ k_{24} & k_{25} & k_{26} & k_{27} \\ k_{34} & k_{35} & k_{36} & k_{37} \\ k_{44} & k_{45} & k_{46} & k_{47} \end{vmatrix} \neq 0 \qquad (5.12)$$

Another rectangular display is useful. This is the array of values of the exponents of the parameters in each π-term:

	e_1	e_2	e_3	e_4	e_5	e_6	e_7
π_1							
π_2							
π_3							

On the left side of this array are displayed the choice of the arbitrary exponents,

	e_1	e_2	e_3	e_4	e_5	e_6	e_7
π_1	1	0	0				
π_2	0	1	0				
π_3	0	0	1				

and the array is completed with the determinations of the exponents e_4, e_5, e_6, and e_7 corresponding to a particular choice of e_1, e_2, and e_3.

This array is called a *matrix of solutions*. The group of π-terms on the left is called a *complete set of dimensionless parameters* for the phenomenon described by x_1, x_2, x_3, ..., x_7. Important to the establishment of a complete set of dimensionless variables is that one parameter (at least) be unique to each π-term. This is assured by composing the left-hand determinant in the matrix of solutions of zeros with a principal diagonal of nonzeros (arbitrarily one).

Look at the dimensional matrix. The nonvanishing determinant (Eq. 5.12) appears as the right-hand determinant in the dimensional matrix. Look at the matrix of solutions. Note that the judicious selection of arbitrary exponents appears as the left-hand determinant and is composed of zeros with a principal diagonal of ones, the identity matrix. If by chance the right-hand determinant

Empiricism

in the dimensional matrix were zero, the columns would have to be interchanged until the right-hand determinant is nonzero, a condition necessary for solutions to the exponential equations to exist. A physically desired parameter within the group $x_1, x_2, x_3, \ldots$ might have to be sacrificed to bring this about.

The *rank* r of the dimensional matrix is the order of the largest nonzero determinant to be found in the matrix. The number of dimensionless groups to be found in the complete set is $(n - r)$, where n is the number of parameters and r is the rank of the dimensional matrix. Note that $r \leq f$.

Should there be no determinant of order four (in this illustration) in the dimensional matrix, then one of order three is placed in the right-hand corner of the matrix of dimensions, and one exponential equation stricken (say the last). The exponential equations become

$$k_{15}e_5 + k_{16}e_6 + k_{17}e_7 = -k_{11}e_1 - k_{12}e_2 - k_{13}e_3 - k_{14}e_4$$

$$k_{25}e_5 + k_{26}e_6 + k_{27}e_7 = -k_{21}e_1 - k_{22}e_2 - k_{23}e_3 - k_{24}e_4$$

$$k_{35}e_5 + k_{36}e_6 + k_{37}e_7 = -k_{31}e_1 - k_{32}e_2 - k_{33}e_3 - k_{34}e_4$$

and four exponents may be taken arbitrarily, e_1, e_2, e_3, and e_4; the matrix of solutions has four rows; and there is *one more π-term in the complete set* than is anticipated by an $(n - f)$ calculation. This is dangerous for it is a serious omission. Always use $(n - r)$ to determine the number of π-terms in the complete set. Now that we have in principle determine the π-terms, it follows that if the function

$$f(x_1, x_2, \ldots, x_n) = 0$$

describes the phenomenon, the function

$$g(\pi_1, \pi_2, \ldots, \pi_{n-r}) = 0$$

will also describe the phenomenon. The advantage of the second formulation can now be assessed. If p points are required to establish a locus satisfactorily, in the case of the dimensionless parameter formulation the number of experiments is

$$N' = p^{n-r-1} \tag{5.13}$$

In the direct approach the number of experiments was

$$N = p^{n-1} \tag{5.14}$$

The ratio of N'/N is of interest

$$\frac{N'}{N} = \frac{p^{n-r-1}}{p^{n-1}} = p^{-r} = \frac{1}{p^r} \tag{5.15}$$

In the case at hand, if $p = 5$ and $r = 4$, then $N'/N = 1/5^4 = 1/625$. The experimentation has been reduced by a factor of 625. The cost has been reduced by a factor of similar order.

It is not that this approach of Buckingham is interesting, or nice, or convenient, or any other pleasant adjective. If a method is available that will dramatically reduce the experimental cost, those who do not use it raise the question of competence.

As a simple example, consider the relationship between the translational kinetic energy of a particle, its mass, and its velocity: $f(E_k, m, v) = 0$, where E_k is the translational kinetic energy, m is the particle mass, and v is the particle velocity (speed). We will use the MLT system of fundamental units. Now

$$\dim(E_k) = FL = ML^2T^{-2}$$

$$\dim(m) = M$$

$$\dim(v) = LT^{-1}$$

and the matrix of dimensions is

	E_k	m	v
M	1	1	0
L	2	0	1
T	-2	0	-1

The rank of the matrix is the order of the largest nonzero determinant to be found within it. Since the matrix is square, the largest determinant is

$$\begin{vmatrix} 1 & 1 & 0 \\ 2 & 0 & 1 \\ -2 & 0 & -1 \end{vmatrix} = 0$$

and therefore the rank is less than three. The 2 by 2 determinant in the upper right-hand corner of the matrix of dimensions is

$$\begin{vmatrix} 1 & 0 \\ 0 & 1 \end{vmatrix} \neq 0$$

and therefore the rank of the matrix of dimensions is two; hence r = 2. The number of fundamental dimensions in the parameter set (E_k, m, v) is three, hence n = 3. The number of dimensionless groups in the complete set is (n - r) = (3 - 2) = 1. The matrix of solutions is

	e_1	e_2	e_3
π_1	1		

We have one exponent that can be chosen arbitrarily. We chose $e_1 = 1$ and write the exponential equations from the dimensional matrix.

$$e_1 + e_2 = 0$$

$$2e_1 + e_3 = 0$$

$$-2e_1 - e_3 = 0$$

Note that the last two exponential equations are identical; the test for the rank of the dimensional matrix tells us this. The number of fundamental dimensions is three and the number of parameters is three; however one exponent can be taken arbitrarily, since there are not enough independent relationships in the exponential equations to determine three exponents. Choosing the first two relations, $e_2 = -e_1$ and $e_3 = -2e_1$, and with e_1 taken arbitrarily as 1, we have $e_2 = -1$ and $e_3 = -2$. The matrix of solution can be completed,

	e_1	e_2	e_3
π_1	1	-1	-2

and the complete set (one π-term) is $\pi_1 = E_k/(mv^2)$. The saving in experimental work to determine $f(E_k/(mv^2)) = 0$ to our satisfaction is

$$N'/N = 1/p^r = 1/p^2$$

If p = 5, the experimental work is cut to 1/25 of direct approach. The fraction depends on the number of points p that satisfactorily establish a curve. If p = 10, the fraction is 1/100. Section 5.8 describes how to reduce the effort even more.

Another set of π-terms may also be a complete set. One way to determine whether another set is complete is to recreate its matrix of solutions. In addition to being dimensionless, sufficiency requires finding within it (n - r) columns that are composed of zeros, with one element nonzero in such a way that the columns could be rearranged to form a determinant of zeros with a principal diagonal of nonzeros. This assures that one parameter (at least) is unique to each π-term.

--Experimental efforts increase exponentially with dimensionality.
--We are indebted to Buckingham for the suggestion of dimensionless parameter grouping, which reduces the dimensionality from n to (n - r).
--The number of π-terms is given by the number of parameters necessary to describe satisfactorily the phenomenon minus the rank of the dimensional matrix.
--Parameters related to the left-hand column of the dimensional matrix are to be purposely chosen for the experimental ease with which these parameters may control the magnitude of the dimensionless group in which they reside.

5.7 USE OF PARTIAL DERIVATIVES

In this section we convert data to multidimensional mathematical curve-fits by using properties of partial derivatives, possibly reducing the experimental effort even further by incorporating an ability to utilize already established relationships. We begin by example, bringing a problem through the first five steps (often called dimensional analysis), and then use the power of the partial derivative to envision the strategy of the experimental effort and to realize the mathematical curve-fit.

Figure 5.30 depicts a right prismatic rod that is stretched by tensile forces whose resultant is coincident with centroids of the cross sections. The stretch δ is expected to be functionally related to the parameters of length ℓ, tensile load P, cross-sectional area A, and the material modulus of elasticity E. Our procedure will be to:

--Identify the set of parameters.
--Establish the dimensional matrix.
--Determine the census of the complete set.
--Establish the exponential equations.
--Establish the matrix of solutions.
--After experimental work, find the functional relationship between the dimensionless variables.

Let us say that our qualitative and quantitative experience with the phe-

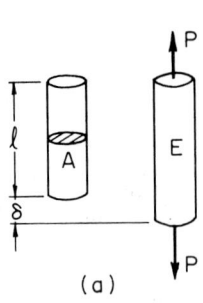
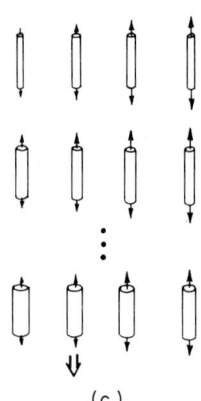
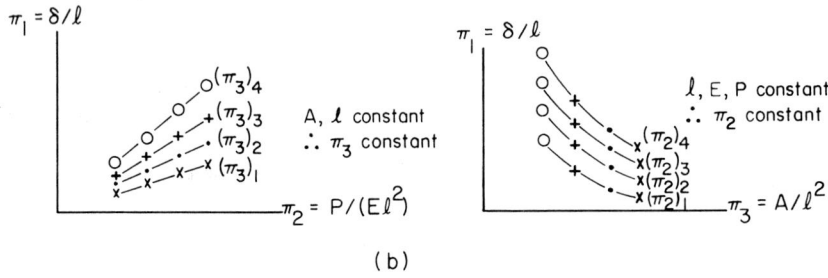

Fig. 5.30. (a) A right circular cylindrical rod stretched by a tensile force. (b) A strategy for empirical determination of the relationship among the π-terms. (c) Specimen sets necessary for the envisioned experimental program.

nomenon convinces us that a single-valued function in some domain exists that can be expressed as

$$f(\delta, P, A, E, \ell) = 0$$

In terms of the fundamental dimensions of force F, length L, time T, temperature θ, and charge Q, we can express the dimensions of all the parameters of the phenomenon.

$$\dim(\delta) = F^0 L^1 T^0 \theta^0 Q^0 = L$$
$$\dim(P) = F$$
$$\dim(E) = FL^{-2}$$
$$\dim(\ell) = L$$
$$\dim(A) = L^2$$

This information is displayed in the dimensional matrix:

	δ	P	A	E	ℓ
F	0	1	0	1	0
L	1	0	2	-2	1

Empiricism

The right-hand determinant of the dimensional matrix is nonzero,

$$\begin{vmatrix} 1 & 0 \\ -2 & 1 \end{vmatrix} = 1 \neq 0$$

and consequently the rank of the matrix is two. Thus Buckingham's equation predicts $\pi = n - r = 5 - 2 = 3$; three dimensionless variables are capable of describing the phenomenon.

The elements of the dimensional matrix are the coefficients of the exponential equations. From the first and second rows of the dimensional matrix, respectively,

$$e_2 + e_4 = 0 \qquad e_1 + 2e_3 - 2e_4 + e_5 = 0$$

where e_1, e_2, e_3, e_4, and e_5 are the exponents of δ, P, A, E, and ℓ, respectively, in any π-term. The exponential equations can be displayed as

$$e_4 = -e_2 \qquad -2e_4 + e_5 = -e_1 - 2e_3$$

The matrix of solutions is established with the left-hand determinant composed of a principal diagonal of ones and all other entries of zero (identity matrix), ensuring that δ and neither P nor A will appear in π_1, P and neither δ nor A will appear in π_2, and A and neither δ nor P will appear in π_3. Expecting π_2 and π_3 to be the independent variables, π_2 can be specified by control of P and π_3 and be specified by control of A.

	e_1	e_2	e_3	e_4	e_5
π_1	1	0	0		
π_2	0	1	0		
π_3	0	0	1		

To complete the row of entries of the solution matrix for π_1, the exponential equations are solved for values of $e_1 = 1$, $e_2 = e_3 = 0$, resulting in $e_4 = 0$ and $e_5 = -1$. To complete the remaining entries for the row corresponding to π_2, the exponential equations, are solved for values of $e_2 = 1$, $e_1 = e_3 = 0$, resulting in $e_4 = -1$ and $e_5 = -2$. The remaining entries for the row corresponding to π_3 are found by solving the exponential equations with $e_1 = e_2 = 0$ and $e_3 = 1$, resulting in $e_4 = 0$ and $e_5 = -2$. The complete matrix of solutions is

	e_1	e_2	e_3	e_4	e_5
π_1	1	0	0	0	-1
π_2	0	1	0	-1	-2
π_3	0	0	1	0	-2

The dimensionless terms are $\pi_1 = \delta/\ell$, $\pi_2 = P/E\ell^2$, and $\pi_3 = A/\ell^2$. The function $g(\pi_1, \pi_2, \pi_3) = 0$ is as capable of representing the quantitative relationships in the stretch of rod phenomena as the original function $f(\delta, P, A, E, \ell) = 0$. Our task now is to discover the functional relationship between the π-terms. If $\pi_1 = \phi(\pi_2, \pi_3)$, it is also true that

$$d\pi_1 = \frac{\partial \pi_1}{\partial \pi_2} d\pi_2 + \frac{\partial \pi_1}{\partial \pi_3} d\pi_3 \qquad (5.16)$$

We note in the graph of the data on the $\pi_1\pi_2$-plot in Fig. 5.30(b) on Cartesian coordinates that the data strings are straight enough and the ordinate intercept sufficiently close to zero to say the function is of a form (a parametric family)

$$\pi_1 = C \pi_2$$

where C is a constant that varies with π_3. The loci on the $\pi_1\pi_3$-plot on Cartesian coordinates are not linear to our satisfaction so that a rectification scheme is sought. The result of the search is that $h(y) = y$ and $g(x) = 1/x$ produces rectification to our satisfaction. The antitransformation produces a parametric family

$$\pi_1 = k(1/\pi_3)$$

where k is a constant that varies with π_2. Now from the $\pi_1\pi_2$-plot (equation family) we can determine $d\pi_1/d\pi_2$, which is $\partial\pi_1/\partial\pi_2$ since the conditions of the plot are constant π_3 (an integral part of the definition of a partial derivative). Hence

$$\frac{\partial \pi_1}{\partial \pi_2} = \frac{d\pi_1}{d\pi_2} = C = \frac{\pi_1}{\pi_2}$$

Likewise

$$\frac{\partial \pi_1}{\partial \pi_3} = \frac{d\pi_1}{d\pi_3} = -k\frac{1}{\pi_3^2} = -\frac{\pi_1}{\pi_3}$$

Substituting the above results of the partial derivative determination into Eq. (5.16), we obtain

$$d\pi_1 = \frac{\pi_1}{\pi_2} d\pi_2 - \frac{\pi_1}{\pi_3} d\pi_3$$

Dividing by π_1 and integrating term by term, we obtain

$$\int \frac{d\pi_1}{\pi_1} = \int \frac{d\pi_2}{\pi_2} - \int \frac{d\pi_3}{\pi_3}$$

$$\ln \pi_1 = \ln \pi_2 - \ln \pi_3 + \ln c$$

$$\pi_1 = c(\pi_2/\pi_3)$$

Substituting the parametric equivalents for the π-terms results in

$$\frac{\delta}{\ell} = c \frac{P}{E\ell^2} \cdot \frac{\ell^2}{A}$$

or $\delta = c(P\ell/AE)$. We have sufficient data (16 observations) to establish $c = 1$ if measurements are in a consistent set of units such as inches and pounds (or metres, newtons, pascals, etc.). Examination of Fig. 5.30(b) shows that all 16 data points are plotted on both graphs. If the tests are destructive we

Empiricism

would use 16 specimens. If tests are nondestructive (elastic limit not exceeded) we can apply varying loads to 4 specimens of different areas but of identical length to establish 16 points. In general this requires p specimens to establish p^2 points, a cost reduction.

The preceding is an example of the approach to a three π-term problem when nothing of the phenomena is known a priori. If we have reason to suspect that $\phi(\pi_2, \pi_3)$ is separable into a product of a function of π_2 and a function of π_3, we may write

$$\pi_1 = \phi_2(\pi_2)\, \phi_3(\pi_3)$$

If ϕ_2 is known a priori, we really have but two dimensionless groups, $\pi_1/[\phi_2(\pi_2)]$ and π_3, related as

$$\frac{\pi_1}{\phi_2(\pi_2)} = \phi_3(\pi_3) \tag{5.17}$$

and our experimental program requires but p points. To be specific, suppose we know that the behavior of a right prismatic body in tension is elastic (i.e., for a given bar, δ is proportional to P), implying that π_1 is proportional to π_2 for a constant π_3. We can incorporate this information into Eq. (5.17) as

$$\pi_1 = \pi_2 \phi_3(\pi_3) \tag{5.18}$$

and our task is to discover an acceptable approximation to ϕ_3. We choose to make a $\pi_1\pi_3$-plot with π_2 held constant. If the tests can be run nondestructively we can repeat the plan of Fig. 5.30, using specimens of the same length and variable area [vertical selections from Fig. 5.30(c)]. Since we no longer have to hold π_3 constant during a $\pi_1\pi_2$ plot, we are free to make the specimens of constant area and variable length. If tests can be conducted nondestructively it is possible to have a single long specimen and slice a little from the end between loadings and stretch measurements. The specimen set creates one data string locus on the $\pi_1\pi_2$-plot.

Figure 5.31(a) shows the data string of p points on a $\pi_1\pi_3$-plot that is not linear to our satisfaction. Rectification is sought as a means to curve fitting. We choose to control π_3 with constant area A and variable length ℓ and keep π_2 constant with constant material E and changing load P to compensate for changing lengths. The equation, as before, is [see Fig. 5.31(b)]

$$\pi_1 = c(1/\pi_3)$$

As before we use the total differential, beginning with our function in the form

$$\pi_1 = \phi_3(\pi_3)\pi_2$$

The partial derivatives are

$$\frac{\partial \pi_1}{\partial \pi_2} = \frac{d\pi_1}{d\pi_2} = \phi_3(\pi_3) = \frac{\pi_1}{\pi_2} \quad \text{(a priori knowledge)}$$

$$\frac{\partial \pi_1}{\partial \pi_3} = \frac{d\pi_1}{d\pi_3} = -c\frac{1}{\pi_3^2} = -\frac{\pi_1}{\pi_3} \quad \text{(from test program)}$$

and as before it follows that

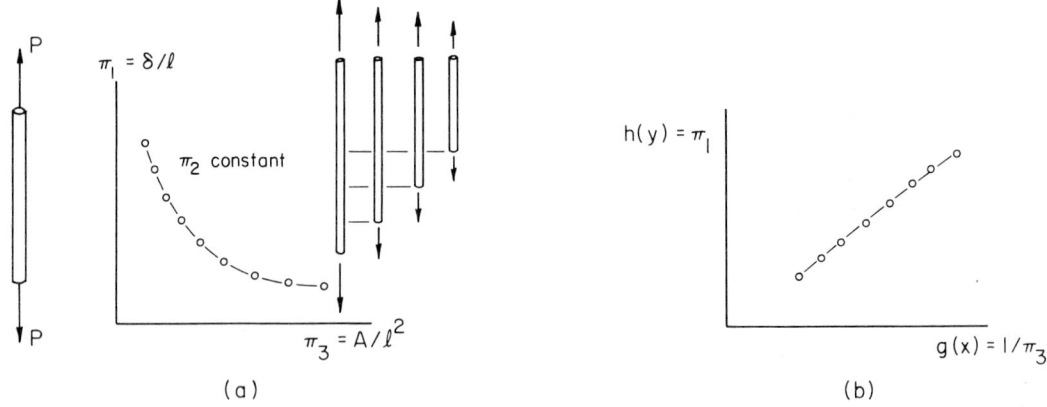

Fig. 5.31. (a) An approach to the stretched rod problem using one rod that is made shorter after each test. Note that the $\pi_1\pi_3$ plot is not linear. (b) Rectification of the data is achieved by a reciprocal transformation applied to π_3.

$$\delta = c\,\frac{P\ell}{AE} = \frac{P\ell}{AE}$$

where c is determined from the p points as unity if units of measurement are consistent such as pounds and inches (or pascals, newtons, and metres). In your strength of materials course you will learn how this equation can be derived from first principle, i.e., how to build a mathematical model for the stretch of a prismatic rod by deduction from first principle and without experiment. Here we demonstrate how the dimensionless groups can be found to form a complete set, how experiments to find functional relations can be envisioned, and how results can be interpreted.

Example 7. Find the complete set of dimensionless variables needed to describe the end deflection of a cantilever beam, incorporating the following a priori information: (1) deflection is directly proportional to load; (2) deflection is inversely proportional to beam width.

Figure 5.32(a) depicts the essential geometry of the cantilever beam. The parameters are the end deflection δ, the load P, the width b, the depth d, the length ℓ, and the material modulus of elasticity E. The engineer, convinced that these parameters are functionally related, writes $f(\delta, P, b, d, \ell, E) = 0$. The dimensions of the parameters are

Fig. 5.32. (a) A rectangular cross-section cantilever beam with an end load and associated transverse end deflection, δ. (b) The only necessary experimental functional determination.

Empiricism

$$\dim(\delta) = L \quad \dim(d) = L$$
$$\dim(P) = F \quad \dim(\ell) = L$$
$$\dim(b) = L \quad \dim(E) = FL^{-2}$$

A dimensional matrix is formed.

	δ	P	b	d	ℓ	E
F	0	1	0	0	0	1
L	1	0	1	1	1	-2

The right-hand determinant in the dimensional matrix is

$$\begin{vmatrix} 0 & 1 \\ 1 & -2 \end{vmatrix} = -1 \neq 0$$

nonvanishing, consequently the rank of the dimensional matrix is 2. The number of π terms is $\pi = n - r = 6 - 2 = 4$. A matrix of solutions is formed as follows:

	e_1	e_2	e_3	e_4	e_5	e_6
π_1	1	0	0	0		
π_2	0	1	0	0		
π_3	0	0	1	0		
π_4	0	0	0	1		

The exponential equations can be written from the dimensional matrix directly as

$$e_2 + e_6 = 0 \qquad e_1 + e_3 + e_4 + e_5 - 2e_6 = 0$$

For completing the matrix of solutions the exponential equations are rearranged with e_5 and e_6 as dependent variables:

$$e_6 = -e_2 \qquad e_5 - 2e_6 = -e_1 - e_3 - e_4$$

By systematically substituting values from the rows of the matrix of solutions, the magnitudes of e_1, e_2, e_3, and e_4 and the rest of the matrix of solutions may be established.

	e_1	e_2	e_3	e_4	e_5	e_6
π_1	1	0	0	0	-1	0
π_2	0	1	0	0	-2	-1
π_3	0	0	1	0	-1	0
π_4	0	0	0	1	-1	0

resulting in

$$\pi_1 = \delta/\ell \qquad \pi_2 = P/E\ell^2 \qquad \pi_3 = b/\ell \qquad \pi_4 = d/\ell$$

The phenomenon can be represented by the function

$$g(\pi_1, \pi_2, \pi_3, \pi_4) = 0 \text{ or } g\left(\frac{\delta}{l}, \frac{P}{El^2}, \frac{b}{l}, \frac{d}{l}\right) = 0$$

Without loss of generality we can write

$$\frac{\delta}{l} = \phi_1\left(\frac{P}{El^2}, \frac{b}{l}, \frac{d}{l}\right)$$

The a priori information that δ was proportional to P leads to the conclusion that the π_2-term is multiplicatively related to the π_1 term as

$$\frac{\delta}{l} = \frac{P}{El^2} \phi_2\left(\frac{b}{l}, \frac{d}{l}\right)$$

The a priori information that δ is inversely related to width b leads to the conclusion that

$$\frac{\delta}{l} = \frac{P}{El^2} \cdot \frac{l}{b} \phi_3\left(\frac{d}{l}\right)$$

This may be written as

$$\frac{\delta E b}{P} = \phi_3\left(\frac{d}{l}\right)$$

Experiments are necessary to establish the functional relationship between $\delta Eb/P$ and d/l. Figure 5.32(b) portrays the nature of the results of such experimentation. It remains to find the appropriate data transformations that rectify the data string so that the functional relationship can be easily determined.

Figure 5.33 depicts a right circular cylinder in pure torsion. The angle of twist β is a function of the applied torque T, the shaft length l, the torsional modulus of elasticity G, and the polar second area moment of the cross section about its centroid J. In this problem one parameter, the angle of twist, is dimensionless. How does the presence of an already dimensionless parameter affect the method of determining a complete set? Given the fundamental dimensions of force F, length L, time t, temperature θ, and charge Q, the dimensions of the problem parameters are expressible as

$$\dim(\beta) = F^0 L^0 t^0 \theta^0 Q^0 = 1$$
$$\dim(T) = FL$$
$$\dim(l) = L$$
$$\dim(G) = FL^{-2}$$
$$\dim(J) = L^4$$

and the functional relationship sought is $f(\beta, T, l, G, J) = 0$. The matrix of dimensions is

	β	T	l	G	J
F	0	1	0	1	0
L	0	1	1	-2	4

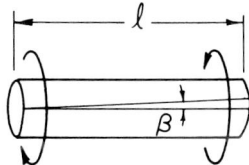

Fig. 5.33. Torsional deflection of a right circular cylinder.

The right-hand determinant of the matrix of dimensions is nonzero; consequently the rank of the dimensional matrix is two and the number of dimensionless groups in the complete set is $\pi = n - r = 5 - 2 = 3$. The exponential equations are written from the matrix of dimensions: $e_2 + e_4 = 0$ and $e_2 + e_3 - 2e_4 + 4e_5 = 0$. These may be more conveniently displayed as

$$e_2 = -e_4$$

$$-2e_4 + 4e_5 = -e_2 - e_3$$

The matrix of solutions is begun by placing an identity determinant on the left:

	e_1	e_2	e_3	e_4	e_5
π_1	1	0	0		
π_2	0	1	0		
π_3	0	0	1		

The first row of the solution matrix is completed by solving the exponential equations with $e_1 = 1$, $e_2 = e_3 = 0$, which results in $e_4 = 0$ and $e_5 = 0$. The second row is completed by solving the exponential equations with $e_1 = 0$, $e_2 = 1$, and $e_3 = 0$, which results in $e_4 = -1$ and $e_5 = -3/4$. The third row is completed by setting $e_1 = e_2 = 0$ and $e_3 = 1$, which results in $e_4 = 0$ and $e_5 = -1/4$. The completed matrix of solutions is

	e_1	e_2	e_3	e_4	e_5
π_1	1	0	0	0	0
π_2	0	1	0	-1	-3/4
π_3	0	0	1	0	-1/4

The complete set consists of

$$\pi_1 = \beta \qquad \pi_2 = \frac{T}{GJ^{3/4}} \qquad \pi_3 = \frac{l}{J^{1/4}}$$

The function to be sought experimentally is $g(\pi_1, \pi_2, \pi_3) = 0$. If, as depicted in Fig. 5.34(a), π_3 is held constant by using a single specimen (consequently holding l and J constant), π_2 may be varied by varying torque T. Different contours may be traced on the $\pi_1\pi_2$-plane with different specimens. The mathematical curve-fit that is adequate in the elastic region is the function

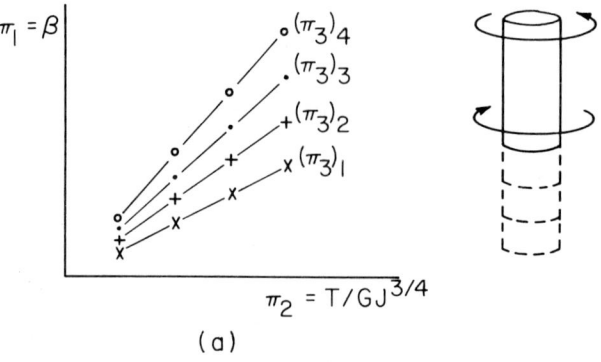

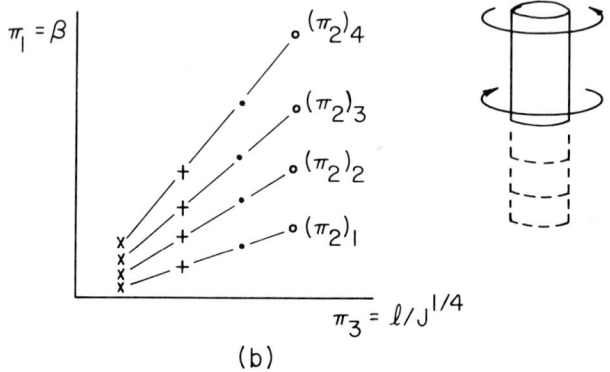

Fig. 5.34. (a) Relating π_1 to π_2. (b) Relating π_1 to π_3.

family parameterized by C_1 and expressed as $\pi_1 = C_1\pi_2$. If, as depicted in Fig. 5.34(b), π_2 is held constant by using a constant torque on specimens of identical cross section, π_3 can be varied by varying the specimen length. Different contours may be traced on the $\pi_1\pi_3$-plane. The mathematical curve-fit that is adequate in the elastic region is the function family parameterized by C_2 and expressed as $\pi_1 = C_2\pi_3$. We are seeking a relation between the π-terms of the form $\pi_1 = \phi(\pi_2, \pi_3)$. The exact differential is

$$d\pi_1 = \frac{\partial \pi_1}{\partial \pi_2} d\pi_2 + \frac{\partial \pi_1}{\partial \pi_3} d\pi_3$$

From the mathematical curve-fits we obtain

$$\frac{\partial \pi_1}{\partial \pi_2} = C_1 = \frac{\pi_1}{\pi_2} \qquad \frac{\partial \pi_1}{\partial \pi_3} = C_2 = \frac{\pi_1}{\pi_3}$$

Substitution in the expression for the exact differential and division by π_1 yields

$$\frac{d\pi_1}{\pi_1} = \frac{d\pi_2}{\pi_2} + \frac{d\pi_3}{\pi_3}$$

This equation is exact and integrable term by term; consequently

Empiricism

$$\ln \pi_1 = \ln \pi_2 + \ln \pi_3 + \ln k$$

or, taking antilogarithms,

$$\pi_1 = k\pi_2\pi_3 = k \cdot \frac{T}{GJ^{3/4}} \cdot \frac{l}{J^{1/4}} = k\frac{Tl}{GJ} = \beta$$

From 16 experimental points, k can be established as unity for a consistent set of units such as inches and pounds (or pascals, newtons, metres) and $\beta = Tl/(GJ)$. This equation can be deduced from first principles in ways to be shown in your strength of materials course. Our purpose here is to indicate how to proceed with a complete set experimentally. Note that the initial presence of a dimensionless term did not frustrate the method. This indicates that a known dimensionless group may be mixed with parameters with dimensions in Buckingham's method to reduce the extent of the experimental program.

Using the exact differential,

$$d\pi_1 = \frac{\partial \pi_1}{\partial \pi_2} d\pi_2 + \frac{\partial \pi_1}{\partial \pi_3} d\pi_3 + \ldots + \frac{\partial \pi_1}{\partial \pi_{n-r}} d\pi_{n-r}$$

requires definition of the functions, to the user's satisfaction,

$$\frac{\partial \pi_1}{\partial \pi_2}, \frac{\partial \pi_1}{\partial \pi_3}, \ldots, \frac{\partial \pi_1}{\partial \pi_{n-r}}$$

or $(n - r - 1)$ functions. If p points will satisfactorily describe a functional relationship, the required number of points N" is

$$N" = (n - r - 1)p$$

For the case of stretch of rod, if p = 5, the brute force approach requires

$$N = p^{n-1} = 4^4 = 256$$

Using dimensionless variables and partial derivatives, the number of points is

$$N' = p^{n-r-1} = 4^2 = 16$$

Using dimensionless variables coupled with partial derivatives and encountering simple functional relationships, we can need as few as

$$N" = (n - r - 1)p = (5 - 2 - 1)4 = 8$$

The case of twist of rod is similar with N = 256, N' = 16, and N" = 8.

Defending any approach other than partial derivatives coupled with Buckingham π-terms is difficult. It allows reduction in dimensionality. Experimental effort increases exponentially with increasing dimensionality. It permits an approach to an experimental effort that may increase linearly with increasing dimensionality.

Example 8. Determine the functional relationship between the π-terms given the data

π_2	π_1	π_2	π_1	π_2	π_1	π_2	π_1
0	0	0	0	0	0	0	0
1	1	1	2	1	3	1	4

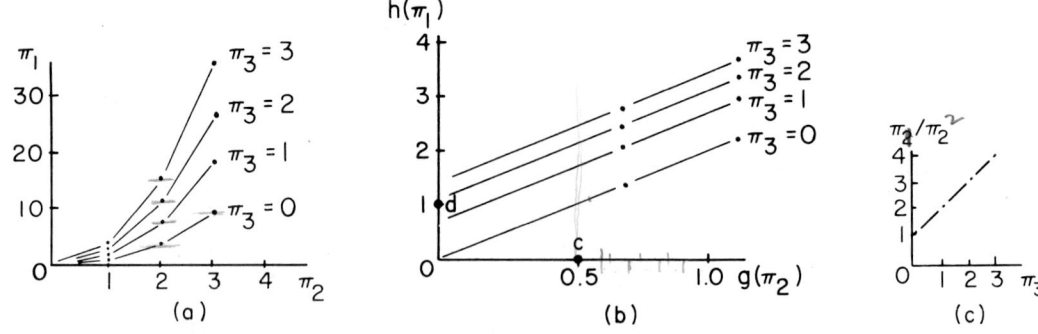

Fig. 5.35. (a) Data strings not rectified on Cartesian coordinates. (b) Data strings rectified by logarithmic transformations to ordinate and abscissa. (c) Data reduced by incorporating results of rectification in (b) is linear on Cartesian coordinates. (Figures are 0.723 actual size.)

π_2	π_1	π_2	π_1	π_2	π_1	π_2	π_1
2	2	2	8	2	12	2	16
3	4	3	18	3	27	3	36
$\pi_3 = 0$		$\pi_3 = 1$		$\pi_3 = 2$		$\pi_3 = 3$	

The raw data are plotted in Fig. 5.35(a) and the data strings are not rectified when plotted on Cartesian coordinates. The next step is to apply a logarithmic transformation to the data:

$\ln \pi_2$	$\ln \pi_1$	$\ln \pi_2$	$\ln \pi_1$	$\ln \pi_2$	$\ln \pi_1$	$\ln \pi_2$	$\ln \pi_1$
$-\infty$	$-\infty$	$-\infty$	$-\infty$	$-\infty$	$-\infty$	$-\infty$	$-\infty$
0	0	0	0.69	0	1.1	0	1.39
0.69	1.39	0.69	2.08	0.69	2.48	0.69	2.77
1.1	2.2	1.1	2.89	1.1	3.30	1.1	3.58
$\pi_3 = 0$		$\pi_3 = 1$		$\pi_3 = 2$		$\pi_3 = 3$	

Figure 5.35(b) shows that when the transformed data are plotted on Cartesian coordinates the data strings are rectified. The scale factor of the abscissa of Fig. 5.35(b) is determined by using point c, its distance of 1 in. from the plotting origin, and its coordinate of 0.5:

$$s_{g(\pi_2)} = X/g(\pi_2) = 1/0.5 = 2 \text{ in.}$$

The scale factor of the ordinate is determined by using point d, its distance of 0.4 in. from the plotting origin, and its coordinate 1:

$$s_{h(\pi_1)} = Y/h(\pi_1) = 0.4/1 = 0.4 \text{ in.}$$

The slope is given by Eq. (5.5) as

$$b = \frac{s_{g(\pi_2)}}{s_{h(\pi_1)}} \left(\frac{\Delta Y}{\Delta X}\right) = \frac{2}{0.4}(0.4) = 2$$

Empiricism

The functional relationship at constant π_3 is

$$h(y) = a + bg(x) \quad \ln \pi_1 = a + 2\ln \pi_2 = a + \ln \pi_2^2$$

Raising e to the $\ln \pi_1$ power gives

$$\exp(\ln \pi_1) = \exp(a + \ln \pi_2^2) = \exp(a)\exp(\ln \pi_2^2)$$

$$\pi_1 = C_1 \pi_2^2 \text{ or } \pi_1/\pi_2^2 = C_1 = \phi(\pi_3)$$

We may proceed in one of two directions. If we redisplay the original data with a π_1/π_2^2 term added, we will have

π_2	π_1	π_1/π_2^2	π_2	π_1	π_1/π_2^2	π_2	π_1	π_1/π_2^2	π_2	π_1	π_1/π_2^2
0	0		0	0		0	0		0	0	
1	1	1	1	2	2	1	3	3	1	4	4
2	4	1	2	8	2	2	12	3	2	16	4
3	9	1	3	18	2	3	27	3	3	36	4
$\pi_3 = 0$			$\pi_3 = 1$			$\pi_3 = 2$			$\pi_3 = 3$		

The next move is to plot π_3 versus π_1/π_2^2, abstracting from the above data.

π_3	π_1/π_2^2
0	1
1	2
2	3
3	4

Figure 5.35(c) shows the plot of these data. Clearly both a and b are 1:

$$\pi_1/\pi_2^2 = a + b\pi_3 = 1 + \pi_3$$

Therefore $\pi_1 = \pi_2^2(1 + \pi_3)$ is the relationship of the original data.

Alternatively we could have rearranged the original data for a π_1 versus π_3 plot at constant π_2:

π_3	π_1	π_3	π_1	π_3	π_1	π_3	π_1
0	0	0	1	0	4	0	9
1	0	1	2	1	8	1	18
2	0	2	3	2	12	2	27
3	0	3	4	3	16	3	36
$\pi_2 = 0$		$\pi_2 = 1$		$\pi_2 = 2$		$\pi_2 = 3$	

These data are plotted in Fig. 5.36. The data strings are rectified. They seem to converge at a common point, $(-1, 0)$. If the plot were carried out with π_1 as ordinate and $1 + \pi_3$ as abscissa, the straight-line family would be a group radiating from point $(0, 0)$. Consequently, $\pi_1 = C_2(1 + \pi_3)$. The total differential is

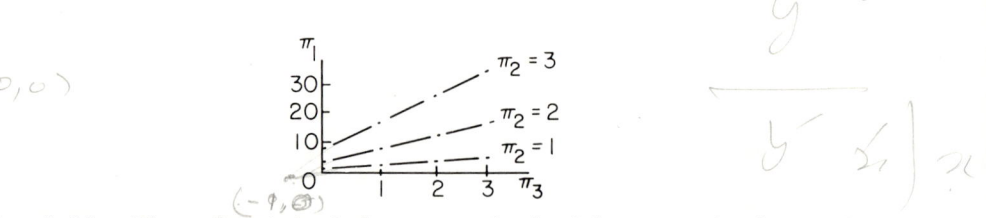

Fig. 5.36. Plot of original data organized with π_3 as abscissa. Data strings satisfactorily rectified on Cartesian coordinates.

$$d\pi_1 = \frac{\partial \pi_1}{\partial \pi_2} d\pi_2 + \frac{\partial \pi_1}{\partial \pi_3} d\pi_3$$

From the previous relation, $\pi_1 = C_1 \pi_2^2$ at constant π_3,

$$\frac{\partial \pi_1}{\partial \pi_2} = \left(\frac{d\pi_1}{d\pi_2}\right)_{\pi_3} = 2C_1 \pi_2 = 2\frac{\pi_1}{\pi_2}$$

From the relation $\pi_1 = C_2(1 + \pi_3)$ at constant π_2,

$$\frac{\partial \pi_1}{\partial \pi_3} = \left(\frac{d\pi_1}{d\pi_3}\right)_{\pi_2} = C_2 = \frac{\pi_1}{1 + \pi_3}$$

Substituting into the total differential,

$$d\pi_1 = 2\frac{\pi_1}{\pi_2} d\pi_2 + \frac{\pi_1}{1 + \pi_3} d\pi_3$$

Dividing through by π_1 yields

$$\frac{d\pi_1}{\pi_1} = 2 \frac{d\pi_2}{\pi_2} + \frac{d\pi_3}{1 + \pi_3}$$

which is exact and integrable term by term. Integrating, we obtain

$$\ln \pi_1 = 2 \ln \pi_2 + \ln(1 + \pi_3) + \ln c \quad \text{or} \quad \pi_1 = c\pi_2^2 (1 + \pi_3)$$

The value of c can be established from

$$c = \pi_1 / [\pi_2^2 (1 + \pi_3)]$$

Sixteen observations of c are available in this case, and in the absence of experimental error, all are 1; consequently the relationship between the π-terms is $\pi_1 = \pi_2^2(1 + \pi_3)$.

5.8 PHYSICAL MODELS AND SIMILITUDE

From Sec. 5.7 we recall that for the stretch of rod example, the functional description of the phenomenon was displayed as $f(\delta, P, A, E, l) = 0$ and was established as $\delta = Pl/(AE)$. Let us consider a course of action if a stretch δ is predicted for a steel rod 100 in. long with a 1 in.² cross-sectional area and a tensile load of 30,000 lbf such as in Fig. 5.37(a). The complicating circumstance is that $\delta = Pl/(AE)$ is not known, and cannot be

Empiricism

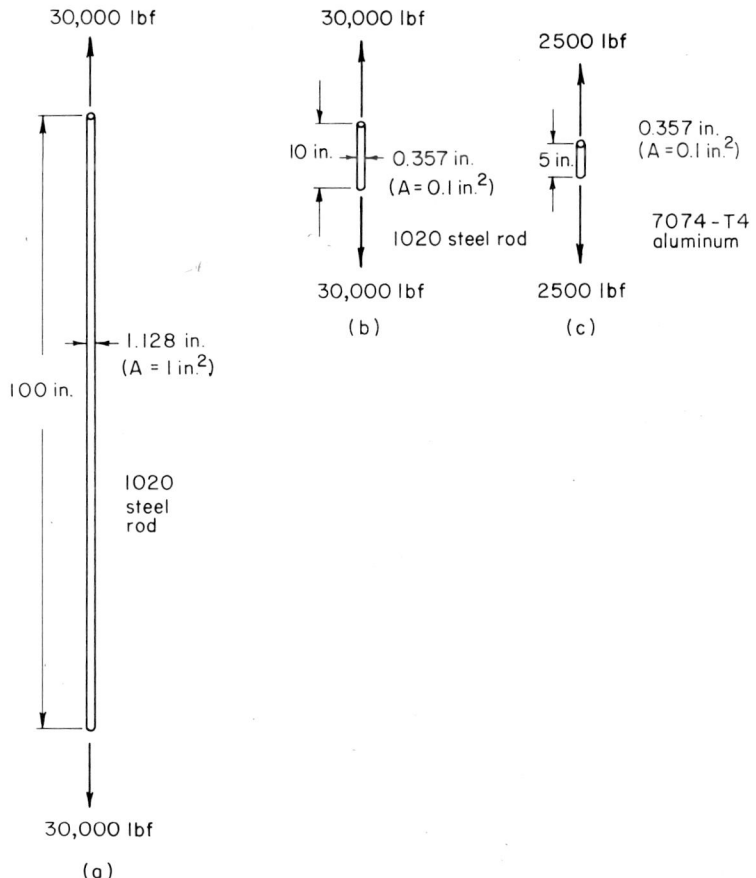

Fig. 5.37. Rods whose stretch under load is desired.

learned in time (or ever). How can one proceed?

Engineers learn by asking questions whose answers suggest alternative approaches to a problem. If we knew that

$$f(0.1, 30{,}000, 1, 30(10^6), 100) = 0$$

satisfied the functional relation between δ, P, A, E, and ℓ, could we draw a schematic picture of the situation? The answer is yes and Fig. 5.37(a) is such a depiction. If

$$f(0.01, 30{,}000, 0.1, 30(10^6), 10) = 0$$

Fig. 5.37(b) depicts the situation. The unique counterpart to

$$f(0.0125, 2500, 0.1, 10(10^6), 5) = 0$$

is depicted in Fig. 5.37(c). Consider the corresponding function of the complete set,

$$g(\pi_1, \pi_2, \pi_3) = 0$$

If $\pi_1 = 0.0025$, $\pi_2 = 10^{-5}$, and $\pi_3 = 250$, then $g(0.0025, 10^{-5}, 250) = 0$. What

$\frac{\delta}{\ell} \qquad \frac{P}{E\ell^2} \qquad \frac{A}{\ell^2}$

does the schematic picture look like? We cannot draw a picture since an infinite set of possibilities of loaded rods that gives rise to $g(0.0025, 10^{-5}, 250) = 0$ is possible. Conversely, all members of this infinite set are represented by a unique point in $\pi_1\pi_2\pi_3$-space. One might suggest that any of these loaded rods is a model of any other in that they share the same $\pi_1\pi_2\pi_3$ description. Furthermore, if we build such a model in which we know π_2 and π_3 and measure π_1 from the model, *this π_1 is unique for all members of the set.* In lieu of using the mathematical model (unknown), we measure an attribute of a physical model (which obeys the true function rather than the approximate curve fit function). In circumstances wherein the π-terms are identical, we say that they are *similar* and that each is a *model* of the other.

Such models may be geometrically similar, i.e., have dimensions in proportion. If the subscript m denotes a model attribute,

$$\delta_m/\delta = s = \text{scale factor} = l_m/l$$

and it follows that $A_m/A = s^2$. The π-terms must be identical; thus

$$\pi_1 = \frac{\delta}{l} = \frac{\delta_m}{l_m} \qquad \pi_2 = \frac{P}{El^2} = \frac{P_m}{E_m l_m^2} \qquad \pi_3 = \frac{A}{l^2} = \frac{A_m}{l_m^2}$$

The load that must be applied to maintain similarity is established from π_2:

$$\pi_2 = \frac{P_m}{E_m l_m^2} = \frac{P}{El^2} \qquad P_m = P \frac{E_m l_m^2}{E l^2} = P \frac{E_m}{E} s^2$$

For a steel model built to 1/10 scale (s = 0.1), the proper load to apply to correspond to a 30,000 lbf prototypical load is

$$P_m = P \frac{E_m}{E} s^2 = 30,000 \frac{30(10^6)}{30(10^6)} 0.1^2 = 300 \text{ lbf}$$

The prediction equation for the prototypical deflection obtained from the observation of the model (measurement) is

$$\delta = \frac{\delta_m}{s} = \frac{\delta_m}{0.1} = 10\delta_m$$

For an aluminum model [E = 10.5(10^6) psi] built to 1/10 size, the proper load is

$$P_m = 30,000[10.5(10^6)/30(10^6)]0.1^2 = 105 \text{ lbf}$$

with a prediction equation of $\delta = 10\delta_m$. For a celluloid model [E = 0.35(10^6) psi] built to 1/10 scale, the proper load is

$$P_m = 30,000[0.35(10^6)/30(10^6)]0.1^2 = 3.5 \text{ lbf}$$

with a prediction equation of $\delta = 10\delta_m$. Note that in addition to geometric similarity between model and prototype there is a loading requirement.

Let us consider a timber beam that has a 6 in. by 12 in. cross section that is simply supported over a span of 144 in. and carries a load of 4800 lbf at 60 in. from the left support. Figure 5.38 depicts the beam in specific and parametric terms. The functional relationship between the parameters is of

Empiricism

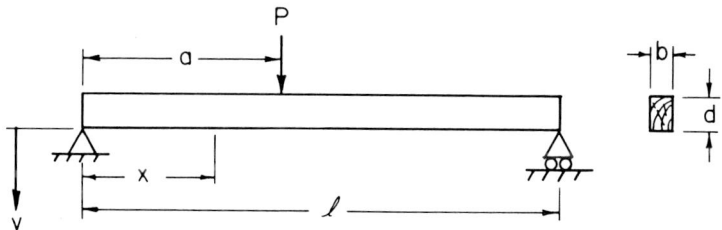

Fig. 5.38. A timber beam whose transverse deflection at location x is desired.

the form

$$f(y, a, b, d, x, P, E, \ell) = 0$$

where y is the deflection under load at position x. It is instructive to discover a complete set of dimensionless variables.

$$\dim(y) = L \quad \dim(x) = L$$
$$\dim(a) = L \quad \dim(P) = F$$
$$\dim(b) = L \quad \dim(E) = FL^{-2}$$
$$\dim(d) = L \quad \dim(\ell) = L$$

The matrix of dimensions is

	y	a	b	d	x	P	E	ℓ
F	0	0	0	0	0	1	1	0
L	1	1	1	1	1	0	-2	1

The right hand determinant is nonzero and consequently the rank of the dimensional matrix is two and the number of dimensionless variables in the complete set is

$$\pi = n - r = 8 - 2 = 6$$

The exponential equations are $e_6 + e_7 = 0$ and $e_1 + e_2 + e_3 + e_4 + e_5 - 2e_7 + e_8 = 0$ or, expressed more conveniently,

$$e_7 = -e_6$$

$$-2e_7 + e_8 = -e_1 - e_2 - e_3 - e_4 - e_5$$

The matrix of solutions appears as

	e_1	e_2	e_3	e_4	e_5	e_6	e_7	e_8
π_1	1	0	0	0	0	0	0	-1
π_2	0	1	0	0	0	0	0	-1
π_3	0	0	1	0	0	0	0	-1
π_4	0	0	0	1	0	0	0	-1
π_5	0	0	0	0	1	0	0	-1
π_6	0	0	0	0	0	1	-1	-2

The complete set is

$$\pi_1 = y/l \qquad \pi_4 = d/l$$
$$\pi_2 = a/l \qquad \pi_5 = x/l$$
$$\pi_3 = b/l \qquad \pi_6 = P/El^2$$

The functional relationship between the parameters may be expressed as

$$g\left(\frac{y}{l}, \frac{a}{l}, \frac{b}{l}, \frac{d}{l}, \frac{x}{l}, \frac{P}{El^2}\right) = 0$$

A study of the above equation (in functional form) suggests that a unique value of any π-term *does not* require a unique value for the parameters that constitute the π-term. Recognition of this suggests the possibility of constructing a physical model of the timber beam in a different size and even of a different material (say steel).

The timber beam with its loading is described by a unique set of π-terms. It is possible to realize the $\pi_1 = y/l$ term with the term y_m/l_m, where the subscript m denotes the model parameter. The set of model π-terms is y_m/l_m, a_m/l_m, b_m/l_m, d_m/l_m, x_m/l_m, $P_m/(E_m l_m^2)$. If the model and prototype π-terms are identical, the modeler is free to decide (1) the linear scale to which the model is built and (2) the material of which it is to be constructed.

The linear model measurements are obtainable from (model dimension) = (scale factor)(prototype dimension). In the case of the deflection y,

$$\frac{y_m}{l_m} = \pi_1 = \frac{y}{l} \qquad y_m = \frac{l_m}{l} y = sy$$

where s is the scale factor. For the other linear dimensions it follows that $a_m = sa$, $b_m = sb$, $d_m = sd$, and $x_m = sx$. The sixth π-term yields

$$P_m = \frac{E_m l_m^2}{El^2} \cdot P = \frac{s^2 P E_m}{E}$$

For a 1/10 size model of steel

$$a_m = 0.1(60) = 6 \text{ in.}$$
$$b_m = 0.1(6) = 0.6 \text{ in.}$$
$$d_m = 0.1(12) = 1.2 \text{ in.}$$

$$P_m = \frac{0.1^2(4800)\,30(10^6)}{1.5(10^6)} = 960 \text{ lbf}$$

The place on the model for the deflection measurement is $x_m = 0.1 x$ and the predicted prototypical (real beam) deflection is $y = y_m/0.1 = 10 y_m$.

5.9 INTERNATIONAL SYSTEM OF UNITS

The International System of Units (Système International d'Unités), officially abbreviated SI, is a modernized version of various metric systems. It was established by international agreement to provide a logical, coherent, and

Empiricism

interconnected network for all measurements in science, industry, and commerce. Of an 1872 international meeting involving 26 countries, 17 signatories in 1875 (including the United States) produced the Metric Convention. They agreed to

--set up metric standards for length and mass
--establish the International Bureau of Weights and Measures (BIPM)
--establish the General Conference of Weights and Measures (CGPM), meeting every six years
--establish an International Committee on Weights and Measures (CIPM), meeting every two years and implementing recommendations of CGPM

By 1960 CGPM consisted of 40 members and it modernized the metric system to create the International System of Units (SI). The candela, a measure of luminous intensity, was adopted in 1971 by CGPM.

The original metric system was created by decree of the National Assembly of France in 1790. It was based on a metre of 1 ten-millionth of the distance from the North Pole to the equator. It was a decimal system with all the conversion advantages that accrue to such a system. The system changed and units such as erg and dyne appeared and the cgs system grew. In 1960 CGPM adopted SI.

We will use SI conventions defined by the National Bureau of Standards henceforth in this book for problems involving metric measures, and abide by as many conventions as possible when problems involve British Engineering units.

See Appendix A for information about and use of the SI measurement system.

The impact of this international agreement will be manifold. Engineers will be required to be bilingual (British Engineering and International System). Soft conversions will be relatively easy; however, many printed informational displays will have to be redone. Diagrams such as PVT surfaces in Figs. 5.14 and 5.15, which are computer-generated contours at preferred values expressed in pounds per square inch, cubic feet per pound, and degrees Fahrenheit, must be reconstructed. The contours do not correspond to convenient or preferred SI values. Recomputation and redrawing will be accomplished by others. However, ordinary graphs may have to be recoordinated locally by the engineer.

Consider the abscissa line graph, Fig. 5.39, which is part of a graph and has coordinates expressed in inches of diameter. The scale factor is

$$s_x = X/x = 1.0 \text{ in.}/2 = 0.5 \text{ in.}$$

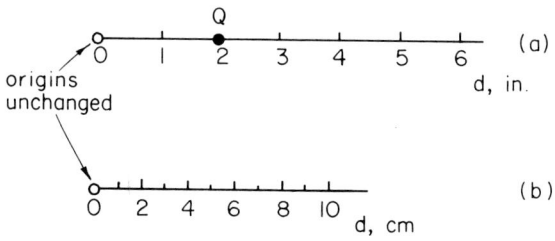

Fig. 5.39. Changing the units of a line graph from inches to centimetres (figure is 0.638 actual size).

If the line graph is to be graduated in centimetres of diameter, the plotting origin is in the same location. The physical reality represented by the graphed point Q is a unique diameter; only the units of measurement differ. The old scale coordinate of Q is 2; on the new scale it is (2)(2.54) = 5.08. The new scale factor is

$(s_x)_{new}$ = X/x = 1.0 in./5.08 = 0.197 in.

The new scale ticks corresponding to 1, 2, 3, etc., are

$X_1 = (s_x)_{new} x$ = 0.197(1) = 0.197 in. = 0.50 cm

X_2 = 0.197(2) = 0.394 in. = 1.00 cm

X_3 = 0.197(3) = 0.591 in. = 1.50 cm

In Fig. P5.4 the line graphs that form the ordinate and abscissa are in British Engineering coordinates. Let us convert them to SI units of megapascals (MPa). The plotting origins of ordinate and abscissa remain in the same locations.

$(s_y)_{old}$ = Y/y = 1.28 in./100 = 0.0128 in.

Had the line graph been prepared with SI units, at the graph of y = 100 kpsi would be a coordinate of 689.5 MPa. The new scale factor is

$(s_y)_{new}$ = Y/y = 1.28 in./689.5 = 0.001 856 in.

The new scale ticks corresponding to 500 and 1000 MPa are

$Y_{500} = (s_y)_{new} y$ = 0.001 856(500) = 0.928 in. = 23.57 mm

Y_{1000} = 0.001 856(1000) = 1.856 in. = 47.14 mm

Figure 5.40 depicts the recoordinated illustration of Fig. P5.4.

The presence of a logarithmic scale as the abscissa line graph of a two-coordinate plot represents a slightly different situation with regard to converting coordinate units. Figure 5.41 depicts the situation. Whether the coordinate is expressed in inches or centimetres, tenfold changes are the same

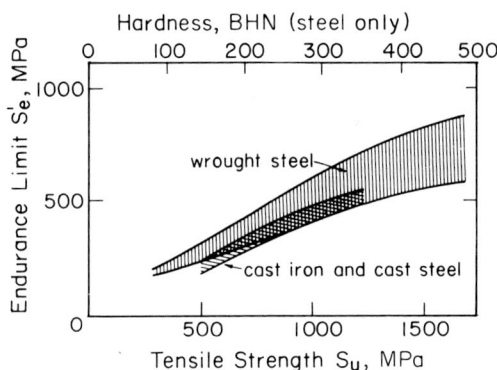

Fig. 5.40. The conversion of Fig. P5.4 from British Engineering to SI units (figure is 0.638 actual size).

Empiricism 167

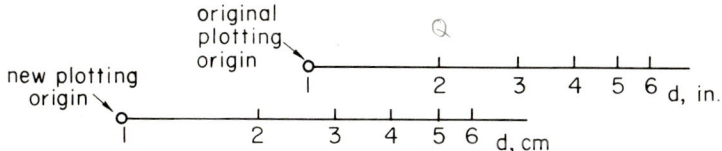

Fig. 5.41. Changing the units on a logarithmic line graph from inches to centimetres (figure is 0.638 actual size).

distance apart on either line graph; hence the invariant in the transformation is taken as the length of a decade. However, since the plotting origin of the original line graph coincides with coordinate 2.54 on the new line graph, the plotting origin has moved.

$$s_{ln\ x} = X/ln\ x = 1.13\ in./ln\ 2 = 1.63\ in.$$

$$X_{10} = s_{ln\ x}\ ln\ 10 = 1.63\ ln\ 10 = 3.78\ in.$$

The scale factor has not changed, i.e., $(s_{ln\ x})_{new} = s_{ln\ x}$. The distance the plotting origin has moved to the left is the same as the distance between the plotting origin and 2.54 on the old line graph; consequently

$$X = s_{ln\ x}\ ln\ x = 1.63\ ln\ 2.54 = 1.52\ in.$$

If we *translate* the old scale 1.52 in. to the left the new line graph is created.

Linear and logarithmic line graphs are treated with different invariants because scale changes are multiplicative (2.54:1) on both linear and logarithmic line graphs. This example should caution the reader to expect that each functional scale type will have a different recipe for carrying out the transformation.

5.10 RECOGNITION OF VARIABILITY

The whole subject matter of statistics is an outgrowth of the recognition of the inherent variability encountered in the empirical process and the search for means of quantitatively communicating information concerning this variability. We encounter variability in the assessment of a measurable attribute of an entity (the breaking strengths of specimens taken from the same bar differ) and from variability in the measurements themselves (the timers of the winner of a race all report different times). We introduce variability into problems by performing calculations using numbers with inherent variability. What can be said of the results? When you calculate sin x using the infinite series,

$$\sin x = x - x^3/3! + x^5/5! - x^7/7! + \ldots$$

and all calculations are subject to truncation errors (as in the case of a digital computer), what can be said of the results?

Continuous (any value is possible) random variables are categorized by *probability density functions* from which many important attributes of the distribution may be deduced, e.g., mean value, standard deviation, skewness, kurtosis, etc. More often, we observe a sample of a population and are anxious to infer the mean of the population and other properties from the sample. We would like to discover the kind of distribution we have and it must be accomplished with some estimate or assurance that the conclusion is "right."

Common distributions are studied because they are useful or they often arise and consequently have been investigated. Names such as *uniform*, *normal* (or *Gaussian*), *log-normal*, *beta*, *gamma*, *exponential*, *extreme value*, and *Weibull* will become important to you, just as they have become familiar to the practicing engineer.

Decision making is the lot of engineers (and others); and they study methodologies that allow rational choices to be made, such as

--Shall we purchase machine A or B?
--Shall we repair the press or buy another?
--Shall we return a lot to the supplier based on a test involving 1 percent of the order?
--Is this outcome the result of pure chance or is a causal agent at work here?

Engineers wish to make inferences concerning entire populations from statistics. If a sample of ten is taken from a large population, what are the confidence bounds on the mean of the population as inferred from the sample? What are the confidence bounds on the standard deviation?

Engineers find themselves testing hypotheses. If 5 lb and 10 lb bags of sugar are unmarked and someone reports weighing (carefully) a bag and states that it weighs 10.04 lbf, there is little doubt that it came from the 10 lb population. But suppose the means were closer; we have 5 lb bags and 81 oz bags and the report was 5.03 lbf. The decision is that the selection came from the 5 lb population. What is the chance that it is wrong? If a sample shows a mean of $\hat{\mu}_1$ and another a mean of $\hat{\mu}_2$, what is the chance that both samples were selected from the same population? One section of a course had one student failure, another two failures, and a third ten failures. What is the chance that the discrepancy is due to chance alone and what is the chance that it is due to a difference in instructor expectation or student performance?

A plot of x and y data shows by least squares curve fitting that $y = 0.97x^{1.02}$. Does this prove that the coefficient is not one or that the exponent is not one? What are the chances that we are confirming a mathematical model $y = x$? If the diameter of bolts exhibits a mean and a standard deviation, what are the mean and the standard deviation of the cross-sectional area? If the diameter of a bearing ball exhibits a mean and a standard deviation, what are the mean and the standard deviation of the ball weight when the density of the steel is likewise variable?

Mathematical preparation for engineering includes many statistical ideas, and imbedded in it is the basis for answers to the questions raised in this section, among others. The deterministic mathematics and arithmetic in which you have been immersed most of your life are the special cases that ensue when all variables and parameters exhibit no variability. Alas, the real world is not deterministic and engineers must use a judicious blend of statistical and deterministic mathematical ideas to build the mathematical models they use to predict the real world response to alternatives they envision.

PROBLEMS

5.1. What is an experiment?
5.2. Criticize the following definitions.
   --Temperature may be considered to be an index of the intensity of molecular activity.
   --The pressure of a system is the force per unit area of its boundaries.
   --Specific volume is the volume occupied by a unit mass of substance.

--A property of a system is any observable characteristic of the system.
--A force is an action exerted by one body on another that tends to change the state of motion of the body acted on.
--Weight is a measure of the attraction of the earth on a body.
--A dyne is the force that will produce an acceleration of 1 cm/s^2 when acting on 1 g mass.
--A force of 1 lbf is defined as a force equal to the force with which the earth attracts a mass of 1 lbm.
--A calorie is the quantity of heat or energy necessary to raise the temperature of 1 g water from 15°C to 16°C.
--The coefficient of expansion is the increase in length per unit length for a 1°C rise in temperature of a body.
--The melting point temperature is the temperature at which a solid changes to a liquid without a change in temperature.
--The heat of fusion of a substance is the number of calories necessary to convert 1 g at the melting point into liquid at the same temperature.

5.3. Define the following entities: electron, proton, neutron, atomic nucleus, chemical element, osmosis, adsorption, absorption, beam, column, clutch, brake, chain, cam, linkage, structure, crank-rocker mechanism.

5.4. The endurance limit of a metal is the magnitude of the largest completely reversed stress that a mirror-polished specimen can endure and exhibit infinite life. For practical purposes 10^6 cycles is taken as infinite life. Figure P5.4 shows the scatter bands enveloping test data on wrought steel and on cast iron and cast steel. Suggest a relationship between endurance limit S_e' and tensile ultimate strength S_u that represents the mean of the wrought steel data. Write a similar expression relating Brinell hardness (Bhn) to endurance limit and to tensile strength in the case of cast iron and cast steel.

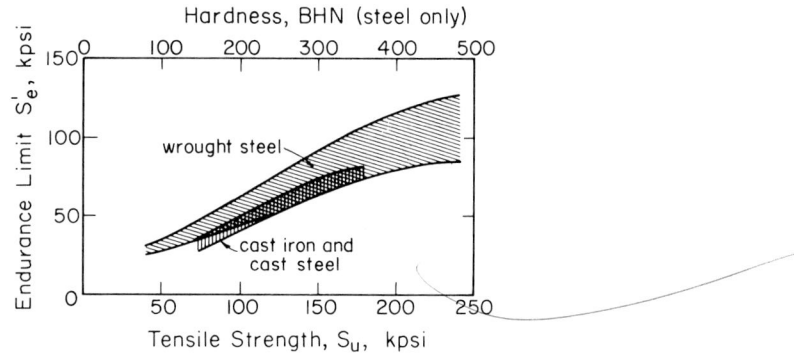

Fig. P5.4. (Shigley, *Mechanical Engineering Design*, Fig. 5-13; and Charles Lipson and Robert C. Juvinall [eds.], *Application of Stress Analysis to Design and Metallurgy*, Univ. Mich. Summer Conf. 1961, Ann Arbor.)

5.5. Figure P5.5 shows fatigue data plotted against an ordinate that is the ratio of endurance limit to tensile strength. The abscissa is the specimen life in cycles. Establish simple approximations to scatter bands. If you were required to establish a locus that separates nonfailures from failures, what would it be?

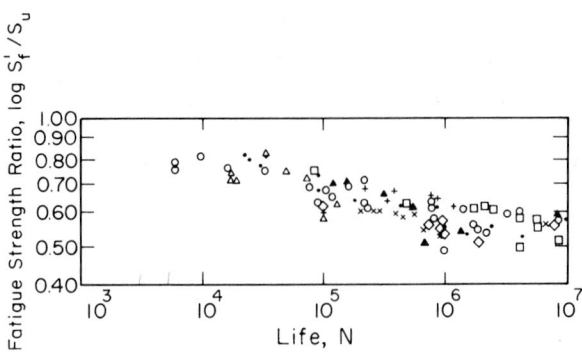

Fig. P5.5. (Shigley, *Mechanical Engineering Design*, Fig. 5-14; and Lee and Juvinall, *Application of Stress Analysis*.)

5.6. Using Fig. P5.5 and Table 2.1, estimate the fatigue strength of a mirror-polished specimen of hot rolled steel, AISI 1018, corresponding to a cycle life of 82 000 cycles. Do the same for AISI 2340 steel, oil quenched, drawn at 400°F.

5.7. The Brinell hardness and tensile strength of wrought aluminum alloys in a soft or annealed condition are tabulated below (see Fig. P5.7).

Brinell Hardness, 500 kg	Tensile Strength, psi	Tensile Strength, MPa
23	13 000	90
28	16 000	110
45	26 000	179
45	27 000	186
45	26 000	179
47	27 000	186
45	27 000	186
26	16 000	110
30	18 000	124
60	33 000	228

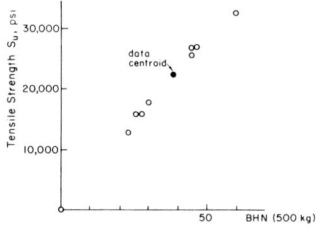

Fig. P5.7

Plot these data and develop an equation for predicting ultimate tensile strength from Brinell hardness using eyeball-best-line and least squares techniques.

5.8. The maximum particle size (μ) that will pass through different mesh numbers (#) is given below (Fig. P5.8).

Empiricism   x   y   171

Mesh #	Maximum Particle Size, μ
20	840
40	420
60	250
80	177
100	149
140	105
200	74
325	44

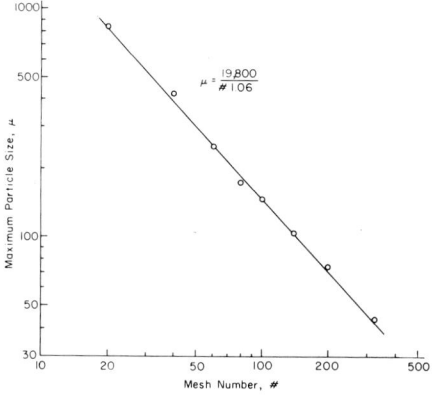

Fig. P5.8

Plot these data and develop an equation predicting particle size from mesh numbers using "eyeball-best-line" and least squares techniques.

5.9. Tabulated below are data of breaking loads of standard hard-drawn copper wire (Fig. P5.9).

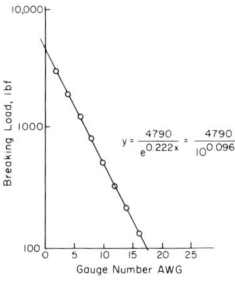

Fig. P5.9

Gauge Number	Breaking Load, lbf	Breaking Load, kN
2	3000	13.3
4	1980	8.8
6	1280	5.7

Gauge Number	Breaking Load, lbf	Breaking Load, kN
8	830	3.7
10	530	2.4
12	340	1.5
14	215	1.0
16	135	0.60

Plot these data and develop an equation for predicting the breaking load from the wire size.

5.10. Electrical resistance of 1 ft lengths of No. 0 AWG standard annealed copper wire at various temperatures results in the data tabulated below (Fig. P5.10).

Temperature, °C	Resistance, Ω
14.0	96.00
19.2	97.76
25.0	100.50
30.0	102.14
36.5	105.00
40.1	105.75
45.0	108.20
52.0	110.60

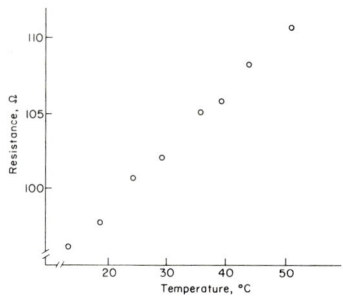

Fig. P5.10

Develop an equation predicting resistance R from knowledge of the temperature T.

5.11. Corrosion tests on specimens of pure magnesium resulted in values of weight increase in pure oxygen at 525°C as tabulated below (Fig. P5.11).

Elapsed Time, hr	Weight Increase, mg/cm^2
0	0
2.0	0.17
4.0	0.36
8.2	0.65
11.5	1.00
20.0	1.68
24.1	2.09
30.0	2.57

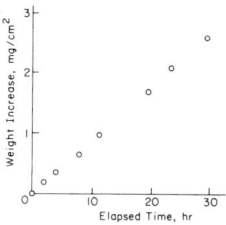

Fig. P5.11

5.12. Develop an equation for predicting weight increase from elapsed time. The approximate discharge rates of water at various heads for 1000 ft of 4 in. pipe are tabulated below (Fig. P5.12).

Head Loss, ft	Discharge, gal/min
1	35.8
2	50.6
4	71.6
6	87.7
9	107.5
12	123.7
16	142.9
20	159.7
25	178.9
30	195.8
40	225.8
50	252.2
75	309.8
100	357.9

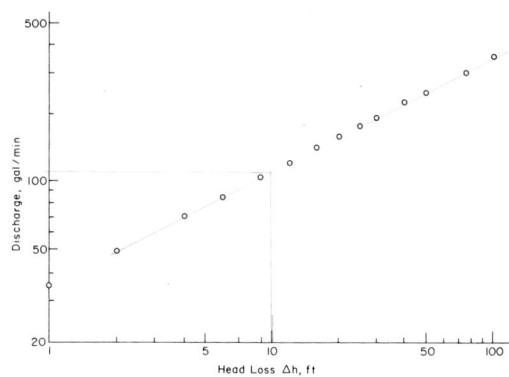

Fig. P5.12

Develop an equation for predicting the discharge from the head loss.

5.13. Steam tables give the specific volume of saturated steam at various absolute pressures as tabulated below (Fig. P5.13).

Absolute Pressure, lbf/in.²	Specific Volume, ft³/lbm
1	333.60
2	173.73
3	118.71
5	73.52
10	38.42
14.7	26.80
20	20.089
25	16.303
30	13.746
40	10.498

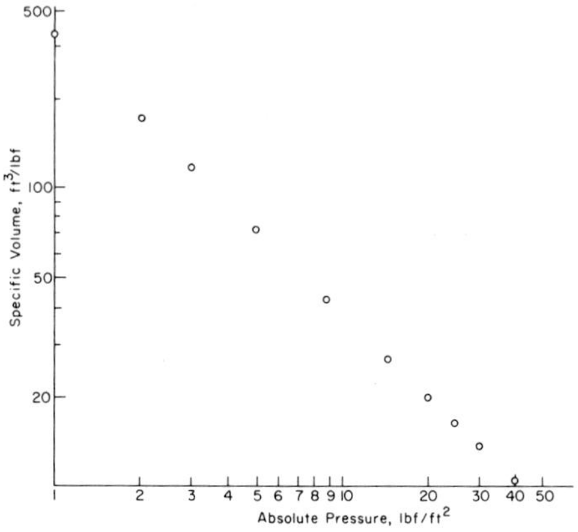

Fig. P5.13

Develop an equation for predicting specific volume from pressure.

5.14. Heat losses from horizontal bare iron hot-water pipes at 180°F in still ambient air at 75°F are tabulated below (Fig. P5.14).

Pipe Diameter, in.	Heat Loss, Btu/ft/day
1	1 896
1 1/4	2 398
1 1/2	2 746
2	3 430
2 1/2	4 140
3	5 054
3 1/2	5 771
4	6 493
4 1/2	7 211
5	8 020
6	9 530
8	12 442
10	15 528
12	18 418
14	20 140
16	23 088
18	25 998

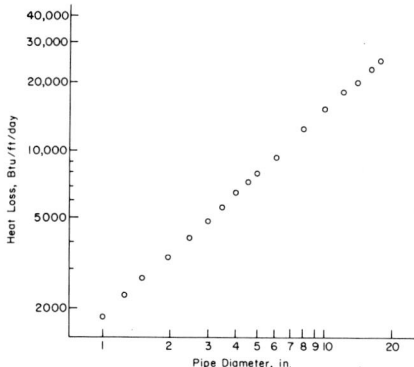

Fig. P5.14

Develop an equation for predicting heat loss from pipe size.

5.15. Creep-strength tests on high chromium (23 percent-27 percent) ferritic steel used in high temperature service resulted in stress levels producing 1 percent deformation in 10 000 operating hours at various temperatures as follows (Fig. P5.15).

Temperature, °F	Stress, psi
1000	6450
1100	2700
1200	1250
1300	570
1400	270

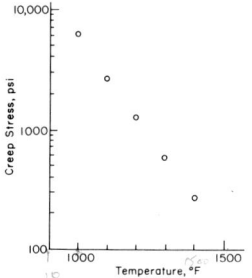

Fig. P5.15

Develop an expression for predicting stress level from temperature.

5.16. Solubility tables give data on solubility of anhydrous potassium alum in water (Fig. P5.16).

Temperature, °C	Solubility $gK_2SO_4Al_2(SO_4)_3/100gH_2O$ for Saturation
0	3.0
10	4.0
20	5.9
30	8.39

Temperature, °C	Solubility $gK_2SO_4Al_2(SO_4)_3/100gH_2O$ for Saturation
40	11.70
50	17.00
60	24.75

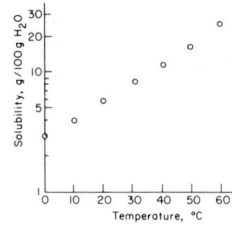

Fig. P5.16

Develop an equation for predicting solubility from temperature.

5.17. The stress s_i (lbf/in.^{2}) at the inner fiber of a curved beam under influence of an external moment M (lbf·in.) is a function of the distance of the inner fiber from the neutral surface h_i (in.), the cross-sectional area of the beam A (in.^{2}), the distance of the neutral surface from the centroid of the cross section e (in.), and the distance of the inner fiber from the center of curvature r_i (in.). Establish a complete set.

5.18. The radius of curvature of a deflected beam ρ (in.) is a function of the bending moment M (lbf·in.), the modulus of elasticity of the beam material E (lbf/in.^{2}), and the second area moment of the cross section about the neutral surface I (in.^{4}). Establish a complete set.

5.19. The strain energy of a rod in tension, U (lbf·in.) is a function of the stress s (lbf/in.^{2}), the length of the rod l (in.), the cross-sectional area of the rod A (in.^{2}), and the modulus of elasticity E (lbf/in.^{2}). Establish a complete set.

5.20. The torque T (lbf·in.) necessary to turn a screw against a load is a function of the axial resisting load F (lbf), the mean thread diameter d_m (in.), the lead of the thread l (in.), and the coefficient of friction μ. Establish a complete set.

5.21. The maximum shear stress in a helical coil spring τ (lbf/in.^{2}) is a function of the axial spring load F (lbf), the mean coil diameter D (in.), and wire diameter d (in.). Establish a complete set.

5.22. The coefficient of friction in a hydrodynamic bearing f is a function of the radius of the journal r (in.), the radial clearance c (in.), the viscosity of the lubricant μ (lb·s/in.^{2}), the journal speed N (s^{-1}), and the nominal loading pressure P (lbf/in.^{2}). Establish a complete set.

5.23. How do we know the significant parameters for a phemomenon?

5.24. Does dimensional analysis tell us anything about the mechanism of operation of a phenomenon, i.e., how it works?

5.25. Can qualitative physical meaning be deduced from π-terms?

5.26. The largest shearing stress τ_{max} (MPa) in a right circular cylindrical bar under torsion is a function of the applied torque T (N·m), the section radius r (mm), and the polar second area moment of the cross sec-

Empiricism 177

tion J (mm^4). Determine the number of dimensionless variables in a complete set and establish the π-terms. Devise a testing scheme to uncover the relationships between the variables.

5.27. The longitudinal (axial) stress induced in a thin-walled right circular cylindrical pressure vessel s (MPa) is a function of the gas pressure p (Pa), the nominal diameter of the pressure vessel D (mm), and the wall thickness t (mm). Determine the number of dimensionless variables in a complete set and establish the π-terms. Devise a testing scheme to uncover the relationship between them.

5.28. The energy U (J) stored as strain energy in a right circular cylindrical bar subject to an applied torque T (N·m), is a function of the length l (mm), the material shear modulus G (MPa), and the second polar area moment of the cross section J (mm^4). Determine the number of dimensionless variables in a complete set and establish the π-terms. Device a testing scheme to uncover the relationship between them.

5.29. The spring rate k (N/mm) of a helical coil spring is a function of the material shearing modulus G (MPa), the wire diameter d (mm), the helix diameter of the coil D (mm), and the number of coils N. Determine the π-terms and devise a scheme of testing that would enable you to discover the functional relationship between the π-terms.

5.30. The torque T (N·m) that can be transmitted through a pair of clutch plates without slipping is a function of the axial force F (N) holding them together, the coefficient of friction between the contacting plates f, the outer diameter of the plates D (mm), and the inner diameter d (mm). Establish the set of π-terms and devise a testing scheme.

5.31. Problems P5.1-P5.16 show it is relatively easy to obtain a good estimating equation when the precision in the physical measurements is high. In many engineering problems this is not the case. For example, in estimating tread wear on commercial tires, the estimation of remaining life may be made on a weight-loss basis or by measurement of depth of the center groove. One would expect a correlation between the two methods.

Weight-Loss Method, x (hundred mi)	Center Groove Method y (hundred mi)
459	357
419	392
375	311
334	281
310	240
305	287
309	259
319	233
304	231
273	237
204	209
245	161
209	199
189	152
137	115
114	112

Estimate, using the eyeball-method, a best line. Determine the centroid of the plotted points and plot an eyeball-best-line through the centroid and the data. Compare these results with Fig. P5.31 in which

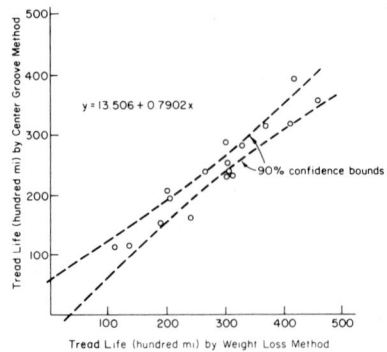

Fig. P5.31

the presumption was made that weight-loss method is error-free with all error in the center groove method determination.

5.32. Data taken in a three π-term empirical curve-fit effort are

π_2	π_1			
	$\pi_3 = 1$	$\pi_3 = 2$	$\pi_3 = 3$	$\pi_3 = 4$
0	0	0	0	0
1	1	0.25	0.11	0.06
2	4	1.00	0.44	0.25
3	9	2.25	1.00	0.56

Using rectification schemes and partial derivative techniques, establish $\pi_1 = \phi(\pi_2, \pi_3)$.

5.33. Data taken in a three π-term empirical curve-fit effort are

π_2	π_1			
	$\pi_3 = 1.1$	$\pi_3 = 1.3$	$\pi_3 = 1.5$	$\pi_3 = 1.7$
0.6	0.57	0.52	0.47	0.42
0.7	0.68	0.63	0.57	0.55
0.8	0.78	0.75	0.72	0.68
0.9	0.89	0.86	0.85	0.84

Using rectification schemes and partial derivative techniques, establish $\pi_1 = \phi(\pi_2, \pi_3)$.

5.34. Data taken in a three π-term empirical curve-fit effort are

π_2	π_1			
	$\pi_3 = 0.8$	$\pi_3 = 0.9$	$\pi_3 = 1.0$	$\pi_3 = 1.1$
0.5	0.49	0.34	0.25	0.19
0.7	0.96	0.67	0.49	0.37
0.9	1.58	1.11	0.81	0.61
1.1	2.36	1.66	1.21	0.91

Using rectification schemes and partial derivative techniques, establish $\pi_1 = \phi(\pi_2, \pi_3)$.

> Newton did not show the cause of the apple falling, but he showed a similitude between the apple and the stars.
>
> Sir D'Arcy Wentworth Thompson

CHAPTER 6

MATHEMATICAL MODELS OF EFFECTS

[handwritten: system boundary ≡ S.b. It bounds all the matter of our interest
surrounding ≡ every thing outside the S.b.]

6.1 INTRODUCTION

Describing a phenomenon and modeling it so that predictions could be made has been reasonably successful, and because of this engineering is possible. When Galileo was performing his crucial experiments, Isaac Newton was unborn. Yet Newton learned from Galileo by absorbing his ideas and reasoning from his writings and from his influence on others. The communication of ideas requires some precision in language and in the delineation of concept. Without this discipline, we would be hopelessly bogged down in communication difficulty. Recognition of reproducibility in cause, effect, and extent led to empiricism's ninth achievement: declarations of mathematical effect statements.

6.2 SYSTEM AND CONTROL SPACE

To model interactions quantitatively between domains of matter, it is necessary to describe events and attributes of matter with precision. To assist in the problems of identifying precisely what we are talking about, three simple definitions are important:

1. A *system* is all matter under observation contained within a prescribed boundary. *[handwritten: e.g. Air in this room, air is system.]*
2. A *boundary* is a fictitious closed surface separating the system from all other matter.
3. The *surroundings* are all matter other than the system, reposing outside the system, that influences system behavior.

The notion of system is fundamental to experimental science. When a system interacts with its surroundings, evidence of this interaction is observable in three places. There are observable changes in the system, in the surroundings, and at the boundary. The experimenter is usually in the surroundings, monitoring the boundary events and the surroundings events. The instrumentation is located in the surroundings. The engineer seeks to observe and report in quantitative terms the events of the system, i.e., describe what is happening to matter under observation from an exterior position.

The engineer is confronted with the necessity of describing phenomena that occur in devices through which substance flows. The engineer is interested in the interaction between the devices and their surroundings. The physical boundaries of the devices suggest suitable monitoring surfaces; but since matter enters and leaves the devices penetrating the monitoring surfaces, the monitoring surface does not enclose a system for more than an instant. The *system* by definition isolates matter and though it may move, distorting the boundary, no matter penetrates the boundary. A monitoring surface that matter penetrates is called a *control surface*, and the enclosure defined by the control surface is a *control region*, *control space*, or *control volume* (or occasionally, *open system*). A control surface is a well-described, closed, continuous surface. As commonly employed, although not essential, the control space is of fixed geometry in some frame of reference.

It should not be surprising that mathematical models describing system behavior are different from mathematical models describing control space behavior.

A few more definitions relating to system description will help us communicate more effectively. Physically and chemically uniform systems are spoken of as *homogeneous* systems. All others are *heterogeneous* systems. *Pure substance* is matter homogeneous in composition with one invariable chemical aggregation. A *phase* is any physically homogeneous aspect of a chemically uniform system. A *property* is any observable characteristic of a system whose change is determined by the end states of a process and whose variation throughout the system is not more than differential order. A *state* of a system is its condition or position as described by its properties. A process is described in terms of its path; the *path* of a process is the sequence of states through which a system passes. An *extensive* property is a property whose magnitude is a function of the mass of the system. An *intensive* property is a property whose magnitude is independent of system mass.

Consider the beam of Fig. 6.1. We will define as system the beam itself, the boundary depicted by the dashed line in Fig. 6.1(b), and the surroundings as the rest of the universe. When we speak of system in this instance, we mean all matter constituting the beam. This matter is within, and remains within, the boundary. If a load is placed on the beam and it deflects, the boundary is distorted but no system or surroundings matter penetrates the boundary. If we are interested in beam behavior (in the large) such as its deflected geometry and the beam is a rolled structural steel section, we can allude to the system as homogeneous; because, in the large, it is physically and chemically uniform.

The beam is a solid throughout and therefore has a physical single phase. A property of the system is the deflection in the center of the span, which is an observable characteristic. We could place a load very carefully on the beam and the beam would sag slowly to its deflected configuration. We could apply the load suddenly and observe the beam deflect to twice its final displacement; with damped oscillation it is independent of the path of the process of loading. We conclude that the deflection is a property of the beam. When the beam is unloaded and essentially straight, we can say it is in state 1 and use the property of center deflection to describe that state. The beam, when statically deflected under the influence of a load, can be described as in state 2 (different from state 1), because its condition as described by the property of center deflection is different. If the beam is at a uniform temperature throughout, temperature is a thermodynamic property of the system. (If the temperature were not uniform temperature would not be a thermodynamic property of the system, since the temperature varies by more than a differential amount within the system.) The temperature, when a property of this system, is an intensive property inasmuch as it is independent of the system mass.

Newton described the interaction between a system (particle) and its surroundings in terms of the acceleration of the system and the system mass. The acceleration must be measured with respect to an inertial (Galilean) reference system. He stated that a system free of acceleration is free of unbalanced force. When the system experiences an acceleration, the surroundings exert a

Mathematical Models of Effects

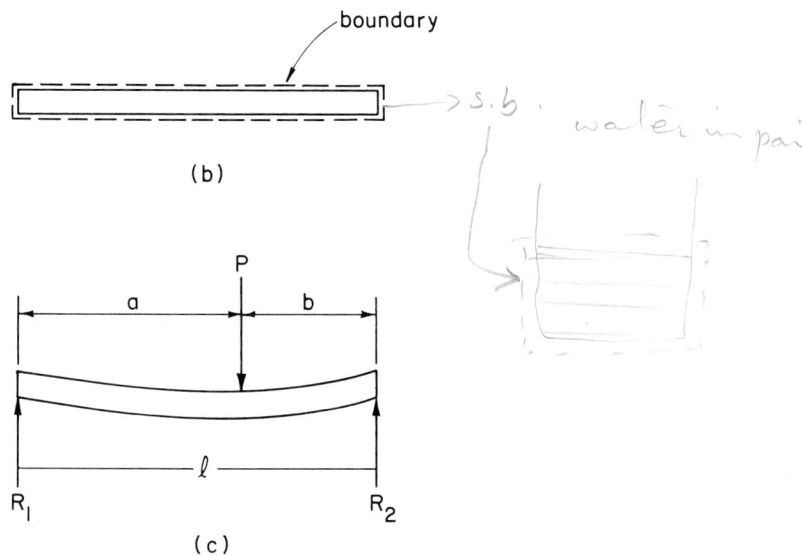

Fig. 6.1. A beam, its boundary, and its parameters.

force on the system (action)--there are equal and opposite vector quantities and the magnitude and direction of the action force is

$$\Sigma \bar{F} = m\bar{a}$$

The symbol for force is written $\Sigma \bar{F}$ to remind us that the force involved here is the vector sum of all tractive influences of the surroundings on the system.

The beam of Fig. 6.1 is at a state of rest (the property of central vertical deflection is time invariant). Newton's law states that for a zero acceleration in an inertial reference frame, the system is free of unbalanced force.

The significant influences of the surroundings on the system of Fig. 6.1 are surface tractions normal to the boundary at each support and at a point under the load. These highly localized normal tractions are represented by their vector sums (equivalents); and the concentrated forces $\bar{R}_1$, $\bar{R}_2$, and $\bar{P}$ replace the significant influences of the surroundings--that is, the significant influences of the surroundings during static deflection of the beam are three parallel forces. Newton's law is the authority for writing the equation

$$\bar{R}_1 + \bar{R}_2 + \bar{P} = \Sigma \bar{F} = m\bar{a} = \hat{0}$$

Since the three forces have parallel lines of action, the equation can be written in scalar form as

$$R_1 + R_2 - P = 0$$

Newton's law for a body of distributed mass relates the couple about the center of mass to the mass moment of inertia and the angular acceleration, and in its simplest form it is

$$\Sigma T = I\alpha$$

Since the system is at rest, the angular acceleration is zero and the sum of the external torques is zero. If the system is moment-free, moments may be summed about any point and set to zero. Summing moments about the left-hand and right-hand reactions, respectively, yields

$$R_2 l - Pa = 0 \quad Pb - R_1 l = 0$$

from which the reactions are determinable as

$$R_1 = Pb/l \quad R_2 = Pa/l$$

Our work above is typical of analyses that engineers perform in the following senses.

    1. It is necessary to define a system or control region, delineate the boundary or control region, and then remove the surroundings from contact with the system or control region.
    2. The matter in the system or currently in the control region will behave identically with actual surroundings present or in the abstraction if the significant influences of the surroundings are identified and simulated.
    3. The significant influences of the surroundings must ·be simulated by mathematical models of sufficient accuracy.
    4. The significant influences of the surroundings must be related to properties of the system (or control region).
    5. The relations established in the preceding step will often be approximate; the error will be made to vanish by a limiting process.
    6. The resulting equations must be solved for appropriate dependent variables. Often the equations are differential equations, and boundary and/or initial conditions must be added to the problem to uniquely determine the solution.

In the preceding problem:

    (a) How did we know that the system should be the beam and not a portion of it?
    (b) How did we know that the significant influences are the three surface tractions cited?
    (c) How did we know that normal surface tractions could be modeled satisfactorily by their vector sum?
    (d) Is the earth an inertial reference? Is Newton's law, as stated, valid?
    (e) Are the reactions calculated accurate within 5 percent? 2 percent? 0.1 percent?

We have opened Pandora's box; in so doing we have shown the topics of interest for the immediate future. We must understand mathematical models of phenomenological events; and hence the problems of reproducibility of cause, effect, and extent of phenomena are of immediate importance. Thus the behavior of systems and identification of systems effects is investigated now. The

Mathematical Models of Effects   183

interrelationships between effects can be understood in terms of notions of
accountability and the manifestations of accountability incorporated in what
are called first principles. Then the problem of deductions and inferences
from first principles can be pursued.

6.3 HEAT AND WORK EFFECTS

   Systems and their surroundings interact. These interactions often have
an identifiable cause or stimulus, the changes occurring in the system and its
surroundings are identifiable in extent, and measures of causal elements can
be related to measures of results. When consistency and reproducibility are
observed, a mathematical model is possible. The change is called an *effect*
and is often named for its discoverer or the principal contributor to its understanding. The equation expressing a quantitative relationship between
cause and effect is called someone's equation (and occasionally and inappropriately, someone's law).

   In describing system behavior, investigators have agreed that when a system changes its state, if the sole effect in the surroundings could have been
the rise of a weight, then the system experiences a *work effect*. Any change
that is not a work effect is called a *heat effect*.

   The quantitative expression for a work effect is given as a matter of
definition (from the rise of a weight in a gravitational field) as

$$dW = \bar{F} \cdot d\bar{s}$$

where the differential of work (a scalar) is the dot product of a force of the
surroundings acting on the system and the differential displacement of the
force. The scalar work is positive when work is done on the surroundings by
the system (a weight rises in the surroundings).

   The heat effect has been interpreted calorimetrically and the definition
is

$$dQ = C \, d\theta$$

where C is the system process heat capacity, θ is the ideal gas-thermometer
temperature of the system, and the differential dQ is positive when heat is
added to the system. A system boundary that allows no heat effect to occur
during a change in state of the system is called an *adiabatic* boundary.
Boundaries that are not adiabatic are called *diathermic* boundaries.

   The work done to accelerate a system is expressible in terms of system
properties.

   Consider the system of Fig. 6.2 at position $\bar{s}$, moving at velocity $\bar{V}$, and
acted on by external force $\bar{F}$ while traversing the interval between $\bar{s}$ and $\bar{s}$ +
$d\bar{s}$. From the definition of work, the work done by the system on the surroundings is

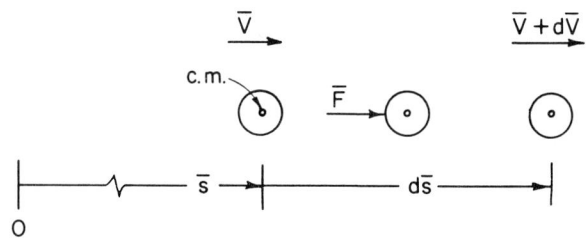

Fig. 6.2. A system undergoing a translational acceleration.

$$dW = -\overline{F} \cdot \overline{ds} = -m\frac{d\overline{V}}{dt}\overline{ds} = -m\,\overline{V} \cdot d\overline{V} = -d[(1/2)mV^2]$$

The external force $\overline{F}$ can be expressed in terms of the system properties of mass and acceleration (after Newton) as $\overline{F} = m\overline{a} = m(d\overline{V}/dt)$. The work done by a system experiencing only a velocity change can be expressed in terms of the negative of the change in the quantity $(1/2)mV^2$. The existence of a work effect (in this instance) can be detected and expressed quantitatively by noting the change in velocity of the system and calculating the change in $(1/2)mV^2$.

When a system is in contact with its surroundings through a diathermic boundary and matter on both sides of the boundary is solid, changes can be observed in the system and surroundings that cannot be made equivalent to the rise or fall of a weight in the surroundings. These changes involve temperature differences between the system and its surroundings. If a temperature gradient (stimulus) exists at a point on the boundary between solid system and solid surroundings, a heat effect is observable calorimetrically, described verbally as the migration of heat or heat transfer. The relationship between the cause, temperature gradient, and the consequences, heat transfer, was shown by Jean Baptiste Biot to be

$$\dot{q} = -k\overline{\nabla}T$$

where: $\dot{q}$ = heat flux in calorimetric units per unit time and area
k = thermal conductivity
T = ideal gas thermometer temperature, any datum
$\overline{\nabla}$ = gradient operator

This relationship is called the *conduction effect*, *heat conduction effect*, or *Biot's heat conduction effect* (and inappropriately, Fourier's conduction equation).

When matter adjacent to a diathermic boundary is solid on the system side and fluid on the surroundings side, heat effects are observable that respond to temperature difference between the fluid well removed from the boundary and the temperature of the system or surroundings at the boundary. The stimulus is the temperature difference between system boundary and fluid surroundings, $T_s - T_\infty$. The consequences are no change of weight position in the surroundings and a calorimetric change, heat migration. Measurements are reproducible and related by the equation,

$$\dot{q} = -h(T_s - T_\infty)$$

where: $\dot{q}$ = heat flux in calorimetric units per unit time and area
h = coefficient of convective heat transfer
T_s = gas thermometer temperature of system boundary, any datum
T_∞ = gas thermometer temperature of fluid surroundings well removed from the boundary, any datum

This heat effect is called a *convection effect*, *heat convection effect*, or the *Newtonian convection effect* (because of Newton's contribution toward its understanding).

When the system at the diathermic boundary is solid and the surroundings approach a vacuum, a heat effect is observable. With the stimulus, system temperature at the boundary, the consequences are no change in elevation of a weight in the surroundings and a calorimetric change, heat migration. Measurements relating temperature and heat flux indicate a relationship of the form

Mathematical Models of Effects

$$\dot{q} \leq -\sigma T^4$$

where: $\dot{q}$ = heat flux in calorimetric units per unit time and area
σ = constant
T = ideal gas thermometer temperature, absolute

This relationship states that the emissive (outgoing) heat flux observed is $\leq -\sigma T^4$. It can be made into an equality by introducing a factor ε, total emissivity of the body, that ranges from 0 to 1. This heat effect is called the *radiation heat effect* and the constant carries the name Stefan-Boltzmann to honor the contributions of Josef Stefan and Ludwig Boltzmann to this complex matter of radiative emission. The problem of incoming radiation and system interaction is a separate matter, and net radiative flux between systems is a complex subject that will be treated in adequate detail in course(s) associated with heat transfer.

6.4 TRACTIVE EFFECTS

When forces are applied to a system by the surroundings, they appear as normal and shearing stresses at the boundary and are spoken of as surface tractions. The effect of tractions on the entire boundary can be represented for some purposes as a force at the center of mass of the system. The result of a surface traction may be the linear and angular acceleration of the system and the distortion of the system geometry. In different circumstances various manifestations are noted and the effects often carry the name of the discoverer or principal contributor.

For example, Hooke's equation (after Robert Hooke) relates stresses to strains in elastic materials. In a wire under tension, a simple linear proportionality exists between the elongation and the tension. To have a unique constant that is independent of the material geometry, Hooke's equation is posed not in terms of elongation and tension but in terms of stress and strain. If we have a wire in tension, the strain in the load direction (x-direction) is

$$\varepsilon_x = s_x/E$$

where ε_x is the strain in the x-direction and E is the modulus of elasticity of the material. If the deformations in the y-direction and z-direction are examined, a lateral contraction is observed. Thus the strains in the other two directions (even though the load is applied in the x-direction) are

$$\varepsilon_y = -\nu(s_x/E) \qquad \varepsilon_z = -\nu(s_x/E)$$

where the constant ν is called Poisson's ratio. The shearing strains are related as

$$\gamma_{xy} = s_{xy}/G \qquad \gamma_{xz} = s_{xz}/G \qquad \gamma_{yz} = s_{yz}/G$$

where G is the modulus of rigidity of the material. The cause of the distortion is the imposition of load (stress) on the system, and the consequences are distortions (strains) in the system. Cause and consequence can be related quantitatively and, at a point in the system,

$$\varepsilon_x = (1/E)(s_x - \nu s_y - \nu s_z) \qquad \gamma_{xy} = s_{xy}/G$$

$$\varepsilon_y = (1/E)(-\nu s_x + s_y - \nu s_z) \qquad \gamma_{xz} = s_{xz}/G$$

$$\varepsilon_z = (1/E)(-\nu s_{sx} - \nu s_y + s_z) \qquad \gamma_{yz} = s_{yz}/G$$

$$G = E/[2(1+\nu)]$$

These equations are attributed to Hooke and are called Hooke's equations relating stress and strain at a point in an elastic body. Elastic distortion as described by these equations is observed in stiff, elastic materials such as steel (but not in rubber) and is described as a *Hookian effect*, the quantitative relations being called Hooke's equations. The Hooke effect is a tractive effect and is reproducible.

If a solid body is in surroundings that are static and fluid and the solid body is taken as a system, the surroundings exert a tractive effect on the system. The surface tractions are normal to the boundary and vary in intensity because of gravitational influence on the fluid. The vector sum of all elemental forces on the boundary can be expressed in terms of the fluid pressure. The expression

$$\overline{f}_s = -\text{grad } p = -\overline{\nabla} p$$

which says that the resultant force on the system (body) per unit of body volume, $\overline{f}_s$, a vector, is equal to the negative of the gradient of the fluid pressure, another vector. The stimulus (cause) is a fluid pressure at the boundary and the consequence is a force applied to the system. It is reproducible, cause (fluid pressure) is measurable, and the consequence (a body force) is measurable; the quantitative relationship between them has been given. This tractive effect is called the *buoyant effect* and the name of Archimedes is often associated with it.

Newton's equation, $\overline{F} = m\overline{a}$, is not for a particular stimulus as is the buoyant force equation but describes the system for unbalanced forces of any origin whatsoever in terms of its property, acceleration; consequently it is far more general. We will treat Newton's laws of motion when we consider principles.

When a fluid boundary of a system is in motion, a "drag" force appears and its magnitude is relatable to properties of the fluid stream at the system boundary. The surface traction that appears is a fluid-induced shearing stress and its magnitude is related to the fluid velocity gradient at the boundary. Again, to obtain simplicity and generality, the stimulus (fluid motion) and the consequence (shearing traction) are related at a point on the boundary, and for a large body of common fluids they are found to be proportional,

$$\tau_x = \mu \frac{\partial u}{\partial y}$$

where: τ_x = shearing stress on a plane perpendicular to the y-direction
$\partial u/\partial y$ = velocity gradient in y-direction
μ = coefficient of viscosity or simply fluid viscosity

This relationship is called the *fluid viscous effect*, and the factor μ is called Newton's viscosity coefficient because of his contribution to the understanding of this effect in quantative terms. The operational definition of viscosity is contained in the equation describing the viscous effect. The viscosity of a fluid is the quotient of the induced shearing stress and the velocity gradient inducing it.

Mathematical Models of Effects

6.5 SURFACE EFFECTS

Forces can be imposed on systems whose boundaries include the free surface of a liquid. The free surface of a liquid exhibits a characteristic behavior that can be explained by visualizing the surface layer of molecules as a membrane under tension. The phenomenon is called surface tension. The "floating" of a dry needle on the trampolinelike surface is a way of balancing body weight against surface tractions induced by the fluid surface.

Capillarity, the rise of a free surface of a liquid in a small bore tube, is another manifestation of surface tension. The difference in air pressure between the inside of a soap bubble and the ambient atmosphere is still another indication of the surface tension phenomenon.

Surface tension, the force per unit length of free surface, is a property of the surface and the liquid. Manifestations of surface tension are forces that influence systems.

6.6 CHARGE EFFECTS

If an electrical resistance is taken as a system and is connected by an open circuit to a battery, we can say that the resistance is the system and all else is the surroundings. If the switch is closed, electrons migrate around the circuit, penetrating the boundary on entrance and on exit. If matter is anything that exhibits inertia, we have to conclude that an electron is matter and that we no longer have a system but a control volume. If we monitor the charge migration rate through the control volume (current) and the potential drop across the control volume, the stimulus (potential, voltage) is related to the consequence (current) measurably and reproducibly by

$$e = iR$$

where R is the proportionality factor, called resistance. The effect is called the *resistive effect*; the equation, referred to as Ohm's equation, may be viewed as the operational definition of electrical resistance.

If the control volume contains a helical coil of vanishing resistance, the relationship between the stimulus (voltage) and the consequence (current) is measurable and reproducible and is given by

$$e = L \frac{di}{dt}$$

where di/dt is the time derivative of current and L is the proportionality factor called inductance. The effect is referred to as the *inductive effect*; the equation, called Henry's equation, may be viewed as the operational definition of inductance.

If the control volume contains parallel plates separated by a dielectric, the stimulus (voltage) and the consequence (current) is measurable and reproducible and is given by

$$e = \frac{1}{C} \int i \, dt$$

where the factor C is called the capacitance. The effect is called the *capacitive effect*, and the equation can be construed as an operational definition of capacitance.

Experimental work has shown that the rise of weight in the surroundings is equivalent to the product of potential drop across the control volume and the amount of charge penetrating the control surface. Thus the work done by the control volume on the surroundings is

$$dW = -e\,dZ = -ei\,dt$$

where Z is the charge. Since the work done involved an ei product, the relationship between e and i depends on whether resistance, inductance, or capacitance is present in the control volume.

$$dW = -i^2 R\,dt = -\frac{e^2}{R}\,dt \quad \text{(resistive)}$$

$$dW = -Li\,di = -\frac{e}{L}(\textstyle\int e\,dt)\,dt \quad \text{(inductive)}$$

$$dW = -\frac{i}{C}(\textstyle\int i\,dt)\,dt = -ci\,de \quad \text{(capacitive)}$$

Forces of attraction exist between charged bodies. The stimuli (the charges on the two bodies) are measurable and reproducible and related by the equation

$$F = \frac{1}{4\pi\varepsilon_0}\frac{q_1 q_2}{r^2}$$

where: q_1 and q_2 = charges on the two bodies, C
r = distance between them, m
ε_0 = permittivity constant, $C^2/N\cdot m^2$
F = force of attraction or repulsion, N

This is Coulomb's *charge attraction effect* expressed for point charges; the equation is Coulomb's equation and may be viewed as the operational definition of permittivity ε_0.

Recent experiments at Princeton show that Coulomb's charge attraction effect is such that the exponent on the distance r is 2 ± 0.000 000 000 001.

6.7 MAGNETIC EFFECTS

Interactions between an electric current in a wire (control volume) and a magnetic field (surroundings) give rise to stimuli (current) and the consequences (forces) are measurable and reproducible and are related by the equation (for a straight wire of infinitesimal length),

$$\overline{F} = i\overline{l} \times \overline{B} \quad \text{or} \quad d\overline{F} = id\overline{l} \times \overline{B}$$

where: i = current in wire
$\overline{l}$ = displacement vector that points along straight wire in direction of current
$\overline{B}$ = magnetic field vector
$\overline{F}$ = force on wire

This effect is the *motor effect* of a magnetic field on a conductor that is carrying current. It is the basis for an electric motor and the effect is exploited to extinguish arcs created by opening switches carrying heavy currents. A blowout coil is placed in the vicinity of the arc, and the arc is moved (and thereby stretched) until it is extinguished.

6.8 OTHER EFFECTS

Other chemical effects require additional thermodynamic information and understanding to express quantitatively. Some chemical effects are means for

Mathematical Models of Effects

monitoring work and heat effects between a system or a control volume and its surroundings.

There are ballistic effects related to nuclear reactions and phenomena. These too express ways in which a system or control volume can experience a heat or work effect.

Another group of effects that may be called *impossible effects* are circumscribed by the second law of thermodynamics; their fascinating exposure is central to a course in thermodynamics.

A word about reproducibility. If you measure the length of a sheet of typewriter paper with an engineer's scale graduated in tenths of an inch, a careful observer reports 11 in. each time. This creates the impression that the sheets *are* 11 in. long. When deviations from the length of 11 in. are so small that the measurement technique (ruler or scale) cannot detect them and, furthermore, lack of detection causes no problems, we can treat the 11 in. as a deterministic quantity--exact, unvarying. If we measure the lengths of sheets of paper with a more precise instrument than a ruler, after we have removed all systematic instrument errors we find that each observation is different. We will try to establish a representative length (mean, median, mode, for example) as well as some measure of the dispersion in the observations (by noting the variance). Exact, unvarying observations are a convenient figment of the human imagination. All observations are directed toward estimation of, say, the mean value and other descriptive attributes of the population of measurables. The declaration of the mean and standard deviation, $N(\hat{\mu}_x, \hat{\sigma}_x)$, is called the result of a measurement, a *couple*.

Since the objective of experiments is to determine the mean of an observed quantity, an instrument may report many different readings for the same quantity. If in spite of the scatter the determination of the mean is good, the observations are regarded as *accurate* though *imprecise*. If the mean is poorly determined but the observations show inconsequential differences, the observations show *precision* and *inaccuracy*. Thus precision error is always present when successive observations of an unchanging quantity yield different numerical values. Accuracy error is always present when the numerical average of successive observations deviates from a known correct value and continues to do so no matter how many successive observations are made. A tachometer is applied to a shaft rotating at 1000 rpm and successive observations are 1050, 950, 1000, 1030, 900, and 980 rpm. The unbiased estimator of the mean value of an infinite number of observations is the sample mean.

$$\hat{\mu} = \frac{\Sigma x_i}{n} = \frac{1050 + 950 + 1000 + 1030 + 900 + 980}{6} = 985 \text{ rpm}$$

$$\hat{\sigma} = \left[\frac{\Sigma (x_i - \hat{\mu})^2}{n - 1}\right]^{1/2} = \sqrt{2990} = 54.68 \text{ rpm}$$

The range $\pm 2\sigma$ is an arbitrary measure of the lack of precision. Figure 6.3 shows the true value plotted on one side of a line and the average reading $\pm 2\sigma$. The range includes the true value, so one may describe the observations as accurate (range includes the mean) but imprecise because the coefficient of variation, $\hat{\sigma}/\hat{\mu} = 0.056$, is large for this physical measurement. A tachometer that reports 950, 952, 948, 951, and 950 rpm has a mean of 950.2 rpm and a standard deviation of $\hat{\sigma} = 1.48$ rpm. Figure 6.4 shows an arbitrary measure of the lack of precision of $\pm 2\sigma$. The range does not include the true value, so one may describe the observations as precise, $\hat{\sigma}/\hat{\mu} = 0.0016$, but inaccurate. The accuracy can be improved by calibration (removing systematic error); then

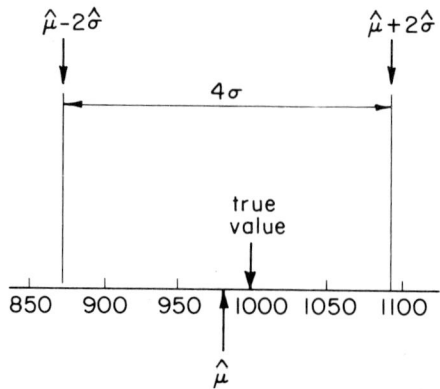

Fig. 6.3. A measurement that is accurate but imprecise.

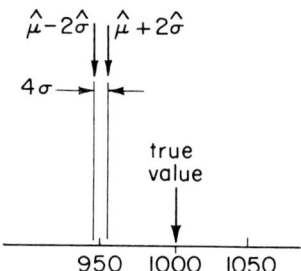

Fig. 6.4. A measurement that is precise but inaccurate (evidence of systematic error).

we can enjoy both accuracy and precision.

In the preceding commentary on effects and their mathematical modeling, remember that the models relate the means of the observations of the measurable attributes of the effect. By a reproducible effect, therefore, we wish to convey (more carefully) that the means of the measurable attributes (stimulus and consequence) are relatable in a consistent fashion. *By an almost reproducible effect we wish to convey the impression that the variation of the measurements is not negligibly small compared to the magnitude of the measurement but is of significant size, and deterministic attitudes do not allow satisfactory descriptions and communications to others nor allow satisfactory predictions.*

PROBLEMS

An engineer must be familiar with our understanding of cause, effect, and extent. Each effect has a name and can be described in a paragraph or less. Illustrations of the utilization of the effect are present in the engineering world. An equation or set of equations quantitatively relate cause and effect and other parameters of the phenomenon. References are available for detailed information for example:

Adsorption

Description: Adsorption is the ability of a material to hold a gas or liquid on its surface. The attraction may be molecular (as with charcoal) or capillary (as with silica gel).

Mathematical Models of Effects

Illustration: Hydrogen adsorbed on platinum is electrically conducting and is used as a hydrogen electrode. Desiccants use silica gel as the active ingredient.

Magnitude: Freundlich's equation is

$$x = AP^n$$

where: x = amount of gas adsorbed per unit area of solid
n = constant, usually between 0 and 1
P = partial pressure of gas in equilibrium
A = surface area of solid

References: H. J. Macintyre, and F. W. Hutchinson, 1950, *Refrigeration Engineering*, Wiley, p. 113.
E. H. MacDougall, 1952, *Physical Chemistry*, 3d ed., Macmillan, pp. 708-12.

You are urged to start a card file with a card for each effect you encounter. For practice, pick ten of the following effects and prepare a card entry for each one.

1. electromagnetic radiation absorption
2. fluid absorption
3. Ampere's force effect
4. Biot-Savart field equation
5. Archimedes' principle
6. Arrhenius' theory of electrolytic dissociation
7. Barkhausen effect
8. Bauschinger effect
9. Bernoulli's theorem
10. Biot's conduction effect
11. Bremstrahlung radiation effect
12. Brewster's law
13. Brownian movement effect
14. capacitance/dielectric effect
15. cavitation effect
16. Christiansen effect
17. Compton effect
18. conservation notions
19. DuLong and Petit law
20. Corbino effect
21. cosmic radiation effect
22. Coulomb's law
23. Curie-Weiss effect
24. Debye frequency effect
25. diffraction effect
26. elastic limit
27. electrocapillarity
28. electrokinetic phenomena
29. Eötvös effect
30. thermal expansion effect
31. Faraday effect
32. Faraday's law of electrolysis
33. Faraday's law of induction
34. Ferranti effect
35. ferroelectric phenomena
36. Fick's law

37. fission effect
38. fluid flow effects
39. friction effects
40. galvanomagnetic and thermomagnetic effects
41. Nernst effect
42. Righi-Leduc effect
43. Gauss's law
44. Geiger-Nuttal rule
45. Gibbs's phase rule
46. Gladstone and Dales's law
47. Graham's law
48. group displacement law
49. Henry's law
50. Hertz effect
51. Hooke's effect
52. ideal gas laws
53. impedance/frequency effect
54. interference effect
55. Johnsen-Rahbek effect
56. Joule's effect
57. Joule-Thomson effect
58. Kelvin's fluid drop principle
59. Kepler's laws
60. Kerr electrostatic effect
61. Kerr magneto-optic effect
62. Kirchhoff's circuit laws
63. Kohlrausch's law
64. Lambert's absorption coefficient
65. LeChatelier's principle
66. Leidenfrot's phenomena
67. liquid crystal phenomena
68. luminescence effect
69. ferromagnetic effect
70. paramagnetic effect
71. diamagnetic effect
72. magnetic dispersion of sound
73. magnetic effects
74. Gauss effect
75. magnetostriction effects
76. crystal memory effect
77. Newtonian convection effect
78. Newton's laws of motion
79. Ohm's resistance effect
80. optical rotary power
81. fluid column resonance effect
82. Paschen's law
83. periodicity of the elements
84. photoelectric effect
85. photomagnetic effect
86. piezoelectric effect
87. pinch effect
88. Poisson's ratio
89. proximity effect
90. Purkinje effect
91. Parkinson's law

Mathematical Models of Effects

92. pyroelectric effect
93. quantum theory
94. radiation pressure phenomena
95. radioactivity phenomena
96. Raoult's law
97. residual voltage effect
98. resistance-strain effect
99. resistance-pressure effect
100. resistance-temperature effect
101. Edison effect
102. Schrödinger's wave equation
103. Schottky effect
104. skin effect
105. Snell's refraction law
106. Stark effect
107. state and change of state effects
108. Stefan-Boltzmann radiation law
109. Stoke's law of fluid drag
110. Stoke's law of fluorescence
111. stress-strain effects
112. surface tension phenomena
113. thermodynamic laws
114. thermoelastic effects
115. thermoelectric effects
116. total reflection phenomena
117. triboelectricity
118. Volta effect
119. Wiedemann-Franz's law
120. Wien effect
121. Wien's displacement law
122. Wood and Ellett effect
123. Zeeman effect
124. Zener effect

> "What's one and one and one and one and one and one and one and one and one and one?"
> "I don't know," said Alice, "I lost count."
> "She can't do addition," said the Red Queen.
>
> **Lewis Carroll**

CHAPTER 7

ENUNCIATION OF FIRST PRINCIPLE

7.1 INTRODUCTION TO ACCOUNTABILITY

Humans first defined a number sequence and formalized counting (enumeration) by a process of summing (adding). If a *part* is defined as countable (enumerable) and the whole is the enumeration of the parts, to say that "the whole equals the sum of its parts" is a matter of definition (axiomatic).

When people learned to count they began a long line of constructs of the intellect that we can call mathematical models of natural phenomena. They learned by experience and a yearning for consistency that a number can be used as a crude description, numbers can be manipulated according to a formalism called logic, and the results are interpretable as an equally crude prediction of consequences of cause in certain natural phenomena. Implicit in this is a sense of time of the "before and after" variety, i.e., recognition of simultaneity and succession in events.

For instance, some early ancestor noted that the complex manifestation called a horse could be described by *one* for the purpose of predicting the consequences of the phenomenon known as "putting horses in the corral." If a group of horses described by *two* were placed in an empty corral (Fig. 7.1) and a group of horses described by *three* were then put in the corral, the formalism of enumeration (adding) yielded *five*, the numerical description of the group now in the corral (available without counting the group in the corral). In other words, the description of the final group *five* could be predicted from the descriptions of the entering groups by using the formalism of addition. The descriptive number *five* said nothing about color, health, breed,

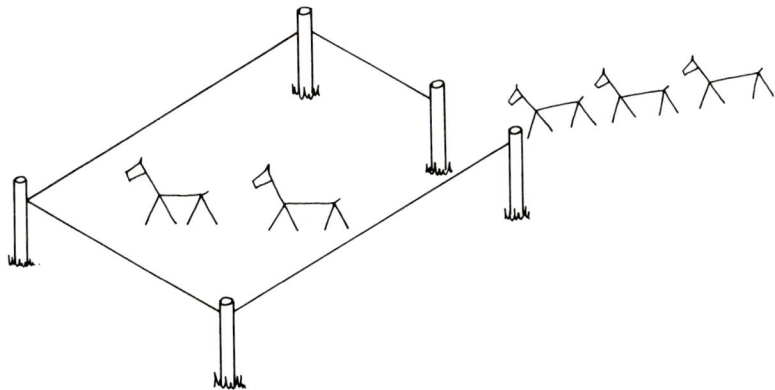

Fig. 7.1. The corral problem.

size, or sex of the horses, nor did the descriptive number *five* change with time, although the horses did (their state of health, for example). The mathematical description is crude but entirely satisfactory for routine daily accounting of the size of the herd. A count of five meant all the horses were in the corral and a count of four meant a horse was missing. This conclusion could be drawn by the tribal chieftain out of sight of the corral, referring only to daily reports of the corral census.

This mathematical description is too crude for the purpose of determining which horse is missing when the count is four or for the purpose of trading five horses for five other horses, because the mathematical model is myopic to matters of size, weight, age, health, training, sex, and breed of the animals. These matters are of great concern to the animal trader.

A mathematical description of a phenomenon, however simple or elegant, after manipulations of logic does not describe the consequences with any greater precision than the original model described the causes. (In fact, a statistical basis exists for saying that manipulation usually will decrease the precision.) A description of matter in the large (macroscopic viewpoint) will say nothing after manipulation about matter in the small (microscopic viewpoint).

Our ancestors, encouraged by initial successes, set about counting everything that could be counted and defined units of measure so that uncountables (continuous rather than discrete things) could be counted. A miller receiving a small oxcart of wheat and a large wagon of wheat (Fig. 7.2) could not pre-

Fig. 7.2. The miller's problem, relating oxcarts, wagonloads, and sacks of flour.

dict the extent of the wheat in the hopper or the number of bags of flour to be milled from the wheat by using the description *one* for the oxcart full of wheat and *one* for the wagonload of wheat. If an attempt were made, the formalism of addition gave $1 + 1 = 2$; yet 50 sacks of flour were milled from the wheat. When confronted with something amiss, people have the resourceful quality of "making it fit." It was found that by defining a unit of measure (such as the bushel) and converting the oxcart load to 20 measures of wheat and the wagonload to 60 measures of wheat, the addition of $20 + 60 = 80$ coincided with the measures of wheat in the hopper. It did not agree with the measures of flour milled, however. After milling there were 50 measures of flour. The formalism of addition could be applied to this case if: (1) one does not use the digits 1 and 1 to describe the incoming loads, but transforms them, through the use of a measure, to bushels; and if (2) one recognizes a law of milling that says the number of bushels of flour is 5/8 the number of bushels of wheat. By being taught this dogma, the miller's apprentice could

First Principle

account in a situation that would confound the wrangler. Thus experience with a phenomenon can suggest ways to preserve prior *modus operandi*.

The subtlety here is that people learn to count, they assume they can account, and they find ways to measure so that previous notions of addition and multiplication enable them to continue to account.

7.2 CONSERVATION OF COUNTABLES

We can add effectiveness to this notion by defining a fictitious closed surface (a boundary or a control surface), Fig. 7.3, observing the matter within, and proceeding to account. The teasing question is how much do we learn about our universe in this way and how much do we learn about ourselves? In other words, is there some of ourselves incorporated in the results of these investigative methods?

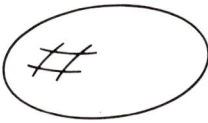

Fig. 7.3. A closed fictitious monitoring surface.

Let us define a fictitious closed surface of fixed geometry as a control surface with the space within (called control space, control region, or control volume) populated with "countables" that are able to pass through (cross) the surface on their way into or out of the control region. If we enumerate countables on two occasions, we define the change in the census of countables, Δc, as

$$\Delta c = \Sigma c_f - \Sigma c_i$$

where Σc_f is the final enumeration of countables and Σc_i is the initial enumeration of countables within the control surface. We will ascribe this statement to an ancestor named Twelve Fingers (who for obvious reasons held the post of tribal mathematician). An ancestor named Eleven Toes (a mathematician of lesser rank) stated that the change in the census of countables can be given by

$$\Delta c = \Sigma c_{in} - \Sigma c_{out}$$

where Σc_{in} is the number of countables crossing the control surface inbound during the accounting interval and Σc_{out} is the number of countables crossing the control surface outbound during the accounting interval.

The statements of Twelve Fingers and Eleven Toes were put to the test of experiment. The control surface was wrapped about an office building, Fig. 7.4. The accounting period was 1 hr. The countables were people. The data were

number of people within control surface at end of test	241
number of people within control surface at beginning of test	238
number of people entering front door	15
number of people entering back door	5
number of people leaving front door	12
number of people leaving back door	4
number of people leaving by window	1

Now, according to Twelve Fingers

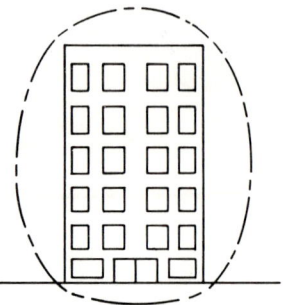

Fig. 7.4. An office building within a control surface.

$$\Delta P = \Sigma P_f - \Sigma P_i = 241 - 238 = 3$$

According to Eleven Toes

$$\Delta P = \Sigma P_{in} - \Sigma P_{out} = (15 + 5) - (12 + 4 + 1) = 3$$

These equations were tried on many office buildings with all kinds of complications, even when people arrived on the roof from adjacent buildings or burgled by tunneling into the basement. In each case ΔP as given by both Twelve Fingers and Eleven Toes was the same. Thus the equation

$$\Sigma P_f - \Sigma P_i = \Sigma P_{in} - \Sigma P_{out}$$

worked flawlessly. With all this experimental evidence it was inevitable that this equation became known as the Finger-Toe law of people for buildings.

7.3 CREATION AND DESTRUCTION OF COUNTABLES

Another researcher in a different part of the world tried to apply the Finger-Toe law to a hospital building and found some discrepancies. In the beginning these were attributed to experimental error (the researcher had to observe the front door and the outpatient clinic entrance simultaneously and a few countables must have slipped by during the count). However, when a large enough staff was assembled to "instrument" the hospital, the discrepancies continued to exist. Concluding it was real, the researcher, mustering courage, stated that the Finger-Toe law in its present form did not apply to hospitals and postulated that as it stood people were not countables unless they could be created and/or destroyed in the hospital. This hypothesis required certain ground rules for observers to agree that people were indeed being created and destroyed.

    1. Dead people arriving by ambulance crossing the control surface were not countables.
    2. Dead people leaving the morgue crossing the control surface in mortuary vehicles were not countables.
    3. Little wiggly, screaming, very small babies were countables.

With these conventions, the researcher stated that

$$[\Sigma P_f - \Sigma P_i] = [\Sigma P_{in} - \Sigma P_{out}] + [\Sigma P_c - \Sigma P_d]$$

where ΣP_c is the number of countables created within the control surface dur-

First Principle

ing the accounting period and ΣP_d is the number of people destroyed in the same period. The term $[\Sigma P_c - \Sigma P_d]$ is the discrepancy discovered in the Finger-Toe law. If all persons who died within the control surface during the accounting period are identified with ΣP_d, it follows from experimental evidence that all persons created within the control surface during the accounting period are babies.

The amended Finger-Toe law as displayed above became known as the "l'Hospital law." When the l'Hospital law contracted to the Finger-Toe law, the countable concerned was said to be *conserved*. The notion of conservation of people is the basis for musters in the armed services and judicial actions based on interpretation of musters.

Some complications threatened the l'Hospital law when data takers were confused on the occasion of fragments of people leaving. If a whole person entered and a person minus an arm or a leg departed, which do we count? It was found that l'Hospital law could be preserved if we agree in the case of less than a whole person that if it wriggled it was a person and if it did not wriggle it was not a countable. Only the existence of the nonconservation term

$$[\Sigma P_f - \Sigma P_i] - [\Sigma P_{in} - \Sigma P_{out}]$$

was taken as evidence of the creation or destruction of countables (people) in the hospital. Thus amputation and disection were not the destruction of people.

Another researcher wondered if l'Hospital law was valid for species of countables (people). A controlled experiment was inaugurated. A male-admit hospital and a female-admit hospital were selected for observation. In a male-admit hospital when the countable was taken as male persons, the l'Hospital law is written as

$$[\Sigma M_f - \Sigma M_i] = [\Sigma M_{in} - \Sigma M_{out}] + [\Sigma M_c - \Sigma M_d]$$

and it was found that in male-admit hospitals ΣM_c was always zero. In a male-admit hospital when the countable was taken as female persons, the l'Hospital law is written as

$$[\Sigma F_f - \Sigma F_i] = [\Sigma F_{in} - \Sigma F_{out}] + [\Sigma F_c - \Sigma F_d]$$

and it was found that in male-admit hospitals ΣF_c was always zero.

In the female-admit hospital, the female accounting is

$$[\Sigma F_f - \Sigma F_i] = [\Sigma F_{in} - \Sigma F_{out}] + [\Sigma F_c - \Sigma F_d]$$

and the male accounting is

$$[\Sigma M_f - \Sigma M_i] = [\Sigma M_{in} - \Sigma M_{out}] + [\Sigma M_c - \Sigma M_d]$$

and it was found that ΣF_c and ΣM_c were not always zero. From this we concluded that the creation of countables (male and female) occurred only in a female-admit hospital and therefore occurred with the female of the species (only women give birth).

When the l'Hospital law is applied to surnames of living people we can, by experiments with city halls and churches as well as hospitals, show that when a person is born, a kind (say Smith), gains one countable and when a person dies, a kind (say Jones), loses one countable. Other ways of creating and destroying Smiths have been observed. If one instruments a city hall one can observe a Smith and Jones entering and two Smiths and zero Jones leaving (marriage) or vice versa (divorce). Also one Smith may enter and one Jones may leave (legal change in name).

What does all this mean? Philosophically we have assumed accountability and made the formal statement,

$$[\Sigma c_f - \Sigma c_i] = [\Sigma c_{in} - \Sigma c_{out}] + \text{discrepancy}$$

A role of the experimentalist is to preserve this equality by identifying the discrepancy terms to be used with each kind of countable and stating rules (conventions) necessary to preserve the validity of the accounting equation. The accountability equation is a formalism reflecting consistency. The rules of logical reasoning are structured to reject contradiction. Many natural phenomena are consistent, i.e., reproducible in cause and effect and extent. It is not surprising that consistent interrelations can be compounded of accountability notions, logical reasoning, and empirical evidence. The resulting mathematical model may be used to *predict*, with often startling accuracy. In a sense then, the accountability equation is not arbitrary, for this amazingly reproducible world of phenomena can be described only in terms of consistent models, of which logic and its quantitative formalism, mathematics, are the foremost candidates as modeling material.

The most fruitful predictive intellectual construct is the mathematical model, and this tool has been embraced as fundamental to engineering. We have shown by an example that

$$[\Sigma c_f - \Sigma c_i] = [\Sigma c_{in} - \Sigma c_{out}] + [\Sigma c_c - \Sigma c_d]$$

can be given useful physical interpretation and quantitative substance. Many subsequent courses will make a point of doing this.

If the countable is taken as the mass of matter called water and a water pump is contained within the control surface (and its leakage is associated with Σc_{out}), it can be shown experimentally that $[\Sigma m_c - \Sigma m_d] = 0$. Then we can say in words that the mass of water is conserved in water pumps.

If we define *steady state* as the condition of time invariance of the countable within the control surface as well as time invariance of countable influx and efflux rates, the accountability equation simplifies to $[\Sigma c_f - \Sigma c_i] = 0$, and it follows, for a steady-state water pump, that

$$[\Sigma m_f - \Sigma m_i] = 0 = [\Sigma m_{in} - \Sigma m_{out}] \quad \Sigma m_{in} = \Sigma m_{out}$$

Within control surfaces through which no countable moves (such as a calorimeter bomb charged only with air), the accountability equation for a mass of matter called air is $[\Sigma(m_a)_f - \Sigma(m_a)_i] = 0$ and we say that the mass of matter called air is conserved.

If the bomb is charged with carbon and oxygen (excess), we find on ignition that (carbon + oxygen → carbon dioxide + less oxygen). When the countable is the mass of matter called graphite, we find that graphite is destroyed. When the countable is the mass of matter called oxygen we find that it is partially destroyed. If the countable is the mass of carbon dioxide we find that it is created. Are chemical phenomena exceptions to the notion of the conservation of matter? The observation is that within our ability to measure, the mass of the reactants is equal to the mass of the products, so conservation of mass in total is preserved by chemical reaction. The conservation of a species is true only if we pick the appropriate species! It has been noted that the mass of chemical *elements* is conserved. In fact, changes in which the masses of the elements are conserved are called (by definition) chemical changes. Thus, for the bomb combustion,

$$\Sigma(m_c)_f = \Sigma(m_c)_i \quad \Sigma(m_o)_f = \Sigma(m_o)_i$$

First Principle   201

and the masses of the elements carbon and oxygen are conserved.
   The lesson is that the uninitiated cannot consistently pick the countables that fit the accountability equation. The wrangler would be lost at the mill, Twelve Fingers and Eleven Toes would be lost at the hospital, and the water pump mechanic would be lost in the calorimeter laboratory. The difference between the initiated and uninitiated is a priori knowledge of countables compatible with accountability notions. This a priori knowledge for the engineer is contained in part in the disciplines of physics and chemistry. The knowledge is taken much further toward applicability in the disciplines of mechanics of solids and fluids; thermodynamics; heat, mass, and momentum transfer; electrical phenomena; materials science; and information theory.

7.4 ACCOUNTABILITY EQUATION FOR DISCRETE COUNTABLES

Our experience with the preceding heuristic approach to the notion of accountability has prepared us for a more formal development of the accountability equation in the case of discrete countables. Figure 7.5 depicts a moving

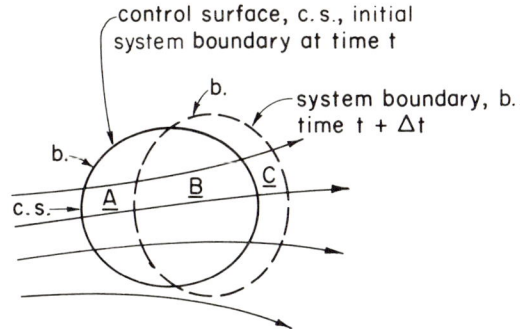

Fig. 7.5. A moving system and its relationship to a control surface.

system whose geometry at the beginning of the accounting interval is congruent with that of the control surface. Domain $\underline{A}$ is occupied by matter that is external to the system and during the accounting interval has penetrated the control surface passing from without to within. Domain $\underline{B}$ is matter associated with the system that has resided within the control surface during the accounting interval. Domain $\underline{C}$ is occupied by matter associated with the system that has moved from within the control surface to without during the accounting interval.

Any increase in the census of countable C within the system during the accounting interval can be calculated by noting the system census of countable C at the end of the accounting interval (t_2) and subtracting the census of C at the beginning (t_1); i.e., $\Delta C = (C \text{ at } t_2) - (C \text{ at } t_1)$. Or

$$\begin{bmatrix} \text{census of countable C} \\ \text{created during account-} \\ \text{ing interval minus census} \\ \text{of countable C destroyed} \\ \text{during accounting interval} \\ \text{within system} \end{bmatrix} = \begin{bmatrix} \text{census of countable} \\ \text{C in domains } \underline{B} \text{ and} \\ \underline{C} \text{ at time } t_2 \end{bmatrix} - \begin{bmatrix} \text{census of countable} \\ \text{C in domains } \underline{A} \text{ and } \underline{B} \\ \text{at time } t_1 \end{bmatrix}$$

In symbols,

$$[\Sigma C_c - \Sigma C_d] = [\Sigma C + \Sigma C]_{t_2} - [\Sigma C + \Sigma C]_{t_1}$$
$$\quad\quad\quad\quad\quad\quad\quad\quad \underline{B}\quad \underline{C}\quad\quad\quad\quad \underline{B}\quad \underline{A}$$

Our objective is to change the system viewpoint of the right side of the above equation to a control region viewpoint. It can be accomplished by adding and subtracting a term that represents the census of countable C in the domain $\underline{A}$ at time t_2 on the right side of the above equation.

$$[\Sigma C_c - \Sigma C_d] = [\Sigma C + \Sigma C]_{t_2} - [\Sigma C + \Sigma C]_{t_1} + \Sigma C\Big|_{\underline{A}\,t_2} - \Sigma C\Big|_{\underline{A}\,t_2}$$
$$\quad\quad\quad\quad\quad\quad\quad\quad \underline{B}\quad \underline{C}\quad\quad\quad\quad \underline{B}\quad \underline{A}\quad\quad\quad \underline{A}$$

$$= [\Sigma C + \Sigma C]_{t_2} - [\Sigma C + \Sigma C]_{t_1} + \Sigma C\Big|_{\underline{C}\,t_2} - \Sigma C\Big|_{\underline{A}\,t_2}$$
$$\quad \underline{A}\quad \underline{B}\quad\quad\quad\quad \underline{A}\quad \underline{B}$$

$$[\Sigma C_c - \Sigma C_d]_{sys} = [\Sigma C_f - \Sigma C_i]_{c.r.} - [\Sigma C_{in} - \Sigma C_{out}]_{c.s.} \quad\quad (7.1)$$

where sys = system, c.r. = control region, and c.s. = control surface. We can show that Eq. (7.1) is not limited to circumstances wherein the system boundary is initially congruent to the control surface. By superposition we can take an accounting interval preceding congruency and a subsequent accounting interval following congruency and write Eq. (7.1) twice and add the equations. This procedure shows that our restriction relaxes to one in which the congruency between system boundary and control surface must occur at some instant during the accounting interval. (See also Prob. 7.39.)

Consider again the widget machine relay inspection problem of Sec. 3.3 and depicted in Fig. 3.6. Let us define a control surface that envelops an inspection and test facility. The countable is to be relays. At the beginning of the working day no relays are in the control region. At the end of the working day the same condition prevails. During the accounting interval of a working day there is an influx of r relays into the control region to the inspection station and three effluxes of relays. There are pfr bad relays from the inspection station, p(1 − f)r bad relays from the test stand, and R good relays placed in widget machines from the test stand. The accountability equation for discrete countables is (or Finger-Toe's equation may be used)

$$[\Sigma C_c - \Sigma C_d]_{sys} = [\Sigma C_f - \Sigma C_i]_{c.r.} - [\Sigma C_{in} - \Sigma C_{out}]_{c.s.}$$

Since relays are neither created nor destroyed within the control region and the control region census of relays is zero at the beginning and the end of the working day, we can write

$$[0 - 0] = [0 - 0] - [r - pfr - p(1 - f)r - R]$$
$$r = pfr + p(1 - f)r + R = pr + R$$

Solving for r and setting R = 1, r = 1/(1 − p), which is the number of relays inspected for each good relay shipped in a widget machine. This less heuristic approach confirms the previous result obtained in Sec. 3.3. It constitutes a much more powerful entree into a problem.

For the discrete countable of linear momentum, you will recall from physics that Newton's law for a particle is expressed as

$$\Sigma \bar{F} = \frac{d(m\bar{V})}{dt} \quad\quad or\ impulse = \frac{Mom}{time}$$

Writing $\Sigma \bar{F}$ as $\bar{F}$, we have $\bar{F} = d(m\bar{V})/dt$. Multiplying both sides by dt and integrating over some interval $t_2 - t_1$, we obtain

$$Impulse = F \cdot t \quad\quad F = ma \quad\quad a = \frac{V_f - V_i}{t} = \frac{V}{t}$$

$$Impulse = \frac{m\bar{V}}{t}$$

First Principle

$$\int_{t_1}^{t_2} \overline{F}\, dt = \int \frac{d(m\overline{V})}{dt}\, dt = \int d(m\overline{V}) = \Delta(m\overline{V})$$

where the integral $\int \overline{F}\, dt = I$, the *linear impulse*. In the form $I = \Delta(m\overline{V})$, the way the momentum of a system is changed is the delivery of a linear impulse to the system by the surroundings in the form of a surface traction at the boundary of some finite duration. In a sense, the delivery of a linear impulse is the way in which system momentum is created or destroyed. In the context of the discrete form of the accountability equation,

$$[(\int F\, dt)_+ - (\int F\, dt)_-]_{sys} = [\Sigma p_c - \Sigma p_d]_{sys}$$
$$= [\Sigma p_f - \Sigma p_i]_{c.r.} - [\Sigma p_{in} - \Sigma p_{out}]_{c.s.}$$

where the plus subscript on the impulse denotes delivery to the system from the surroundings and the minus subscript vice versa. If, during an accounting interval $t_2 - t_1$, no surface traction exists delivering a linear impulse to the system, the creation and destruction of momentum term $(\Sigma p_c - \Sigma p_d)_{sys}$ vanishes and momentum is conserved in the system during the accounting interval. In the control region that is congruent with the system at some instant during the accounting interval, the difference between final and initial linear momentum is accounted for by the linear momentum influx and efflux during the accounting interval.

As an example of the application of the discrete accountability equation with linear momentum as the countable, consider the linear inelastic collision between two particles as depicted in Fig. 7.6. The accounting interval is arbitrary as we shall see, but to describe the event we call a collision it is necessary that the accounting interval include the event. Let the accounting interval be from t_1 to t_4 and the countable be the linear momentum in the x-direction. We can write for such a countable

$$[\Sigma p_c - \Sigma p_d]_{sys} = [\Sigma p_f - \Sigma p_i]_{c.r.} - [\Sigma p_{in} - \Sigma p_{out}]_{c.s.}$$

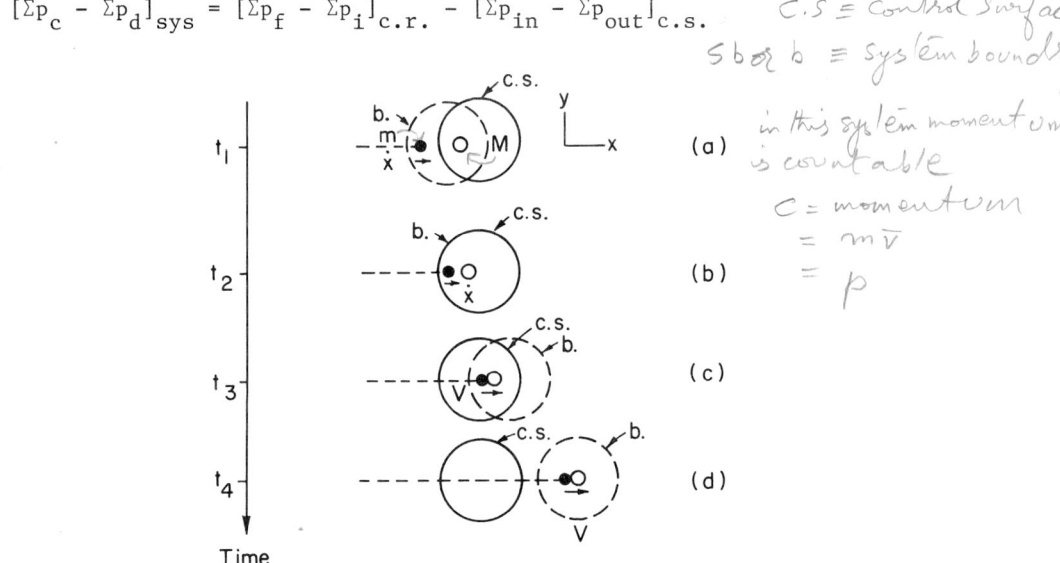

Fig. 7.6. A linear inelastic collision between two particles.

and for the accounting interval $t_4 - t_1$

$$[\Sigma p_c - \Sigma p_d] = [0 - 0] - [m\dot{x} - (m+M)V] = (m+M)V - m\dot{x}$$

For the accounting interval $t_3 - t_2$ we can write

$$[\Sigma p_c - \Sigma p_d] = [(m+M)V - m\dot{x}] - [0 - 0] = (m+M)V - m\dot{x}$$

giving the same expression. For the accounting interval t_1 to t_3

$$[\Sigma p_c - \Sigma p_d] = [(m+M)V - 0] - [m\dot{x} - 0] = (m+M)V - m\dot{x}$$

For the accounting interval $t_4 - t_2$ we can write

$$[\Sigma p_c - \Sigma p_d] = [0 - m\dot{x}] - [0 - (m+M)V] = (m+M)V - m\dot{x}$$

All accounting intervals that straddle the collision event give rise to the same equation. For the system we know that the countable momentum cannot be created or destroyed, since no impulse was delivered to the system by the surroundings or vice versa. Thus momentum is conserved and

$$[(\int_{t_1}^{t_4} F\, dt)_+ - (\int_{t_1}^{t_4} F\, dt)_-]_{sys} = [0 - 0] = 0$$

where the subscript + on the impulse denotes delivery to the system and − denotes delivery to the surroundings. An expression for V can be written as

$$V = \frac{m}{m+M}\dot{x} \tag{7.2}$$

Choosing the countable as linear kinetic energy and the accounting interval as t_2 to t_4, we write

$$[\Sigma KE_c - \Sigma KE_d]_{sys} = [\Sigma KE_f - \Sigma KE_i]_{c.r.} - [\Sigma KE_{in} - \Sigma KE_{out}]_{c.s.}$$

$$= [0 - \tfrac{1}{2} m\dot{x}^2] - [0 - \tfrac{1}{2}(m+M)V^2] = \tfrac{1}{2}(m+M)V^2 - \tfrac{1}{2} m\dot{x}^2$$

Substituting for final velocity V from Eq. (7.2), we obtain

$$[\Sigma KE_c - \Sigma KE_d]_{sys} = \tfrac{1}{2} m\dot{x}^2 \left(\frac{m}{m+M} - 1\right)$$

This equation indicates that linear kinetic energy of translation is destroyed in the collision, since the right-hand side of the above equation is negative for finite material bodies.

For a collision other than completely inelastic, Fig. 7.7 depicts the two different trajectories of the bodies. In this circumstance the countable momentum is a vector quantity that can be handled by components. We write

$$[\Sigma p_c - \Sigma p_d]_x = [\Sigma p_f - \Sigma p_i]_x - [\Sigma p_{in} - \Sigma p_{out}]_x$$

$$= [0 - m\dot{x}] - [0 - (mv\cos\theta - MV\cos\phi)] = 0$$

$$[\Sigma p_c - \Sigma p_d]_y = [\Sigma p_f - \Sigma p_i]_y - [\Sigma p_{in} - \Sigma p_{out}]_y$$

$$= [0 - 0] - [0 - (mv\sin\theta - MV\sin\phi)] = 0$$

We have two scalar simultaneous equations in mv and MV:

$$mv\cos\theta - MV\cos\phi = m\dot{x}$$

First Principle

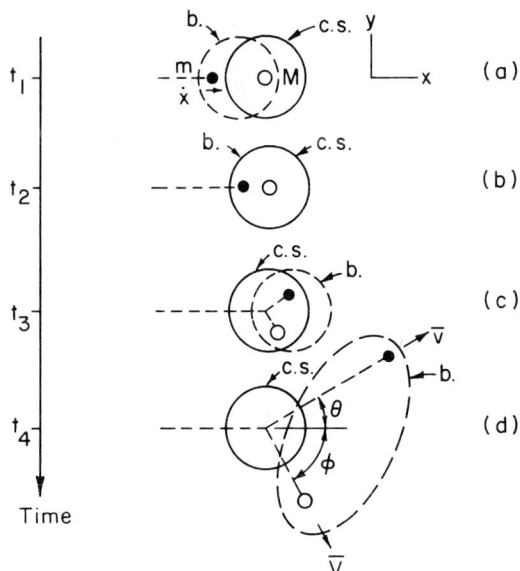

Fig. 7.7. A collision between two particles.

$mv \sin \theta - MV \sin \phi = 0$

Consequently,

$$mv = \frac{\begin{vmatrix} m\dot{x} & \cos\phi \\ 0 & -\sin\phi \end{vmatrix}}{\begin{vmatrix} \cos\theta & \cos\phi \\ \sin\theta & \sin\phi \end{vmatrix}} = \frac{m\dot{x} \sin\phi}{\sin(\theta + \phi)} \quad (\theta \neq 0, \pi;\ \phi \neq 0, \pi)$$

$$v = \dot{x} \frac{\sin\phi}{\sin(\theta + \phi)} \tag{7.3}$$

$$V = \frac{m}{M} \dot{x} \frac{\sin\theta}{\sin(\theta + \phi)} \quad (\theta \neq 0, \pi;\ \phi \neq 0, \pi) \tag{7.4}$$

Linear kinetic energy is a scalar countable, and in the accountability equation (7.1) the countable C may be designated linear kinetic energy. We write for the accounting interval $t_4 - t_1$ of Fig. 7.7:

$$[\Sigma KE_c - \Sigma KE_d]_{sys} = [\Sigma KE_f - \Sigma KE_i]_{c.r.} - [\Sigma KE_{in} - \Sigma KE_{out}]_{c.s.}$$

$$= [0 - 0] - \left[\tfrac{1}{2} m\dot{x}^2 - \left(\tfrac{1}{2} mv_2^2 + \tfrac{1}{2} MV^2 \right) \right]$$

$$= \tfrac{1}{2} mv_2^2 + \tfrac{1}{2} MV^2 - \tfrac{1}{2} m\dot{x}^2$$

We may substitute for v_2 from Eq. (7.3) and for V from Eq. (7.4) and obtain

$$= \tfrac{1}{2} m \frac{\dot{x}^2 \sin^2\phi}{\sin^2(\theta + \phi)} + \tfrac{1}{2} M \left(\frac{m}{M}\right)^2 \dot{x}^2 \frac{\sin^2\theta}{\sin^2(\theta + \phi)} - \tfrac{1}{2} m\dot{x}^2$$

$$[\Sigma KE_c - \Sigma KE_d] = \frac{1}{2} m\dot{x}\left[\frac{\sin^2\phi}{\sin^2(\theta+\phi)} + \frac{m}{M}\frac{\sin^2\theta}{\sin^2(\theta+\phi)} - 1\right] \quad (\theta \neq 0, \pi; \phi \neq 0, \pi)$$

In general, the right-hand brackets will be less than zero, indicating that in general, kinetic energy of translation will be destroyed in a ballistic collision.

The power of Eq. (7.1) can be further exemplified by an example requiring more involved accounting. Our inability to answer the question exactly is not shown as directly by other methods. In fall 1965, 179 freshmen students registered in fence-post engineering. During the fall quarter 33 dropped out, during winter quarter 25 dropped out, during spring quarter 30 dropped out, and during the summer 3 dropped out. The fall 1966 enrollment of sophomores was 100. Losses were 20 fall, 14 winter, 14 spring, and 3 summer. The 1967 junior enrollment was 79 with losses of fall 4, winter 9, spring 9, and summer 2. No transfer students entered as seniors. Assess the probability of a student becoming a senior in fence-post engineering if the student enters the university as a freshman. Figure 7.8 shows a control surface placed about the university campus. Let

S_u = university freshmen who become seniors
S_t = transfer students who become seniors
F_u = university freshmen who leave the university (the department)
F_t = transfer students who leave the university (the department)

We will recognize people as countables of group U, of group T, and of group (U + T). In Fig. 7.8(a) the countable is people of group U (university beginners). At t = 3 yr, the system contains all people of group U but some lie within and some without the control surface.

$$[\Sigma C_c - \Sigma C_d] = [\Sigma C_f - \Sigma C_i] - [\Sigma C_{in} - \Sigma C_{out}]$$

$$[\Sigma U_c - \Sigma U_d] = [\Sigma U_f - \Sigma U_i] - [\Sigma U_{in} - \Sigma U_{out}]$$

Since people of group U are conserved, $[\Sigma U_c - \Sigma U_d] = 0$. The control region census of people of group U at the end of the accounting interval is S_u. The control region census of people of group U at the beginning of the accounting interval is 179. The influx of people of group U through the control surface is zero and efflux is F_u. Substitution into the above equation yields

$$[0] = [S_u - 179] - [0 - F_u] \quad S_u + F_u = 179$$

Figure 7.8(b) considers the countables as people of group T (transfer beginners). Similar substitution into the accountability equation yields

$$[0] = [S_t - 42] - [0 - F_t] \quad S_t + F_t = 42$$

The number of seniors can be deduced to be $79 - 4 - 9 - 9 - 2 = 55$ and the number of dropouts to be 166. Thus we can write

$$S_u + S_t = 55 \quad F_u + F_t = 166$$

We have a set of four equations and four unknowns, and the equations may be written as

$$S_u + 0 + F_u + 0 = 179$$

$$0 + S_t + 0 + F_t = 42$$

First Principle

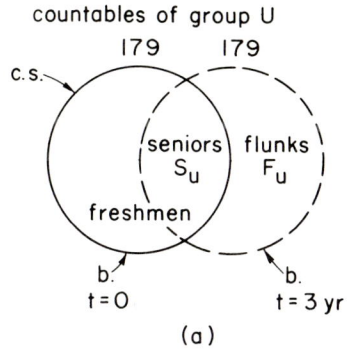

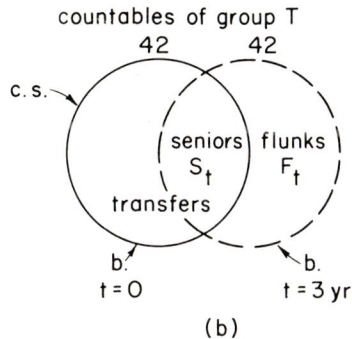

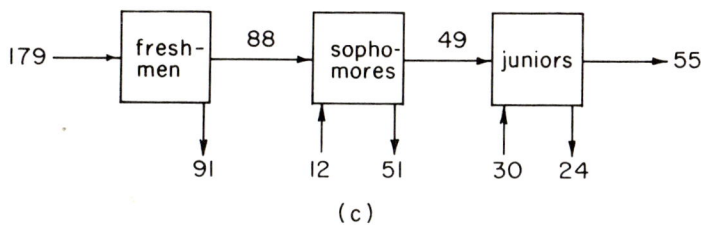

Fig. 7.8. (a) A system of 179 university beginners shown three years apart and the control surface (campus fence). (b) A system of 42 university transfers shown three years apart. (c) The input and output of people U and T to freshman, sophomore, junior, and senior classifications.

$$S_u + S_t + 0 + 0 = 55$$

$$0 + 0 + F_u + F_t = 166$$

A solution of the above equations, using Cramer's rule, results in the indeterminant form 0/0. Insufficient information exists to establish S_u or S_t uniquely. The next best thing is to establish bounds on S_u.

Figure 7.8(c) indicates the input and output of people of group U and group T to freshman, sophomore, junior, and senior classifications. It is clear that the 55 seniors are of group U or T. The sum of 55 can be the results of the sum of the smallest and largest values of S_u and S_t:

$$55 = S_u + S_t = (S_u)_{max} + (S_t)_{min} = (S_u)_{min} + (S_t)_{max}$$

The smallest possible value of S_t exists when (30 - 24) or 6 transfers are successful in becoming seniors; $(S_t)_{min} = 6$. The largest possible value of S_t exists when the (12 + 30) or 42 transfers are successful in becoming seniors; $(S_t)_{max} = 42$. Substitution into the above equation results in

$$55 = (S_u)_{max} + 6 = (S_u)_{min} + 42$$

Solutions for $(S_u)_{max}$ and $(S_u)_{min}$ are

$$(S_u)_{max} = 55 - 6 = 49$$

$(S_u)_{min} = 55 - 42 = 13$

If one began in 1965 as a freshman at the university, the probability of becoming a senior in fence-post engineering is expressible as

$(S_u)_{min}/179 \le P \le (S_u)_{max}/179$

$13/179 \le P \le 49/179$

$0.073 \le P \le 0.274$

7.5 ACCOUNTABILITY EQUATION FOR A CONTINUUM OF COUNTABLES

For engineering purposes another form of Eq. (7.1) is also generally used:

$$[\Sigma C_c - \Sigma C_d]_{sys} = [\Sigma C_f - \Sigma C_i]_{c.r.} - [\Sigma C_{in} - \Sigma C_{out}]_{c.s.} \tag{7.5a}$$

The left side of Eq. (7.5a) represents the change in census of the countables in the system (note subscript sys) due to the process of creation or destruction of countables during the accounting interval. A handier notation for this term is $(\Delta C_{cd})_{sys}$, the change in census of countables is the system of initial congruence due to creation and destruction of countables. The first of the two right-hand terms represents the change in the census of countables within the control region during the accounting interval Δt, and a more convenient notation is

$$C_{c.r.}\Big|_{t+\Delta t} - C_{c.r.}\Big|_{t}$$

where t is the instant of the initiation of the accounting interval. The second term on the right represents the difference in the enumeration of countables crossing the control surface during the accounting interval, and a better notation is $\Delta C_{c.s.}$. Substituting these notational changes into Eq. (7.5a) we have

$$(\Delta C_{cd})_{sys} = C_{c.r.}\Big|_{t+\Delta t} - C_{c.r.}\Big|_{t} - \Delta C_{c.s.}$$

Let us divide each term by the accounting interval Δt, and take the limit as the accounting interval vanishes:

$$\lim_{\Delta t \to 0} \frac{(\Delta C_{cd})_{sys}}{\Delta t} = \lim_{\Delta t \to 0} \frac{C_{c.r.}\Big|_{t+\Delta t} - C_{c.r.}\Big|_{t}}{\Delta t} - \lim_{\Delta t \to 0} \frac{\Delta C_{c.s.}}{\Delta t}$$

In the limit

$$\dot{C}_{sys} = \frac{\partial}{\partial t}(C_{c.r.}) - \dot{C}_{c.s.}$$

The control surface and the system boundary are congruent for the entire instantaneous accounting interval, a help in sketching a diagram for a problem.

Let us give some geometric interpretation of this equation. Figure 7.5 shows a system whose boundary is changing shape with time. In the beginning of the accounting interval, the matter in the system is within the boundary drawn as a solid line. At the end of the accounting interval all the system matter is still resident within its boundary, but the boundary is in a different location and its geometry has changed. A control region is depicted with its control surface shown as the solid line in Fig. 7.5. The geometry of the control surface is invariant, as are its location and orientation.

First Principle

Domain A is a domain in which countables identified with the surroundings of the system have penetrated the control surface and entered the control region, and at instant t + Δt these countables are within the control region. Domain B is a domain in which matter is resident within the system and the control region during the entire accounting interval Δt. Domain C is a domain where matter identified with the system has penetrated the control surface; left the control region; and at instant t + Δt, is outside the control region. The term ΔC is net creation of countables within the control region. As the accounting interval gets smaller, the geometry of the system and of the control region approach coincidence; and at the limit, $\dot{C}$ is the rate of creation of countables in *either* the control region or the system since their domains are identical at the instant t.

The form

$$\dot{C}_{sys} = \frac{\partial}{\partial t}(C_{c.r.}) - \dot{C}_{c.s.}$$

is much more useful as an accountability equation since it relates $\dot{C}_{sys}$ to attributes of the control region of coincident geometry at instant t. If C is a census of countables and c is the census per unit mass, c = dC/dm and the above equation can be written,

$$\dot{C}_{sys} = \frac{\partial}{\partial t} \oiiint c\, dm - \dot{C}_{c.s.}$$

where the symbol $\oiiint$ denotes integration over the control region. The net rate of influx of countables through the control surface can be shown to be (Fig. 7.9)

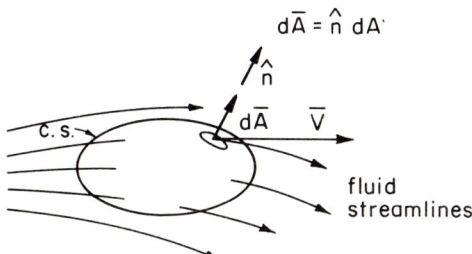

Fig. 7.9. Monitoring the net rate of influx of countables through the control surface.

$$\dot{C}_{c.s.} = -\oiint c\rho \overline{V}_r \cdot d\overline{A}$$

where ρ is the mass density of influxes and effluxes, $\overline{V}_r$ is the velocity *relative to the control surface* and $d\overline{A}$ is an elementary area of the control surface. The accountability equation can be expressed in the form

$$\dot{C}_{sys} = \frac{\partial}{\partial t} \oiiint c\, dm + \oiint c(\rho \overline{V}_r \cdot d\overline{A}) \qquad (7.5b)$$

where: ρ = density of matter within control region at point on control surface
$\overline{V}_r$ = velocity of matter at point on control surface relative to control surface
$d\overline{A}$ = element of area of control surface, vector $\hat{n}\, dA$ that is outwardly drawn

210   Chapter 7

$(\bar{V}_r \cdot \overline{dA})$ = component of matter relative velocity at control surface normal to control surface

$(\rho\bar{V}_r \cdot \overline{dA})$ = rate of mass efflux through control surface element dA

$c(\rho\bar{V}_r \cdot \overline{dA})$ = rate at which countable c is leaving control surface element dA

An alternate development of the accountability equation in a continuum of countables is as follows. In Fig. 7.5, we focus our attention on the matter in the system during the accounting interval Δt. Any increase in the countable C within the system during the accounting interval can be calculated by noting the system census of countable C at the end of the accounting interval and subtracting the system census of the countable C at the beginning of the accounting interval. In algebraic terms,

$$\begin{bmatrix}\text{increase in census}\\ \text{of countable C with-}\\ \text{in system during ac-}\\ \text{counting interval }\Delta t\end{bmatrix} = \begin{bmatrix}\text{system census of}\\ \text{countable C at end}\\ \text{of accounting inter-}\\ \text{val at time }t + \Delta t\end{bmatrix} - \begin{bmatrix}\text{system census of counta-}\\ \text{ble C at beginning of}\\ \text{accounting interval at}\\ \text{time }t\end{bmatrix}$$

We can abbreviate somewhat as follows,

$$\Delta C = \begin{bmatrix}\text{census of C}\\ \text{in domains }\underline{B}\\ \text{and }\underline{C}\text{ at }t + \Delta t\end{bmatrix} - \begin{bmatrix}\text{census of C}\\ \text{in domains }\underline{B}\\ \text{and }\underline{A}\text{ at }t\end{bmatrix}$$

or in terms of summation notation,

$$\Delta C = \left[\sum_{\underline{B}} C + \sum_{\underline{C}} C\right]_{t+\Delta t} - \left[\sum_{\underline{B}} C + \sum_{\underline{A}} C\right]_t$$

Our objective now is to change the system viewpoint of the right side of the above equation to a control surface monitoring viewpoint. This change can be accomplished by adding and subtracting the term $\sum_{\underline{A}} C$ evaluated at $t + \Delta t$ to the right side of the above equation:

$$\Delta C = \left[\sum_{\underline{B}} C + \sum_{\underline{A}} C\right]_{t+\Delta t} - \left[\sum_{\underline{B}} C + \sum_{\underline{A}} C\right]_t + \sum_{\underline{C}} C\bigg|_{t+\Delta t} - \sum_{\underline{A}} C\bigg|_{t+\Delta t}$$

Expressing the above equation in words is helpful.

$$\Delta C = \begin{bmatrix}\text{control region}\\ \text{census at }t + \Delta t\end{bmatrix} - \begin{bmatrix}\text{control region}\\ \text{census at }t\end{bmatrix} + \begin{bmatrix}\text{net influx of C}\\ \text{during }\Delta t\end{bmatrix}$$

(annotated: efflux)

We now divide each term by the accounting interval Δt and approach the limit as $\Delta t \to 0$. The left side becomes the time rate of increase in the countable C in the system that has the same geometry and contents at instant t that the control region has at instant t. The first two right-hand terms represent the change in control region census in the limiting process when divided by Δt and the time rate of increase in control region census of countable C. The extreme right-hand term in the limit as Δt approaches zero represents the net time rate of influx of countable C through the control surface. In words, as the result of the limiting process, the above equation becomes,

$$\begin{bmatrix}\text{time rate of increase}\\ \text{in census of countable}\\ \text{C in system with same}\\ \text{geometry of boundary}\\ \text{as control surface at}\\ \text{instant }t\end{bmatrix} = \begin{bmatrix}\text{time rate of increase}\\ \text{in census of countable}\\ \text{C in control region at}\\ \text{instant }t\end{bmatrix} + \begin{bmatrix}\text{net time rate of}\\ \text{influx of countable}\\ \text{C through control}\\ \text{surface at instant }t\end{bmatrix}$$

(annotated: efflux)

(7.5c)

First Principle   211

In words this is the accountability equation for a continuum of countables. It is one of the most important equations in engineering and you will encounter it innumerable times in your engineering work.

7.6 ENUNCIATION OF FIRST PRINCIPLE

First principles or *laws* are statements of inductions of sweeping and general validity concerning our environment. They cannot be derived, deduced, or argued from nonempirical axiomatic statements. All that logic can provide is consistency and protection from contradiction. True and false in the logical sense only means consistent with or inconsistent with given or axiomatic statements.

Many attributes of our environment are reproducible and consistent. If humans could build into some axiomatic statements the essence of their environment, they could make consistent deductions that would have a counterpart in nature. Where does one obtain such axiomatic statements that are permeated with the essence of reality?

People observe their environment often, with millions of experiments and thousands of investigators. Then a genius performs an exceedingly astute act of induction, looking at the morass of data and confusion, even apparent contradiction. Either through good fortune or penetrating insight (usually a fortuitous blend of both), the genius captures and enunciates an essence of the environment. Such a statement is offered as an axiom from which deductions can be made that explain (predict) a broad range of consequences. In due time it is called a law, a principle, or first principle because of its great power (utility).

When a glass rod is rubbed with silk the rod exhibits a positive charge. A negative charge appears on the cloth. The facts that the rod charge is positive, the cloth charge is negative, and they are equal in magnitude (within our ability to measure) suggests not the creation of charge but the disturbance of the neutrality of charge on each body by a mechanical redistribution process. The assertion, *electrical charge is conserved*, began as an axiomatic statement. Review of all experimental work, small scale (atomic and nuclear) and large scale, has failed to produce a single exception or contradiction. If there are n experiments, the probability of the next experiment producing a contradiction is 1/n. Since n is an extremely large number, the probability of an exception is extremely small (but finite). In an engineering project, the success of which affects life, property, corporate survival, or the destiny of a nation, which way should an engineer decide--accept the axiomatic statement (*charge is conserved*) or reject it?

If the statement *electric charge is conserved* contains an essence of our universe, we can believe that logical deductions made from that statement will still contain that essence. Here is the basis for the very existence of engineering. The reader, at least mentally, should preface the statement with *matter behaves as if* electric charge is conserved. After all, charge is a human mental construct and it is not reasonable that nature "knows" what charge is and takes steps to conserve it. The statement, *matter behaves as if charge is conserved*, is a powerful axiom that allows deductions that predict a broad range of consequences. Because of its power (great utility), the conservation of charge statement is called a *law*. Note that it is a conservation statement because our accounting approach is partly a search for invariants. A conservation statement allows someone to say dogmatically (with the authority of great experimental effort and a chance of being wrong of the order 1/n) that before and after a system-surroundings event, the census of the charge in the system plus surroundings will be the same before and after the event. If charge census can be expressed quantitatively, there exists an *equation* valid before and after the event.

It has been the human experience that matter is conserved, and it was not until the work of Einstein and others that the blending of the conservation of matter principle and the conservation of energy principle was brought about. In circumstances where interconversion of matter and energy are not involved, the separate principles are still used, saying that "in this circumstance matter is conserved" or "in this circumstance energy is conserved." The conservation of matter idea is a statement for a *system*. All the matter enveloped by the closed boundary is present within the boundary at all times and hence *is* the system now or later or at any time whatever. For a quantitative statement for a system, we might write: If m is the mass of the system (measure of the extent of matter), dm/dt = 0, or more simply,

$$\dot{m} = 0$$

where $\dot{m}$ is the first time derivative of m with respect to t. This equation is simplicity itself. What is the corresponding statement for a control region? For this we must utilize the accountability relation for discrete or continuous countables as appropriate to our problem. What we discover is an important disappointment for engineers--there is no unique statement to be made such as the one displayed above. Each and every control region has a conservation of matter statement that is unique but the statement differs from one control region to another. This means, much to our disappointment, that the appropriate conservation of matter statement for the control region at hand must be tailor-made by the engineer at the time. In subsequent courses, you will be continually, routinely doing just that, and you will become very adept at it from practice and from necessity.

As an example consider the reducing elbow depicted in Fig. 7.10. Equation (7.5c) is used with the countable mass. The law of conservation of mass for any system whatsoever states that $\dot{m} = 0$; therefore,

$$\begin{bmatrix} \text{time rate of increase in census of countable mass in system} \\ \text{with same boundary geometry as control surface at t} \end{bmatrix} = \dot{m} = 0$$

If we have steady flow of fluid (all successive snapshots of the control region are identical and indistinguishable), there is always the same census of mass in the control region; consequently, the time rate of change of this constant must be zero, or

$$\begin{bmatrix} \text{time rate of increase in census of} \\ \text{countable mass in control region at t} \end{bmatrix} = 0$$

Given the uniform velocity field at entrance, the rate of influx of mass into the control region from the left is (density)(velocity)(area):

$$\text{time rate of mass entrance from left} = \rho_1 V_1 A_1$$

$$\text{time rate of mass exiting from right} = \rho_2 V_2 A_2$$

$$\begin{bmatrix} \text{net time rate of influx of countable} \\ \text{mass through control surface at t} \end{bmatrix} = \rho_1 V_1 A_1 - \rho_2 V_2 A_2$$

Substituting in Eq. (7.5c), we obtain

$$0 = 0 - (\rho_1 V_1 A_1 - \rho_2 V_2 A_2)$$

$$\rho_1 V_1 A_1 = \rho_2 V_2 A_2$$

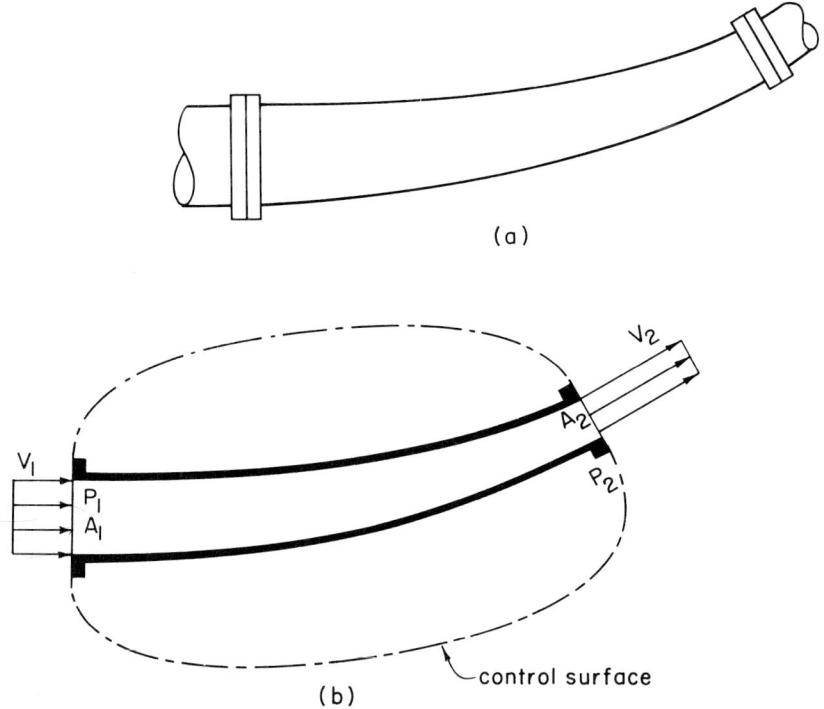

Fig. 7.10. (a) A reducing elbow; (b) control surface used in analyzing the reducing elbow.

for this control region. Remember that the conservation of mass statement is a system statement with a unique mathematical expression, but the conservation of mass statement for a control region is unique only to that region and the engineer must deduce or derive such a statement for each control region of interest.

Conservation of energy is a very important conservation principle, and to understand it we spend a good deal of time in subsequent courses probing the subtleties of the statement and its consequences. Understand again that we are speaking of events that do not involve matter-energy interconversion.

Innumerable experiments led to the statement called the *first law of thermodynamics*. The integration of the differential of work dW around the closed path representing the sequence of states of a system that returned to its original state led to the conclusion that $dW \neq 0$. Similar integration of the differential of heat dQ around the closed path representing the sequence of states of a system that returned to its original state led to the conclusion that

$$dQ \neq 0$$

However, as closely as could be measured, the integral of the difference of the differentials, $\oint(dQ - dW)$ was zero, consistently, without plausible exception. Thus

$$\oint(dQ - dW) = 0$$

This led to the statement:

The cyclic integral of heat addition to a system executing a complete cycle is proportional to the cyclic integral of the work done by the system during the cycle.

It follows from the definition of property that the change (dQ - dW) is a change in a system property, the $\int(dQ - dW)$ between two state points is independent of path, and $(Q_{12} - W_{12})$ is a property of the system. The name given to this property is *energy*. It is therefore clear that the only way to change the energy of a system is to cause it to experience a work effect, a heat effect, or both. A working statement of the first law of thermodynamics for a system is

$$Q = \Delta E + W \text{ or } dQ = dE + dW$$

where Q is the heat effect during a process, W is the work effect during a process, and ΔE is the change in energy of the system during the process. If a system is prohibited from entertaining both a heat effect and a work effect, then $\Delta E = 0$ and the energy of the system is conserved. If heat effects and work effects are allowed, then $\Delta E = Q - W$ and the energy change is accounted for. If Q is regarded as energy added to the system and W is regarded as energy leaving the system, then

$$\Delta E + W - Q = 0$$

always, and to some, this is the conservation of energy statement. A preferred view is that when the energy of a system changes, a calorimetric effect and a work effect are observable in the surroundings, and quantitatively, the difference is the energy change. Perhaps we should say that the difference between the measures of the heat effect and the work effect is used to infer the system energy change. The idea that something flows between system and surroundings during heat and work effects is a carry-over from caloric fluid notions. It may be a convenient tactical fiction, but it is philosophically disturbing because no experimental evidence shows anything "flowing" between system and surroundings. The finer understanding of these problems and viewpoints is left to your studies in thermodynamics.

The statement of the first law of thermodynamics for a system is

$$\Delta E = Q - W$$

For a control region we must use the accountability relation, Eq. (7.5c), the countable to be energy E. The first law of thermodynamics for any system is often expressed as

$$Q = \Delta E + W \text{ or } dQ = dE + dW \text{ or } \dot{Q} = \dot{E} + \dot{W}$$

where: Q = heat effect between system and surroundings during a change of state, the positive sign associated with a heat addition to the system
W = work effect between system and surroundings during a change of state, the positive sign associated with work done by the system on its surroundings
E = energy of the system, a thermodynamic property

Energy is the property of a system whose change is given by the difference between the heat effect and the work effect. We recognize that the energy of a system can be changed in many ways. Empirical evidence shows that the state

First Principle

of pure substance may be established by specification of three independent properties, at least one of which must be extensive. Thermodynamicists identify system energy changes that occur in the absence of gravitation, motion, charge, magnetism, and capillarity as changes in system *internal* energy, ΔU;

$$Q = \Delta U + W$$

The energy of a system can be changed in many ways--by changing the elevation z, the velocity V, the angular velocity ω, the magnetic flux density β, the liquid surface area A, and the charge Z, to name several. The energy differential dE can be written

$$dE = dU + \frac{\partial E}{\partial z} dz + \frac{\partial E}{\partial V} dV + \frac{\partial E}{\partial \omega} d\omega + \frac{\partial E}{\partial Z} dZ + \frac{\partial E}{\partial \beta} d\beta + \frac{\partial E}{\partial A} dA + \ldots$$

Since E is a thermodynamic property, it satisfies the mathematical requirements imposed on an exact differential. The previous equation is open-ended, gaining terms as scientists discover ways and means of changing the energy (heat or work effects) of a system.

The structure of the differential dE suggests experiments in which the partial derivatives may be given acceptable mathematical form. For instance, the experimenter who maintains at zero all differentials except one can (in principle) uncover a useful functional relationship between one parameter and the energy change. If we limit our considerations to systems that undergo changes of state in the absence of charge, magnetism, and capillarity, we can write

$$dE = dU + \frac{\partial E}{\partial z} dz + \frac{\partial E}{\partial V} dV + \frac{\partial E}{\partial \omega} d\omega$$

where $(\partial E/\partial z)dz$ is the change in potential energy of elevation and the last two terms are the change in kinetic energy of translation and rotation.

For illustration let us take an appropriate set of three independent properties to establish U as mass m, temperature T, and pressure p. The internal energy of the system is expressed as U(p, T, m).

Let us apply the preceding equation under circumstances wherein the changes of state occur isothermally and isobarically and the rest mass is the significant mass. Then $dU = 0$ and

$$dE = dE_p + dE_k$$

where E_p is the potential energy of elevation and E_k is the kinetic energy of translation and rotation. Substituting this relationship into the first law statement for a system, we have

$$dQ = dE + dW = dE_p + dE_k + dW$$

Under changes of state in an adiabatic system ($dQ = 0$) without work effects with the surroundings ($dW = 0$), the first law simplifies to

$$dE_p + dE_k = 0$$

Integration of this equation produces the common form of the statement of the principle of the conservation of the mechanical energy of a system:

$$E_p + E_k = \text{constant}$$

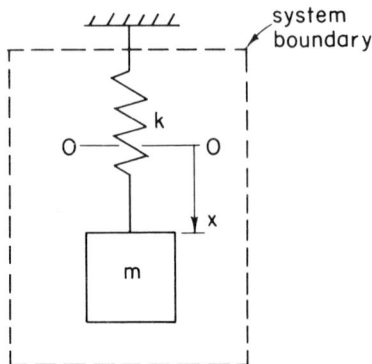

Fig. 7.11. A spring and block in vertical oscillation.

Remember that this equation is valid for systems of pure substance in which there is no heat or work effect, two independent thermodynamic properties (such as temperature and pressure) are invariant, nonrelativistic speeds are involved, and gravitation and motion are present; however, charge, magnetism, and capillarity are excluded. The constant in the above equation can sometimes be conveniently evaluated when a mechanical system is oscillatory so that both E_p and E_k become zero at some different times, and evaluation of one component is possible when the other is zero.

$$E_p + E_k = \text{constant} = E_p \Big|_{E_k = 0} = E_k \Big|_{E_p = 0}$$

Depicted in Fig. 7.11 is a block at the end of a spring that has oscillating vertical movement. The system is defined as the block and the spring. The datum for block position coordinate is the free length block position. Such potential energy as the system may possess is in two forms: (1) potential energy of elevation due to the displacement of the block from the datum of x and (2) potential energy of strain of the spring. If the spring rate is k lbf/in., at a displacement x the potential energy of the system is

$$(E_p)_{spring} = \int_0^x F_x dx = \int_0^x kx\, dx = (1/2)kx^2$$

$$(E_p)_{block} = \int_0^x F_x dx = \int_0^x -mg\, dx = -mgx$$

$$E_p = -mgx + (1/2)kx^2$$

The kinetic energy of the system at displacement x is the kinetic energy of the block and the kinetic energy of the spring (neglected):

$$(E_k)_{block} = \int_0^x F_x dx = \int_0^x m \frac{dv}{dt} dx = \int_0^v mv\, dv = (1/2)mv^2$$

$$E_k = (1/2)m\dot{x}^2$$

We do not know the constant from the conservation of mechanical energy equation, and we do not wish to limit our results to a definite displacement extreme of x_0; therefore we use the differential form,

$$dE_p + dE_k = 0$$

First Principle

$$\frac{dE_p}{dt} + \frac{dE_k}{dt} = 0$$

$$\frac{d}{dt}[-mgx + (1/2)kx^2] + \frac{d}{dt}[(1/2)m\dot{x}^2] = 0$$

Carrying out the differentiation yields

$$-mg\dot{x} + (1/2)(2)kx\dot{x} + (1/2)(2)m\dot{x}\ddot{x} = 0 \text{ or } m\ddot{x} + kx = mg$$

This expression is a second order nonhomogeneous differential equation with constant coefficients. Its solution contains a complementary portion with two arbitrary constants and a particular integral.

$$x = C_1\cos(\sqrt{k/m}\,t) + C_2\sin(\sqrt{k/m}\,t) + mg/k$$

The initial conditions establish C_1 and C_2. Note that if the initial displacement is mg/k and the initial velocity is zero, the solution becomes $x = mg/k$, which is independent of time as well as constant. The system is consequently static, motionless.

This condition gives rise to another useful concept. The system is in its static equilibrium position (configuration). To change coordinate x in either positive or negative sense, it is necessary for the surroundings to work on the system. The first law statement is $dE = -dW$. The work done by the system on the surroundings is negative, making dE a positive quantity whether movement occurs in the positive or negative x-direction. If dE is always positive when moving from the static equilibrium configuration, the system energy must be at a minimum at the static equilibrium configuration.

This concept can be applied to the preceding problem in the following manner. Let the static equilibrium configuration be required. The energy of the system at static equilibrium is of two forms: (1) potential energy of elevation due to the displacement of the block from the datum of x and (2) potential energy of strain in the spring. Consequently the energy of the system is expressible as

$$E = (1/2)kx^2 - mgx$$

Inasmuch as the system energy is at a minimum when the position coordinate x has the value x_s, we can differentiate the preceding expression with respect to x, equate to zero, and call the established and necessary value of x to be the static equilibrium position x_s.

$$\frac{dE}{dx} = (1/2)(2)kx - mg = 0$$

from which it follows that $x_s = mg/k$.

Building a conservation of energy statement for a control region using Eq. (7.5c) with energy as the countable is sufficiently complicated to be distracting to the purposes of this course. The decision to omit the evolution is made reluctantly, for it is a beautiful development. It is made in appropriate detail and introspection early in the first thermodynamics course.

Newton's three laws of motion can be stated:

1. Every body (system) persists in its state of rest or uniform motion in a straight line unless it is compelled to change that state by forces impressed on it.
2. The change in momentum is proportional to the net force, i.e.,

$$\Sigma \overline{F} = \frac{d}{dt}(m\overline{V})$$

3. To every action there is always an equal reaction, or the mutual actions of two bodies on each other are always equal and oppositely directed along the same straight line.

The word *body* is comparable to the use of the word *system*. Under circumstances where a system is free of force, i.e., the net resultant force from external matter on the system is zero, the time rate of change of momentum is zero and the system momentum is time-invariant. Under these circumstances it is commonly said that *linear momentum is conserved* (or matter behaves as if linear momentum is conserved).

Newton's second law may be written for the matter within a control region by using the accountability equation, Eq. (7.5c):

$$\begin{bmatrix} \text{time rate of increase} \\ \text{in census of countable} \\ \text{momentum in system} \\ \text{with the same geometry} \\ \text{of boundary as control} \\ \text{surface at t} \end{bmatrix}_x = \begin{bmatrix} \text{time rate of increase} \\ \text{in census of counta-} \\ \text{ble momentum in con-} \\ \text{trol region at t} \end{bmatrix}_x - \begin{bmatrix} \text{net time rate of} \\ \text{influx of countable} \\ \text{momentum through} \\ \text{control surface at t} \end{bmatrix}_x$$

This equation is of necessity a vector equation, since the countable is a vector quantity and the subtraction involved is a vector subtraction. If we take components in some direction, say x, we have a scalar equation. As an example, consider the conveyor belt carrying granular material away from a hopper at uniform speed V, as depicted in Fig. 7.12. The material enters the control region penetrating the control surface at a uniform rate of $\dot{w}$ mass/s. The influxing material has no velocity component in the x-direction. The effluxing material leaves with an x-component of velocity equal to $\overline{V}$. The sand in free fall exhibits no internal pressure; exit sand has negligible internal pres-

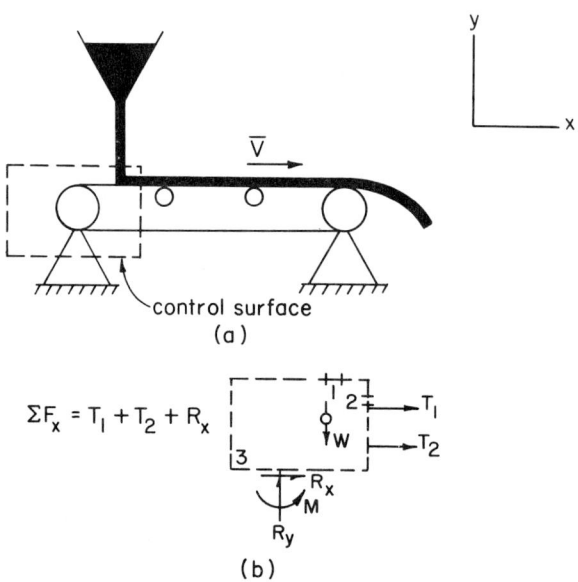

Fig. 7.12. A sand conveyor belt.

First Principle

sure. Thus areas 1 and 2 of Fig. 7.5(b) are free of normal and shearing tractions. Belt tensions are normal tractions; support tractions are normal and shearing with couple M. If the countable chosen is the x-component of the momentum mV_x, the time rate of change of momentum for the system with the same geometry as the control region at instant t is

$$\begin{bmatrix} \text{time rate of increase} \\ \text{in census of countable} \\ mV_x \text{ in system with same} \\ \text{geometry of boundary as} \\ \text{control surface at t} \end{bmatrix} = \frac{d}{dt}(mV_x) = \Sigma F_x = T_1 + T_2 + R_x$$

$$\begin{bmatrix} \text{time rate of increase} \\ \text{in census of countable} \\ mV_x \text{ in control region} \\ \text{at t} \end{bmatrix} = 0 \qquad \begin{bmatrix} \text{net time rate of} \\ \text{influx of countable} \\ mV_x \text{ through control surface at t} \end{bmatrix} = 0 - V_x \dot{w}$$

Substitution into the accountability equation yields

$$\Sigma F_x = 0 - (0 - V_x \dot{w}) = V_x \dot{w}$$

For the second belt of the cascading conveyors of Fig. 7.13, using the countable of the x-component of the linear momentum mV_x for the control region depicted in the figure,

$$\begin{bmatrix} \text{time rate of increase} \\ \text{in census of countable} \\ mV_x \text{ in system with same} \\ \text{geometry of boundary as} \\ \text{control surface at t} \end{bmatrix} = \frac{d}{dt}(mV_x) = \Sigma F_x$$

$$\begin{bmatrix} \text{time rate of increase} \\ \text{in census of countable} \\ mV_x \text{ in control region} \\ \text{at t} \end{bmatrix} = 0 \qquad \begin{bmatrix} \text{net time rate of} \\ \text{influx of countable} \\ mV_x \text{ through control surface at t} \end{bmatrix} = u_x \dot{w} - V_x \dot{w}$$

Substitution into the accountability equation yields

$$\Sigma F_x = 0 - (u_x \dot{w} - V_x \dot{w}) = \dot{w}(V_x - u_x)$$

The power requirement to keep the belt moving uniformly is $\Sigma F_x V_x$ or

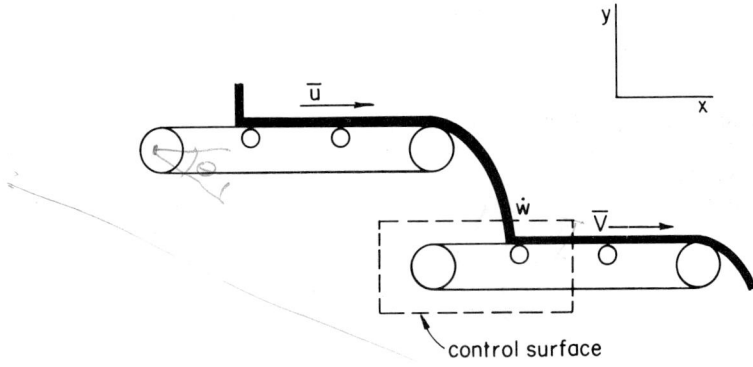

Fig. 7.13. Two cascading sand conveyor belts.

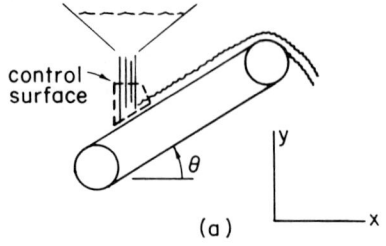

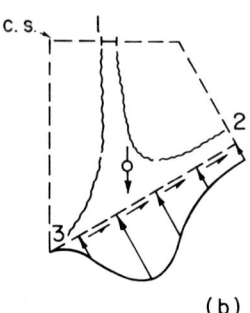

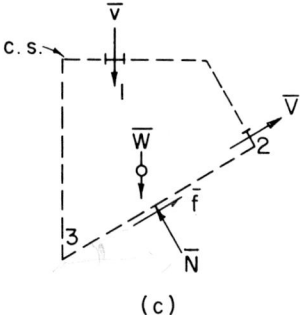

Fig. 7.14. (a) An inclined sand conveyor belt; (b) control surface of fixed geometry used in the analysis; (c) modeling decisions for the influence of the surroundings on the matter in the control region.

$$\text{power} = \dot{w} V_x (V_x - u_x)$$

When $u_x = 0$, the solution contracts to the previous simpler circumstance.

In another example, consider the conveyor belt, depicted in Fig. 7.14(a), that is inclined at an angle of θ to the horizontal with the granular material flowing uphill. A control surface of fixed geometry is established, Fig. 7.14(b). The significant influences of the surroundings on the matter within the control surface are recognized as follows. The effect of atmospheric pressure on the control surface effectively vanishes; it is modeled as free of normal and shearing tractions due to atmospheric fluid. Sand enters at 1 in a condition of free fall and consequently exerts no normal or shearing tractions on the control surface. Sand leaves through the control surface at 2. The pressure in the sand is trivial and the decision to model as free from normal

First Principle

or shearing tractions is made. The bottom of the control surface experiences distributed normal and shearing tractions of unknown distribution. The distributions are related, however, through the Coulomb friction effect. A body force influences the matter within the control surface because of the proximity of the planet earth. The remainder of the control surface, 3, is the portion that is neither 1 nor 2.

Figure 7.14(c) shows the decision to model the control surface as experiencing surface tractions. The normal surface traction $\bar{N}$ is the vector sum of the continuously distributed normal traction on the bottom of the control surface, which represents part of the influence of the belt. The shearing traction $\bar{f}$ is the vector sum of the distributed shearing traction, which represents the remainder of the influence of the belt on the matter in the control space. The concentrated body force $\bar{W}$ is placed at the center of mass of the matter within the control surface and directed toward the earth. The sand penetrates the control surface in area 1 with an average velocity $\bar{v}$ and leaves through area 2 with an average velocity $\bar{V}$. For the countable $(mV)_x$, noting that $\Sigma F_x = [\bar{f} + \bar{N} + \bar{W}]_x$,

$$[\bar{f} + \bar{N} + \bar{W}]_x = [0] - [0 - \dot{w}V \cos \theta]$$

$$f \cos \theta - N \sin \theta = \dot{w}V \cos \theta$$

For countable $(mV)_y$, we have

$$[\bar{f} + \bar{N} + \bar{W}]_y = [0] - [-\dot{w}v - \dot{w}V \sin \theta]$$

$$f \sin \theta + N \cos \theta - W = \dot{w}v + V\dot{w} \sin \theta$$

It is convenient to arrange the equation as simultaneous equations in f and N,

$$f \cos \theta - N \sin \theta = \dot{w}V \cos \theta$$

$$f \sin \theta + N \cos \theta = W + \dot{w}(V \sin \theta + v)$$

Solving simultaneously for f and N with Cramer's rule, we obtain

$$f = \dot{w}(V + v \sin \theta) + W \sin \theta$$

$$N = (W + \dot{w}v) \cos \theta$$

When $\theta = 0$, the shearing surface traction supplied by the belt is $f = \dot{w}V$ as before, and $N = W + \dot{w}v$. The normal surface traction supplied by the belt is the weight of the material in the control region plus the inertia force due to the incoming sand. When $\theta = -\phi$ (the belt goes downhill),

$$f = \dot{w}(V - v \sin \phi) - W \sin \phi$$

$$N = (W + wv) \cos \phi$$

The force f can be zero when $\dot{w}(V - v \sin \phi) - W \sin \phi = 0$ and an angle can be found where no force is required to sustain uniform belt motion. When f becomes negative a force has to be supplied in a direction opposite to the motion to prevent acceleration. In this range of θ we have a sand motor. When $\dot{w} = 0$, then $f = -W \sin \phi$, indicating that the belt force alone fights a component of the weight of the material in the control region. The same is true of the force $N = W \cos \phi$.

The force between any two particles having masses m_1 and m_2 and separated by a distance r is an attraction along the line joining the particles; it has the magnitude

$$F = G(m_1 m_2 / r^2)$$

where G is a universal constant. The force is independent of presence of other bodies and properties of the intervening space. Note that the significant property of both particles, mass, is precisely the same significant property as for resistance to acceleration. Uniform spheres may be represented by concentrated masses at their respective centers of mass. Other bodies require integration methods to arrive at forces and lines of action.

Because the significant property for matter in gravitation and in motion is the same, we experimentally determine mass of a body by a weighing procedure rather than by the more difficult acceleration procedure.

Gravitational forces manifest themselves in systems and control volumes as *body forces* applied through the center of mass and along a line connecting the center of mass of the system with the center of mass of the earth. These body forces are influences due to surroundings but do not manifest themselves at the boundary or control surface and are therefore distinguished from *surface tractions*.

7.7 DELINEATION OF SYSTEM AND CONTROL REGION

The ideas behind system and control volume allow the definition of either to be a powerful tool in the analysis and resolution of the problem. Understanding is available from the reasoning process when axiomatic statements incorporate the essences of the behavior of matter. The stratagem used in analysis is to require our knowledge of effects and first principles to teach us what the significant parameters are in a given circumstance and to express the relationships between them.

We will require that a system defined be removed from its surroundings, and in our mind's eye the surroundings are missing. Clearly, in this circumstance the behaviors of the systems resident in the universe and resident in our consciousness will be different. In one circumstance the remainder of the universe is capable of influencing the system, while in the other no influence is possible (the surroundings are nonexistent). Thus we begin with a system that cannot behave in our mind's eye as does the system in the real world. We then close the discrepancy between the real world behavior and the abstract behavior by *imitating* the influences of the surroundings. These imitations are chosen from our reservoir of understanding of cause and effect. The better the imitation, the closer the approximation to real world behavior.

In the consideration of an aluminum ladder resting against a house, the influences of the real surroundings on the system (ladder) are manifold. A heat effect is present, since the garage air temperature is different from the outdoor air temperature. The ladder is now in the sun and solar radiation is augmenting the convective effect between the ladder and its ambient air environment. The ground temperature is different from the ladder temperature and a conductive effect exists between the ladder and the ground and the house. There are the tractive influences of the ground on the ladder and the house on the ladder. The ladder is changing in length, the locality of its contact with the house is changing, and the spatial distribution of the surface traction is changing. Charge is migrating within the ladder in sympathy with nearby radio broadcasting stations, and the voltage between the ground and house ends of the ladder is changing. The wind exerts a drag force on the ladder. The air exerts a buoyant force on the ladder. The expanding ladder is having a work interaction with its surroundings. Indeed, the aluminum is reacting with surrounding fluid and our system is really a control volume, since matter of various species is combining chemically with the metal. Some of the aluminum is sublimating--atoms of aluminum are escaping into the atmosphere. Many more effects are in progress from the partial absorption of per-

First Principle   *Inertial frame of ref.*

meating X rays to the ionization of some aluminum atoms from cosmic radiation.

What hope is there of understanding what is going on when the ladder leans against the wall? Only a few effects of the environment have a significant role in determining parameters. For example, if we are interested in the forces exerted on the ladder by the surroundings (ground and wall), the body force due to the proximity of the earth (weight) is significant, and the body force due to the buoyant effect is insignificant and can be neglected. The influence of the changing geometry of the ladder with time is insignificant. The only significant influences are the initial geometry of ladder orientation and the weight. The error we make in showing a force at the top of the ladder due to the wall, a force at the bottom of the ladder due to the ground, and a body force at the center of mass due to the ladder's weight, incorporating the wall-ground-ladder geometry, is exceedingly small.

It is apparent that skill is important to an engineer's ability to analyze. (1) The first element of that skill is to identify the system or control region in such a way that meaningful parameters are exposed. (2) The significant influences of the surroundings must be identified and the insignificant ones suppressed in the interests of simplicity. (3) The model of the influences that is sufficient in accuracy has to be identified and utilized.

If we are interested in whether the ladder will fall by itself, we have to be clever enough to realize that the system that will help us understand the stability condition is the entire ladder, not part of it and not the ladder plus the house. If we understand nature to the point that we realize the tractive influences of the surroundings on the ladder are instrumental in determining the stability condition, it will dawn on us that taking the ladder as a system will force the tractive influences to reveal themselves. *Without this realization we are blocked.*

The engineer must develop some savvy as to how the universe behaves. The significant influences on the ladder are the surface tractions, but we do not know just how these are distributed as pressures over the zone of contact between ladder and house or ground. If we take the vector sum of these tractions and call them reactions, what are their lines of action? To a priori state that we know them is foolishness. To say that if we arbitrarily place them at the center of the geometric contact is to introduce error but not significant error (if understanding of the stability condition is our objective). To know if we can do this or not is a *skill* on which useful analysis depends.

Is Newton's law for a system fixed with respect to the earth valid? Of course not, for the earth is a noninertial reference, but the discrepancy between the summation of external forces of that which is and that which $\bar{F} = ma$ predicts is small enough to be neglected in the case of the ladder. How do we know that? Savvy, sense of proportion, experience, insight, or recognition of the magnitude of terms in Newton's law written for a noninertial reference. All these are good words for alternative moves that can produce meaningful results. An engineer has to be skillful at this too.

Let us say that the significant influences of the surroundings are identifiable. We are, for example, interested in the pressure of a stagnant atmosphere at 1 km altitude above the earth's surface. Most students know that the pressure at the bottom of a column of fluid is related to the weight of the column of fluid above it. A column a kilometre high and 1 m^2 in cross section could be identified as a system. The significant influences are the tractive effects of the surroundings on the system. We have the lateral pressures on the sides, the pressures at the top and bottom, and the fluid weight. The fluid weight is expressible in terms of the density and the volume. Ah, there's the rub! Density varies from sample to sample taken from the system by more than an infinitesimal amount and thus is not a property of the system,

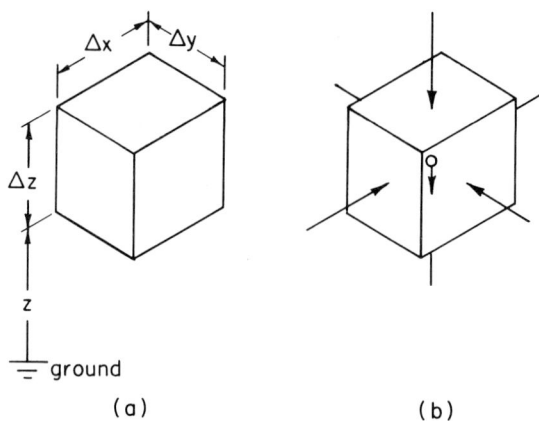

Fig. 7.15. A infinitesimal system in a column of stagnant atmosphere, and identification of the significant influences of the surroundings.

since it violates the definition of property. Since the pressure variation and the density variation are intimately related, it is clear that we cannot obtain a meaningful relationship from a kilometre-high system.

We are forced to take as a system an element of the fluid column across which the density does not vary by more than an infinitesimal amount, restricting the height of the system to a differential height. Figure 7.15 shows the system in isolation with the significant influences of the surroundings identified. A body force exists directed toward the earth due to the proximity of the earth. The net effect of the air above is a downward normal traction; the net effect of the lateral surroundings is equal normal surface tractions on the lateral faces; the net effect of the air below is an upward normal traction. Note the absence of any quantitative statement or expression. The role of this step is to identify the significant influences of the surroundings. We can argue that the tractive influences on the surfaces of the rectangular parallelepiped are purely normal, since we are dealing with a static fluid. A fluid is incapable of sustaining a shearing stress unless continuous deformation prevails.

As a matter of tactics, a good engineer begins with a diagram that delineates the geometry of the system, as in Fig. 7.15(a). A second figure shows that system with the surroundings removed and the significant influences of the surroundings declared, as in Fig. 7.15(b). This figure is the basis for all future analysis. If it is wrong, all else is wrong. If it is correct there is hope for correct deduction. Since it is the basis for the solution and displays the engineer's understanding of the phenomenon, a good engineer will display the second sketch, which will form the basis for further analysis. False steps will retreat to this point, so it is important to have a record of the fundamental decisions of what *is* significant. The ladder problem recently discussed will result in sketches such as shown in Fig. 7.16.

When we are interested in what goes on within a beam such that is resists loading, deforms somewhat, and does not break, we can choose as a system a portion of the beam, as depicted in Fig. 7.17(a) and (b). The fact that the boundary cuts the beam itself forces us to consider how the internal resistances are related to the external loading and the beam geometry.

Consider a thermocouple sensor in the fluid stream that is connected through various instrumentation to a temperature-time recorder. If there are no instrumental errors, would the temperature-time locus be accurate or in er-

First Principle   225

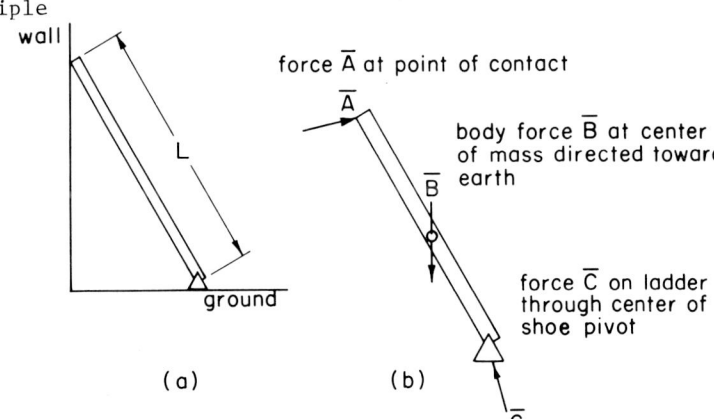

Fig. 7.16. A ladder as a system and identification of significant influences of the surroundings.

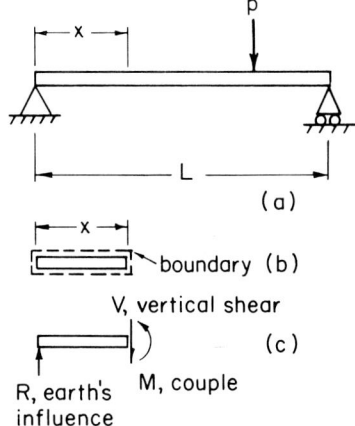

Fig. 7.17. A portion (b) of a beam (a) as a system and identification (c) of significant influences of the surroundings.

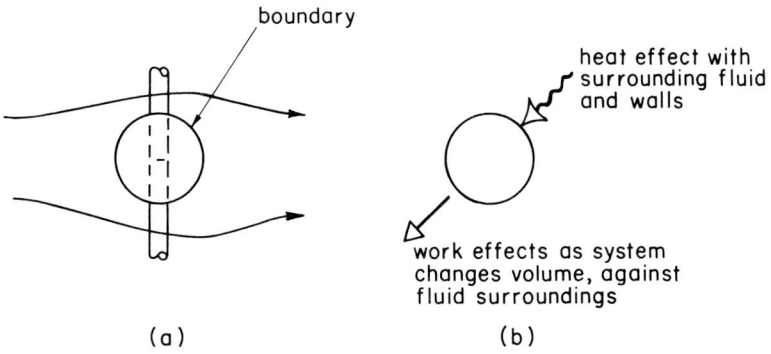

Fig. 7.18. A thermocouple bead as a system and identification of significant influences of the surroundings.

226   Chapter 7

ror? The heart of this operation is in the thermocouple bead, so the system
is taken as the thermocouple bead as indicated in Fig. 7.18. The significant
influences are heat and work effects with the surroundings. The engineer
draws Figs. 7.18(a) and (b).

A temperature distribution exists in a solid because of imposed boundary
conditions. Is the temperature distribution predictable from knowledge of the
boundary conditions, solid geometry, and material properties? An equation for
the prediction of temperature will be the result of statements that can be
made about a system of which temperature is a property; and since temperature
is varying from point to point, it is necessary that the system be infinitesimal in extent. Figure 7.19 shows the engineer's representation of the significant influences of the surroundings for the purpose of describing the tem-

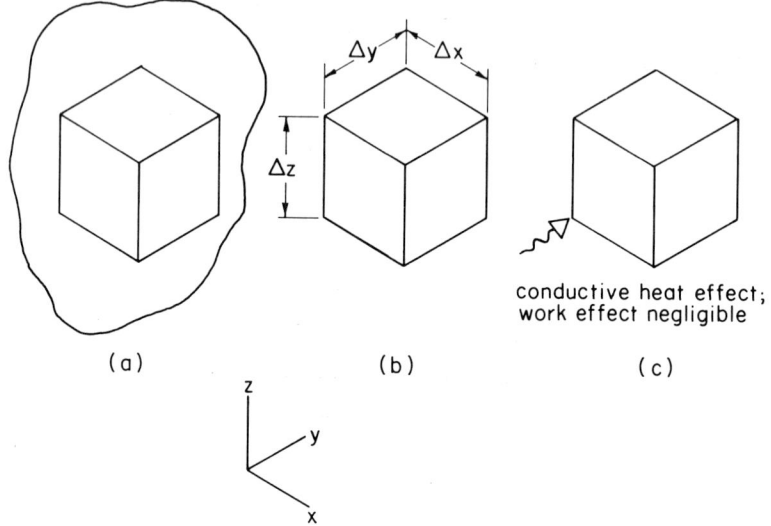

Fig. 7.19. A prismatical element of a solid body as a system and identification of significant influences of the surroundings.

perature distribution. In a solid the only heat effect that can be present at
the boundary of the infinitesimal system is a conductive effect. The only
work effect is tractive interaction with the surroundings, and its magnitude
can be so small in comparison with the heat effect that its presence is neglected, introducing trivial error.

An external brake shoe is in contact with a brake drum. The tractive effects of the surroundings are to be idealized into a resultant force equivalent to the shearing tractions $\bar{x}$ and a resultant force equivalent to the normal tractions y; see Fig. 7.20(a). The shoe is isolated as the system. The
significant influences are the surface tractions due to the brake drum and the
pivot pin, as shown in Fig. 7.20(b).

In seeking the reactions on a reducing elbow, a control volume is defined
that includes the matter presently in the elbow (Fig. 7.21). The significant
influence of the downstream fluid on the control surface is a distributed normal traction. The significant influence of the upstream fluid is another distributed normal traction. The significant influence of the pipe wall on the
control surface is resolvable into a distributed normal traction and a distributed shearing traction about the periphery of the control surface. The

First Principle

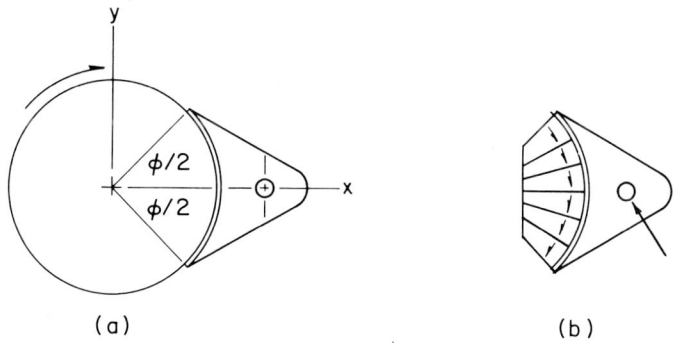

Fig. 7.20. An external brake shoe as a system and identification of the significant influences of the surroundings.

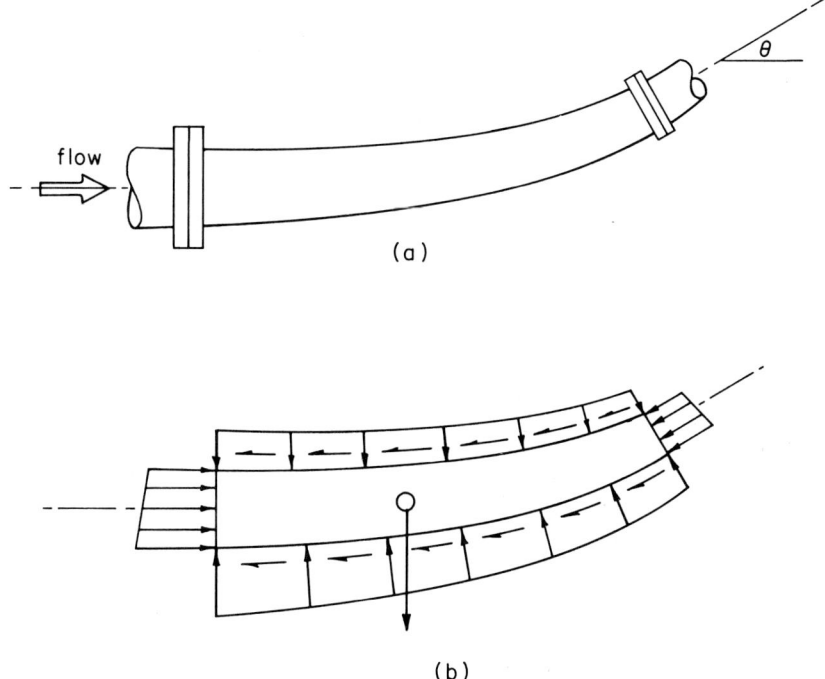

Fig. 7.21. The interior of a reducing elbow as a control region and identification of the significant influences of the surroundings.

influence of the proximate earth on the matter within the control region is a body force acting through the center of mass of the control region directed toward the center of the earth.

The delineation of what is surroundings and what is system (or control region) is accomplished by describing the boundary (or control surface). It must be done clearly for the analyst and for the reader of the work. Description in words and a concise diagram help to describe precisely what the analyst is talking about.

Deductive; Reasoning from Known Principle to an unknown.
Deterministic; To reach decision after thought & investigation.
assertive: unduly confidant; positive, claiming to affirm.
Probabilistic: Likelihood, chance.

Chapter 7

7.8 PREDICTION WITH MATHEMATICAL MODELS

Mathematical models of physical phenomena can be classified as (1) deterministic deductive, (2) assertive deductive, and (3) probabilistic deductive. All these are created by a deductive process from given information and known effects and first principles. The deterministic deductive model uses deterministic formulations of significant influences. The assertive model usually interjects one or more effect statements that are conjecture, supposition, or a best guess from available experience. The assertive model in its final form cannot be used directly, for it must be tested against reality. If it does not conform, the assertive element is untrue. If it does conform, the assertive information is not proven; all we can say is that nature acts as if the assertion were valid. The probabilistic deductive model uses one or more nondeterministic variates in the formulation, rendering the final formulation nondeterministic, and care must be exercised regarding inferences to be drawn and the marshaling of evidence to test or "prove" the final formulation.

We devote most of our attention to the deterministic deductive model and its derivation, inasmuch as most of the student's early studies employ it almost exclusively. An assertive deductive model is discussed to show its nature. The probabilistic model usually requires a reasonable knowledge of statistics and consequently will be discussed only briefly.

The following steps are usually present in the formulation of a deterministic deductive mathematical model incorporating first principle(s).

0. *Declare the system.*
1. *Isolate* a finite or infinitesimal system or control region.
2. *Identify* the significant influences of the surroundings.
3. *Qualify* significant influences with mathematical models of effects.
4. *Relate* influences to system or control space behavior, using first principle(s). *(form equations)*
5. *Limit* as Δx, Δy, Δz, Δt, etc., approach zero, if necessary.
6. *Solve* resulting equation(s) for variable(s) of interest.

As an example of these steps and to show consistency despite varying viewpoints, a block capable of horizontal oscillation will be analyzed and a mathematical model of its behavior will be deduced. Figure 7.22 depicts a smooth block carried on smooth rollers and restrained by a spring. The problem is to predict the position of the block at any time.

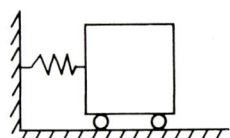

Fig. 7.22. A smooth block on smooth rollers constrained by a spring.

System approach: the system is the block. Figure 7.23(a) shows the block in its natural environment and declares the coordinate of position x. We declare the system to be the block and isolate the system (block) from its surroundings. In Fig. 7.23(b) and (c) we identify and give a mathematical model to the significant influences of the surroundings. The decision is that the normal tractive influences due to the rollers (N_1 and N_2), the normal tractive influence due to the spring (kx), and the body force due to the proximity of the earth ($\overline{W}$) are all the significant influences. Any error of omission at this step propagates to the conclusion. When we say the static equi-

force of friction pull

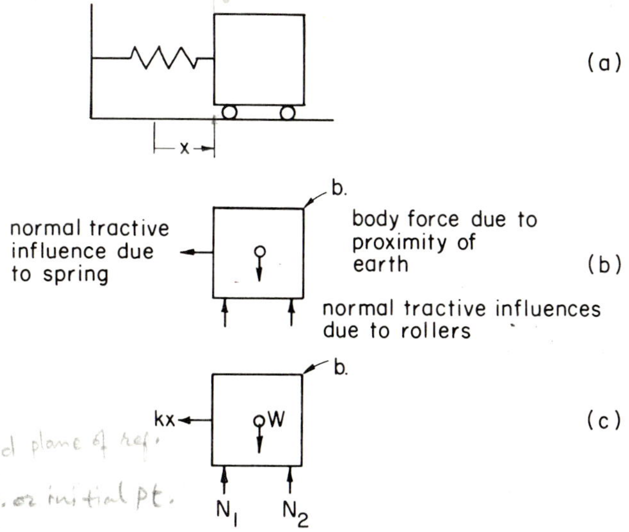

First Principle

Fig. 7.23. The block displaced a distance x. Significant influences of the surroundings are identified and modeling decisions are made.

(a) isolate
(b) identify
(c) qualify

librium position of the left face of the block represents the datum for the coordinate x and the spring's tractive influence is kx, we are declaring that a linear spring model is satisfactory, i.e., our conclusion will not be distorted by a noticeable amount. When we model the tractive influences of the rollers as force vectors normal to the block surface, we are saying that there is no friction of consequence and that the inertias of the rollers are of no consequence. The body force due to the proximity of the earth is a force vector whose line of action contains the center of mass of the system (block) and is directed perpendicular to the horizontal. We relate the significant influences using first principle and we choose Newton's second law of motion for a particle in an inertial reference frame. Since the xyz-reference frame is fixed in the earth and for this purpose it is inertial, we write

$$\Sigma \bar{F} = m\bar{a}$$

The statement for components in the x-direction is

$$\Sigma F_x = ma_x$$

The summation of surface tractions in the x-direction is $-kx$, therefore

$$-kx = m\ddot{x}$$

and it follows that $m\ddot{x} + kx = 0$, which is a differential equation of motion. We will not pursue it further as its solution is straightforward, given two initial conditions. The limit step (5) is unnecessary since we have a finite system. The solve step (6) is omitted as self-evident.

Control region approach: no matter penetrates control surface, inertial coordinates used. Figure 7.24(a) depicts the control surface established so that at no position during motion does the block penetrate the control surface. Figure 7.24(b) indicates the isolated control region and the significant influences of the surroundings. Again using the static equilibrium position of the left face of the block as datum for the coordinate x, we adopt a linear spring model and use kx as the spring force. The xyz-reference frame

230   Chapter 7

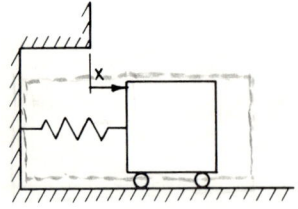

(a) Isolate

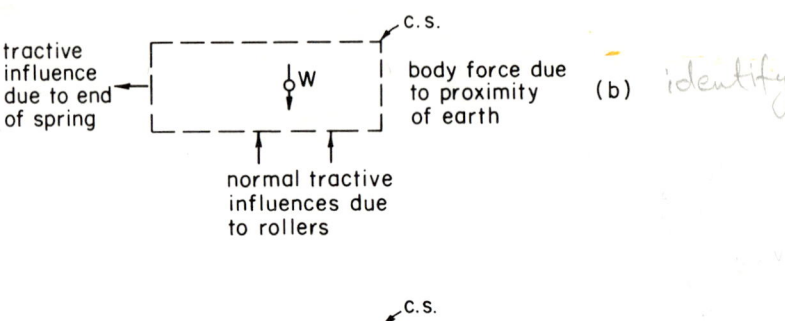

(b) identify

m = mass of block
m_s = mass of spring
$\dot{x}$ = velocity
$m\dot{x}$ = mom. of block
$m_s \dot{x}$ = mom. of spring

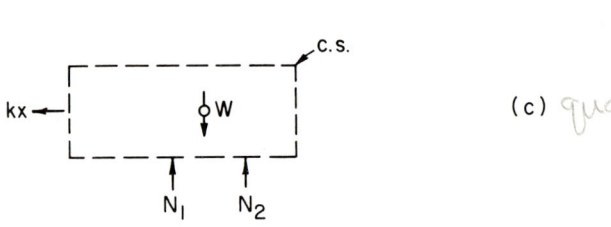

(c) qualify

Fig. 7.24. The block, a control surface containing the block and spring, and the significant influences of the surroundings on the matter in the control region and the qualification decisions.

is fixed in the earth and for this problem it is an inertial reference. The control surface is stationary in the xyz frame. We note that for the chosen countable mV_x

$x \equiv$ distance

$\begin{bmatrix} \text{time rate of increase} \\ \text{in census of countable} \\ mV_x \text{ in system with} \\ \text{same geometry of bound-} \\ \text{ary as control surface} \\ \text{at } t \end{bmatrix} = \dfrac{d}{dt}(mV_x) = \Sigma F_x = -kx$

$\dfrac{dx}{dt} = \dot{x} =$ velocity

$\dfrac{d^2x}{dt^2} = \ddot{x} =$ accel.

$\begin{bmatrix} \text{time rate of increase} \\ \text{in census of countable} \\ mV_x \text{ in control region} \\ \text{at } t \end{bmatrix} = \dfrac{d}{dt}[\dot{x}m + (1/2)\dot{x}m_s]; \quad \begin{bmatrix} \text{net time rate of} \\ \text{influx of countable} \\ mV_x \text{ through control} \\ \text{surface at } t \end{bmatrix} = 0$

Substitution into the accountability equation yields

$-kx = \dfrac{d}{dt}[\dot{x}m + (1/2)\dot{x}m_s] - 0$

Simplification gives

or $-kx = \ddot{x}m + \dfrac{1}{2}\ddot{x}m_s$

or $-kx = \ddot{x}m\left(1 + \dfrac{1}{2}\left(\dfrac{m_s}{m}\right)\right)$

or $kx + \ddot{x}m\left[1 + (1/2)\left(\dfrac{m_s}{m}\right)\right]$

First Principle   231

$$m\ddot{x} + \frac{k}{1 + (1/2)(m_s/m)} x = 0$$

Note that this approach provides an opportunity to discover the influence of the spring mass m_s. In the system approach, the mathematical model of the spring tension at the right end of the spring was taken as kx and in the control region approach the tension in the left end of the spring was taken as kx. Spring tension is constant throughout a spring in a static situation, but in a dynamic situation it is an approximation. The control region approach to this problem, requiring a momentum census throughout the control region, forces us to think about whether the spring contributes any momentum to this census. If the left end of the spring is stationary and the right end of the spring is moving, clearly the spring must have momentum. The question is, Is it significant to the problem? *assume*

Control region approach: *matter penetrates control surface, inertial coordinates used.* Figure 7.25(a) depicts a constant geometry control surface chosen so that the block is partially emergent by the amount x. Let

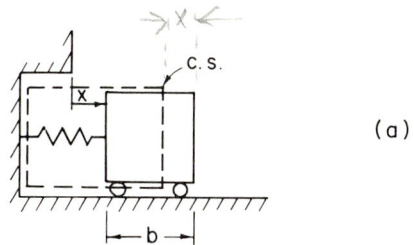

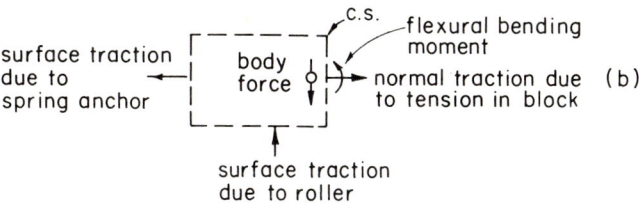

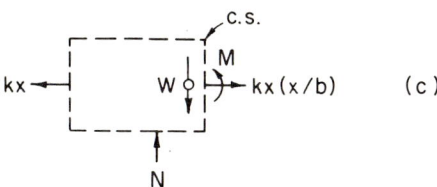

Fig. 7.25. The block and spring and a control surface through which the block emerges a distance x, identification of significant influences of surroundings on matter within the control region, and the qualification decisions.

232   Chapter 7

　　　b = horizontal length of the block
　　　m = mass of the block
　　　mx/b = mass of the block portion that has penetrated control surface and
　　　　　　 is part of the surroundings
　　　m(1 - x/b) = mass of portion of block within control region
　　　mẋ/b = time rate of mass penetration (efflux) of control surface

Figure 7.25(b) indicates the significant influences of the surroundings on the matter in the control region and their mathematical modeling. Note the normal traction on the control surface due to tension in the block. It is left to readers to satisfy themselves that this tension can be modeled as kx^2/b. Figure 7.25(c) depicts the mathematical modeling of the significant influences. In preparation for substituting in the accountability equation, we note

$$\begin{bmatrix} \text{time rate of increase} \\ \text{in census of countable} \\ mV_x \text{ in system with} \\ \text{same geometry of bound-} \\ \text{ary as control surface} \\ \text{at t} \end{bmatrix} = \frac{d}{dt}(mV_x) = \Sigma F_x = -kx + kx(x/b)$$

The momentum census within the control region is given by $\dot{x}$ times the mass within the control surface, namely, $m(1 - x/b)$.

$$\begin{bmatrix} \text{time rate of increase} \\ \text{in census of countable} \\ mV_x \text{ in control region} \\ \text{at t} \end{bmatrix} = \frac{d}{dt}[\dot{x}m(1 - x/b)]$$

The rate of momentum efflux through the control surface is the velocity $\dot{x}$ times the time rate of mass efflux, $m\dot{x}/b$. We write

$$\begin{bmatrix} \text{net time rate of influx} \\ \text{of countable } mV_x \text{ through} \\ \text{control surface at t} \end{bmatrix} = -\dot{x}(m\dot{x}/b)$$

Substitution in the accountability equation yields

$$[-kx + kx(x/b)] = \frac{d}{dt}[\dot{x}m(1 - x/b)] - [-\dot{x}(m\dot{x}/b)]$$

$$-kx(1 - x/b) = m\ddot{x} - m\dot{x}^2/b - mx\ddot{x}/b + \dot{x}m\dot{x}/b$$

$$-kx(1 - x/b) = m\ddot{x}(1 - x/b)$$

and if $(1 - x/b) \neq 0$, then $m\ddot{x} + kx = 0$ as before.

Many viewpoints are possible in either the system or control region approach to the formulation of deterministic deductive mathematical models. In this recitation of alternative approaches, we omitted the system viewpoint using a noninertial frame of reference that would come about if we define block matter as the system and attach the coordinate system to the block. We could take a constant geometry control region with the coordinates fixed in the block (noninertial) or allow the geometry of the control surface to vary as a function of time. The presentation here illuminates the steps of isolate, identify, qualify, relate, limit, and solve and shows that the results are independent of viewpoint. It should suggest to the reader ways to check a derivation. Also, in certain problems, particular geometries are natural and simple and the evaluation of terms is easier. The experienced analyst exploits

First Principle

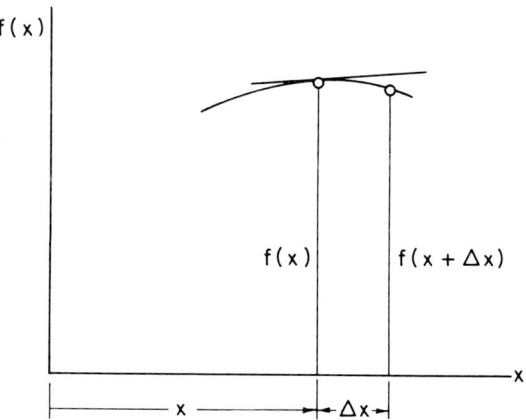

Fig. 7.26. A function of x evaluated at x and at $x + \Delta x$.

this whenever possible, and experience is often the best guide to an initial approach. Since we have used derivatives in the preceding derivations, a caution is in order. Figure 7.26 graphically depicts a functional relationship between $f(x)$ and x. An ordinate to the curve is erected at abscissa x and at the incremented abscissa position $x + \Delta x$. The ordinates to the curve are $f(x)$ and $f(x + \Delta x)$, respectively. The definition of a derivative of a function of x is

$$\left.\frac{dy}{dx}\right|_x = \lim_{x \to 0} \frac{\Delta y}{\Delta x} = \lim_{\Delta x \to 0} \frac{f(x + \Delta x) - f(x)}{\Delta x}$$

It follows that

$$\left.\frac{dy}{dx}\right|_x = \frac{f(x + \Delta x) - f(x)}{\Delta x}$$

Prediction of the ordinate at $x + \Delta x$ can be made from the approximation

$$f(x + \Delta x) = f(x) + \left.\frac{dy}{dx}\right|_x \Delta x$$

where: Δx = positive increment in abscissa x

$\left.\frac{dy}{dx}\right|_x$ = slope of the curve at abscissa x; its sign, + or -, reflects whether $f(x + \Delta x)$ is larger or smaller than $f(x)$

A negative slope will make the quantitative determination of

$$\left.\frac{dy}{dx}\right|_x \Delta x$$

negative. In Fig. 7.26, the slope of the curve is the tangent to the curve at abscissa x. It intercepts the ordinate erected at abscissa $x + \Delta x$ in such a fashion as to predict the magnitude of $f(x + \Delta x)$. The prediction is too large. The error in the prediction is smaller as the increment Δx becomes smaller. In the limit as Δx approaches zero, the prediction is exact. We may use the expression

$$f(x) + \left.\frac{dy}{dx}\right|_x \Delta x$$

to predict $f(x + \Delta x)$ precisely if we will pass to the limit as $\Delta x \to 0$. Figure

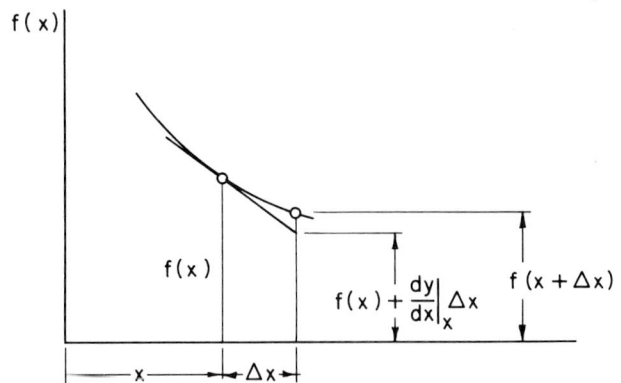

Fig. 7.27. A function of x with negative slope in the neighborhood of x and x + Δx.

7.27 shows the essential geometry in the case where the slope of the curve is negative in the region of interest.

Figure 7.28 shows a cross section of the system for the atmospheric pressure variation problem. The tractive influences of the surroundings on the system are (in the x-direction) normal tractions on the bottom and the top of the system and a body force due to the proximity of the earth acting at the center of mass. Figure 7.28(a) defines the geometry of the system, and (b) depicts the significant surface tractions in the x-direction as well as the body force; (c) replaces the surface tractions with their equivalent vector sums.

It is possible to relate these influences using a first principle (Newton's law for a system at rest) by stating that the system is at rest, persists at rest, and in our universe the vector sum of the external forces on the system is null. That is, $\Sigma \mathbf{F} = \hat{\mathbf{0}}$; therefore, $\Sigma F_x = 0$ and

$$p\Delta y \Delta z - (p + \frac{dp}{dx}\Delta x)\Delta y \Delta z - \rho \Delta x \Delta y \Delta z = 0$$

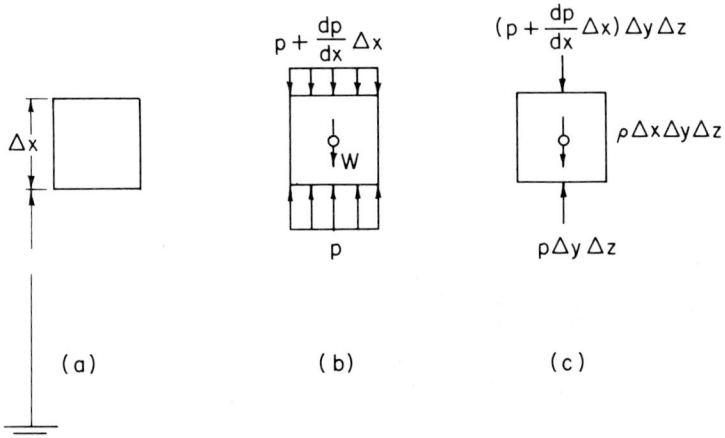

Fig. 7.28. The infinitesimal system of the stagnant atmosphere problem with the pressures displayed in (b) and the forces in (c).

First Principle

Consequently, $dp/dx = -\rho$.

It is clear that dp/dx is negative and the pressure decreases with increasing altitude. This *conclusion* (not assumed, presumed, or necessarily known in advance) is a consequence of Newton's law expressed for a static fluid.

Suppose we had initially recognized that the pressure decreases with altitude and had written the upper surface normal force as $(p - \Delta p)\Delta y \Delta z$; then the equation summing the forces in the x-direction to zero would have been

$$p\Delta y \Delta z - (p - \Delta p)\Delta y \Delta z - \rho \Delta x \Delta y \Delta z = 0$$

We obtain $\Delta p/\Delta x = \rho$, and in the limit as Δx approaches zero we have

$$\frac{dp}{dx} = \rho$$

This equation formed the limit as Δx approached zero and therefore "must be right." It clearly predicts something contrary to nature and therefore not observable in nature. What have we done that is inadvisable or inconsistent? We have defined a derivative contrary to the way everyone else does (contrary to definition), yet used the conventional notation for a derivative. Is there a tag we can place on these "homegrown" derivatives to avoid confusing ourselves and others and possible misinterpretations of the predictions made by first principle? There is no tag, so adhere to the universal definition of a derivative (unless there are good reasons not to), and then *be careful*.

Thus, in an atmospheric pressure distribution problem, if we conclude that the atmospheric pressure p is a function of the altitude x, then at the altitude x we must assign a pressure of p. At the altitude $x + \Delta x$ we must assign the pressure $p + \Delta p$, i.e., the pressure at position x *plus* the change in pressure as we move an incremental distance Δx in altitude. This procedure ensures that derivatives taken by differentiation rules will agree in sign with derivatives inadvertently constructed in the process of writing equations and taking limits. In the case of the stagnant atmosphere, the imposition of Newton's law of motion on the system whose surroundings are modeled by three colinear forces led to the equation $dp/dx = -\rho$.

The equation modeling the pressure variation in the atmosphere has been shown to be

$$\frac{dp}{dx} + \rho = 0$$

which is a differential equation and must be "solved." If ρ is constant, or so nearly so that it may be modeled as a constant, integration term by term is possible, specifically,

$$p + \rho x = C$$

where C is a constant of integration. If the pressure is known where the altitude origin is located, $p = p_0$ when $x = 0$ and substitution in the solution above yields

$$p_0 + 0 = C$$

and substitution of p_0 for C in the same solution results in the expression

$$p - p_0 = -\rho x \tag{7.6}$$

If ρ is not a constant but is a function of altitude, the equation that must be solved is

$$\frac{dp}{dx} + \rho(x) = 0$$

If the fluid is a gas or a mixture of gases, such as air, whose states are well described by the equation of state for an ideal gas and particularized for air, then

$$p/\rho = RT_0 \qquad (7.7)$$

where R for ideal air is 53.35 ft·lbf/(lbm°R). Substituting for ρ in the differential equation, we obtain

$$\frac{dp}{dx} + \frac{p}{RT_0} = 0$$

If the range of use of the final equation is sufficiently small in altitude that the temperature T may be treated as a constant, T_0, separation of variables results in

$$\frac{dp}{p} + \frac{dx}{RT_0} = 0$$

each term of which is exact and integrable. The integration is

$$\ln p + x/(RT_0) = \ln C$$

where C is a constant of integration. Taking antilogarithms gives

$$p/C = e^{-x/(RT_0)}$$

If $p = p_0$ when $x = 0$, it follows that $C = p_0$ and the equation becomes

$$p/p_0 = e^{-x/(RT_0)} \qquad (7.8)$$

If the temperature variation in the fluid cannot be ignored, its nature must be taken into account. Soundings of the atmosphere indicate that an expression of the form

$$T = T_0(1 - ax)$$

is reasonable for altitudes below 36 000 ft (see Fig. 7.29). The differential equation

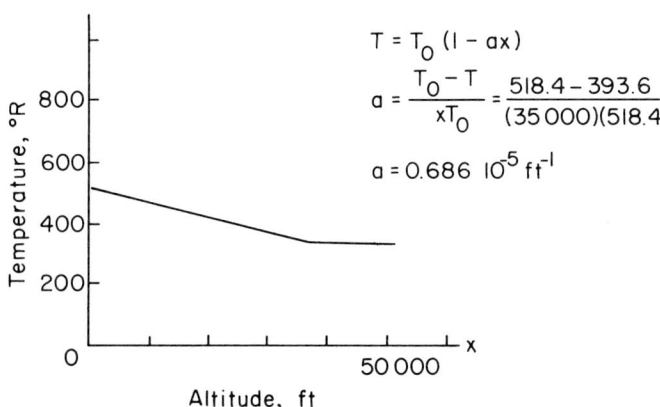

Fig. 7.29. The variation of temperature with altitude in the earth's atmosphere.

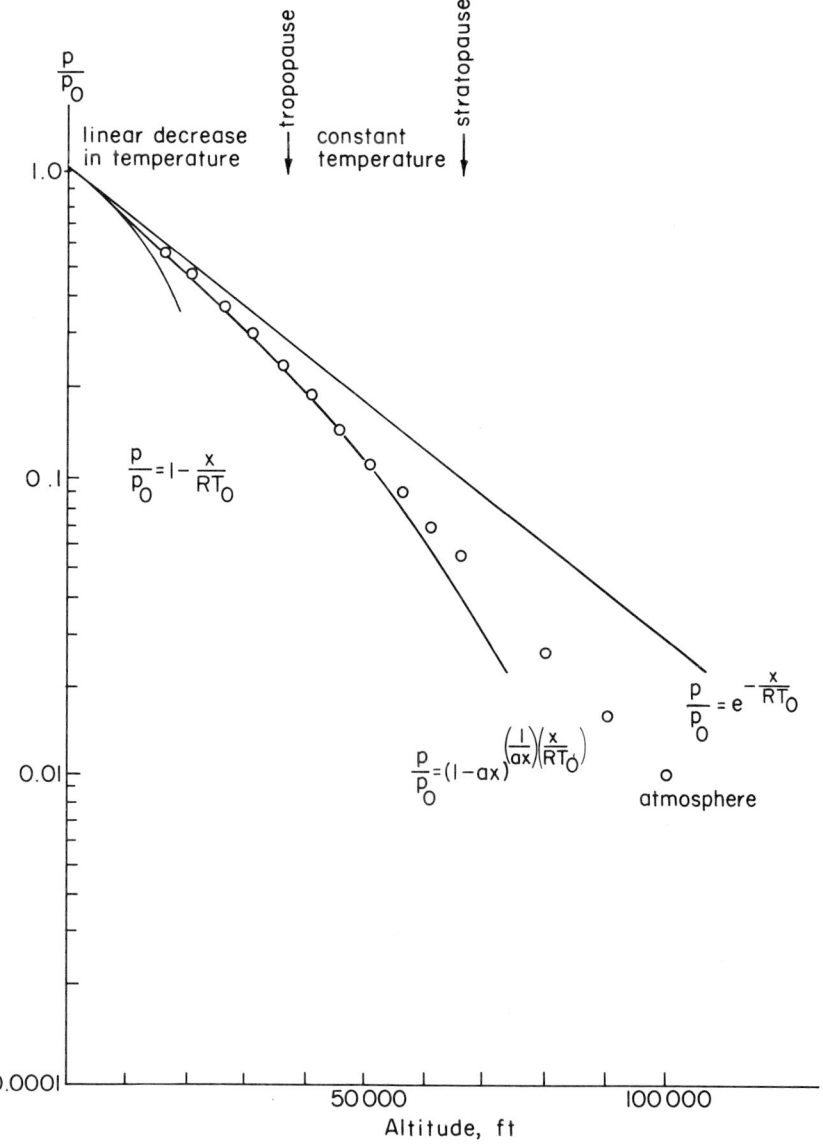

Fig. 7.30. The three models of pressure variation with altitude, with data points from the earth's atmosphere.

$$\frac{dp}{dx} + \frac{p}{RT(x)} = 0 \text{ or } \frac{dp}{dx} + \frac{p}{RT_0(1-ax)} = 0$$

may have variables separated as follows:

$$RT_0 \frac{dp}{p} = -\frac{dx}{(1-ax)}$$

Integration between p_0 and p involves the corresponding limits of 0 and x:

$$RT_0 \int_{p_0}^{p} \frac{dp}{p} = -\int_0^x \frac{dx}{(1-ax)}$$

In the right-hand integral let $u = 1 - ax$ and $du = -a\, dx$. Then

$$-\int_0^x \frac{dx}{(1-ax)} = \frac{1}{a}\int_1^{1-ax} \frac{du}{u} = \frac{1}{a}\ln(1-ax)$$

It follows that

$$RT_0 \ln(p/p_0) = (1/a)\ln(1-ax)$$

Taking antilogarithms and solving for p/p_0 yields

$$p/p_0 = (1-ax)^{1/(aRT_0)} \tag{7.9}$$

The three relations, Eqs. (7.6) with (7.7) substituted, (7.8), and (7.9),

$p/p_0 = 1 - x/(RT_0)$ (unvarying density and temperature, ideal gas)

$p/p_0 = e^{-x/(RT_0)}$ (unvarying temperature, ideal gas)

$p/p_0 = (1-ax)^{1/(aRT_0)}$ (linearly varying temperature, ideal gas)

all come from different system models. They are all consistent with Newton's law of motion, however. Examine Fig. 7.30 for the prediction of these different mathematical models and their correlation with real atmospheric observations.

7.9 CHECKING VALIDITY

In solving a problem an engineer might have to perform a hundred elementary tasks. If one's average reliability in being correct is 0.99 on each task, what is the probability of obtaining the correct answer?

$$\text{Prob} = 0.99^{100} = 0.366$$

Frightening, isn't it? If it reminds you of your experience with some college course, you understand the problem firsthand. How can you improve the performance? Increase the average reliability. How do you do that? By checking!

If a hundred steps are concatenated to reach a result, where in the chain of events is a mistake most wasteful of effort that must be repeated? Early! When do most people think of checking? At the end. A good engineer checks each step as it is made, checks groups of steps, and checks results. As an example of some useful kinds of checking applied to the same result, consider the equations developed in Sec. 7.6 for tangential and normal forces on the conveyer belt:

$f = \dot{w}(V + v\sin\theta) + W\sin\theta$

$N = (W + wv)\cos\theta$

Limiting case check. We examine the effect on the solution of allowing variables to range from the point of vanishing to increasing without upper bound. Does the effect make sense or confirm a result obtained from other sources? For example, let the angle the conveyer belt makes with the horizon vanish, i.e., $\theta = 0$. Then

First Principle

$$f = \dot{w}(V + 0) + 0 = \dot{w}V$$

agreeing with our previous result for a horizontal belt. When the sand flow rate is vanishingly small we have

$$f = 0(V + v \sin \theta) + W \sin \theta = W \sin \theta$$
$$N = (W + 0) \cos \theta = W \cos \theta$$

which makes sense.

Dimensional check. Equations must be dimensionally homogeneous. Now

$$\dim(f) = F \qquad \dim(W) = F$$
$$\dim(\dot{w}) = MT^{-1} = FTL^{-1} \qquad \dim(N) = F$$
$$\dim(V) = LT^{-1} \qquad \dim(\sin \theta) = 1$$
$$\dim(v) = LT^{-1} \qquad \dim(\cos \theta) = 1$$

and the F unit equation is

$$[F] = FTL^{-1}[LT^{-1} + LT^{-1}(1)] + [F(1)] = [F] + [F] + [F]$$

which is dimensionally homogeneous.

Symmetry. We expect that the angle θ changing from plus to minus will affect f, decreasing it; and we do not expect a change of θ from plus to minus to affect N at all:

$$f = \dot{w}[V + v \sin(-\theta)] + W \sin(-\theta) = \dot{w}(V - v \sin \theta) - W \sin \theta$$
$$N = (W + \dot{w}v)\cos(-\theta) = (W + \dot{w}v)\cos \theta$$

confirming our expectation.

Experience. We expect the belt speed sustaining force f to vary directly with the sand flow rate $\dot{w}$, belt speed V, sand speed v, belt angles less than 90° (θ), and weight of the sand resident in the control region. An examination of the equation for f confirms this. In other words we are saying that in our experience the gross properties of the equation "make sense." We expect the belt normal force N to vary directly with the sand weight W, sand flow rate $\dot{w}$, and sand speed v and to vary indirectly with the belt angle to the horizon, which is confirmed by inspection. Again the results do not violate or contradict our previous physical experience.

Validity of assumptions. For assumptions made during derivation (say, the implicit decision that atmospheric drag on sand and belt is negligible) we can inquire:

--Was this a necessary assumption?
--Has embracing it hidden an important influence of the surroundings on matter in the control region?
--Did we qualify our result with an explicit statement of this assumption?
--Is the assumption defendable at all belt speeds V?
--Did this assumption make the model incongruent with nature?

The answers to these questions can help uncover shaky work. The word *assumption* shouldn't be used in engineering because ignoring atmospheric drag in the modeling process is a *decision* on the engineer's part that atmospheric drag is not significant; and engineers must be responsible for all decisions they make, whether made by commission or omission. It is a good idea to list these decisions in any development.

Experiment. We can subject the results to experimental verification.

Alternate method of derivation. Many times this is wishful thinking because many engineering solutions strain our ability to get there; otherwise the problem would already have a recorded solution. For many of the simpler steps in a solution however, this is a feasible method of check. In Chap. 2, remember the differing results of two models of the bolts?

Have a colleague check your work. This is sometimes possible and is especially effective when two different engineers are asked to work on an aspect of a problem independent of one another.

The insufficiency of checking methods. Methods of check are often, in the mathematical sense, directed to verification of matters of necessity, not sufficiency. In the equation

$$f = (1/2)\dot{w}(V + v \sin \theta) + W \sin \theta \quad \text{(incorrect)}$$

the limiting case check, dimensional check, symmetry check, experience check, assumption check will not uncover the erroneous 1/2. The experimental check *can*, a colleague check *may*, and an alternate method check *might* detect it.

Nothing has been said about numerical precision, since we have been discussing an algebraic result, but establishing the range of error, the significant figure content, or the dispersion of the result of a series of calculations is an interesting, instructive, and somewhat scary enterprise.

Nothing has been said about goofs in transcription, interchange of digits in numbers, reading the wrong line from a table, or punching the wrong button on a calculator (and obtaining the wrong answer to 8 significant figures). Watch experienced (surviving) engineers. They read tabular entries from the edge of an *opaque* straightedge for a reason and immediately check by entering the table in another order. They check computational results by doing the calculation again in a different order (for a reason).

Methods of check detect troubles; they do not rectify troubles. They are not infallible, but all engineers check, check, and check again. If your spouse did it you might be tempted to categorize it as nagging. It is all to the purpose of increasing the reliability of each step. If your overall reliability in reaching a yes-no decision with a complex sequence of steps is less than 0.5, you can be replaced by a coin at the capital cost of one cent and the loss of return on that investment.

7.10 ASSERTIVE MODEL OF AN EPIDEMIC

We recall that deductive deterministic mathematical models combine established effects with conservation statements (first principles) to reach a useful predictive formulation consistent with nature. An assertive deterministic mathematical model combines one (or more) axiomatic effect statement(s) with established effects and first principle(s) to reach a formulation. This mathematical model must be tested against nature for consistency. Agreement verifies a useful predictive mathematical model but does not prove the quantitative expression used for the axiomatic effect statement. The steps isolate, identify, qualify, relate, limit, and solve continue to be effective.

As an example of the formulation of an assertive model, we examine the phenomenon called a disease epidemic. Sometimes we are surrounded by cases of disease, but the number who are eventually infected is relatively minor. Other times the disease snowballs until it appears as if "everyone" has contracted it. Let us build a model to predict the numbers of a population that will contract a disease. We have a population of persons susceptible to the dis-

First Principle   241

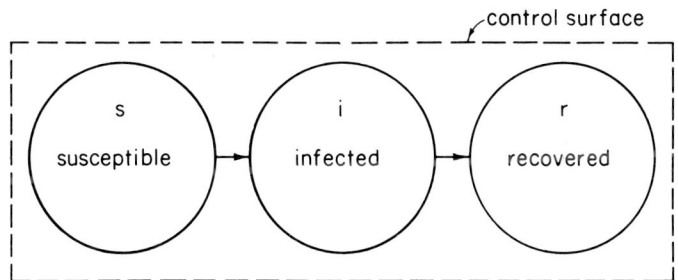

Fig. 7.31. A control surface encapsulating susceptible, infected, and recovered persons in the epidemic problem.

ease. The actual population may be greater because some persons are immune. We shall designate those who are not immune as susceptibles. We will allow a single case of the disease to infect a susceptible. To simulate the contagion process, we will stipulate that every person in the population shake hands with every other person at 8:00 every morning in the town square. At any time the number of *susceptibles* is denoted as s, the number of *infecteds* is denoted as i, and the number of *recovereds* is denoted as r. The identification of susceptibles, infecteds, and recovereds as the only categories present is a simplification and is found in all modeling. The unique evolution from susceptible to infected to recovered is also a simplification. The identification of people as countables and the recognition of their conservation comes from understanding. The accountability equation for discrete countables is useful. Figure 7.31 places a control surface about all people involved and denotes their unique path of change if change occurs. For the control region of Fig. 7.31, we can write

$$[\Sigma C_c - \Sigma C_d]_s = [\Sigma C_f - \Sigma C_i]_{c.r.} - [\Sigma C_{in} - \Sigma C_{out}]_{c.s.}$$

where the subscripts c = created, d = destroyed, f = final, and i = initial. The term on the left side of the equation is zero, since people are conserved. The term on the extreme right is zero, since no countables (people) cross the control surface. It follows that

$$[\Sigma C_f - \Sigma C_i]_{c.r.} = 0 \text{ or } \Sigma C_f = \Sigma C_i = s + i + r = n$$

where the total population is n by definition. The equation

$$s + i + r = n$$

is a *deterministic deductive* model. Figure 7.32 depicts the enclosure of the

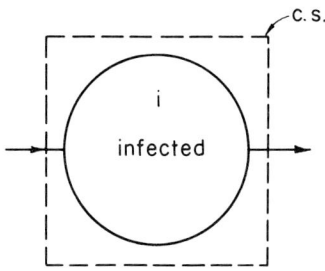

Fig. 7.32. A control surface encapsulating infecteds.

infected population within a control surface. The significant influences of the surroundings are the influx of people who were susceptibles and have become infecteds and the efflux of people who were infecteds and have become recovereds.

The number of susceptibles who penetrate the control surface and become infecteds is (by assertion) proportional to the number of infecteds and to the number of susceptibles. In the time interval Δt (an accounting interval) this may be represented as (ais Δt), where a is a constant. This is an assertion. It seems reasonable but has no foundation in fact. The number of infecteds who cross the control surface to become recovered is (by assertion) proportional to the number of infecteds and may be represented as (bi Δt), where b is a constant. This seems reasonable but has no foundation in fact. The accountability equation for this control space is

$$[\Sigma C_c - \Sigma C_d]_s = [\Sigma C_f - \Sigma C_i]_{c.r.} - [\Sigma C_{in} - \Sigma C_{out}]_{c.s.}$$

$$0 = [i + \Delta i - i] - [ais \, \Delta t - bi \, \Delta t]$$

$$\Delta i / \Delta t = ais - bi$$

This equation blends a conservation statement with two effects whose mathematical form is not based on experiment but on "reasonableness," i.e., assumption or presumption, axiomatically and hypothetically. Thus the above equation is an *assertive deterministic* model. Figure 7.33 passes a control surface around the susceptibles. In a similar fashion we can write

$$[\Sigma C_c - \Sigma C_d]_s = [\Sigma C_f - \Sigma C_i]_{c.r.} - [\Sigma C_{in} - \Sigma C_{out}]_{c.s.}$$

$$0 = [s + \Delta s - s] - [0 - ais \, \Delta t]$$

$$\Delta s / \Delta t = -ais$$

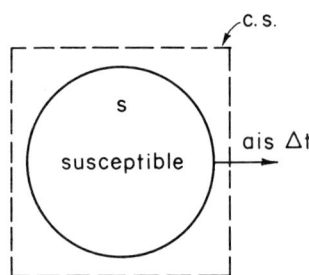

Fig. 7.33. A control surface encapsulating susceptibles.

Similarly Fig. 7.34 passes a control surface around the recovereds and we can write

$$[\Sigma C_c - \Sigma C_d]_s = [\Sigma C_f - \Sigma C_i]_{c.r.} - [\Sigma C_{in} - \Sigma C_{out}]_{c.s.}$$

$$0 = [r + \Delta r - r] - [bi \, \Delta t - 0]$$

$$\Delta r / \Delta t = bi$$

Of the four equations that constitute the model of an epidemic as previously described,

$$\Delta s/\Delta t = -ais \quad \Delta r/\Delta t = bi \quad \Delta i/\Delta t = ais - bi \quad n = s + i + r$$

three incorporate assertive effect statements. Consequently, this set of

First Principle   243

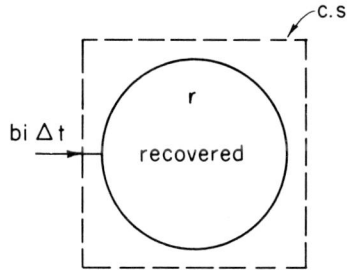

Fig. 7.34. A control surface encapsulating recovereds.

equations constitute an assertive deterministic model of an epidemic. We must check its predictions against experience (nature) before we can determine whether it would be useful.

We will use a continuum model rather than limit the variables to integer values. Integration of these equations is not simple because of the (i × s). The populations are functions of time, i.e., i(t), r(t), and s(t). Numerical integration is possible by saying

$$s = n - 1 \quad i = 1 \quad r = 0 \quad t = 0$$

initially, selecting a magnitude for Δt, and calculating

$$\Delta i = (ais - bi)\Delta t \quad \Delta r = bi\,\Delta t \quad \Delta s = -ais\,\Delta t$$

Then at instant $t + \Delta t$, the current values of t, i, r, and s are upgraded

$$t \to t + \Delta t \quad i \to i + \Delta i \quad r \to r + \Delta r \quad s \to s + \Delta s$$

To study graphs of i(t), s(t), and r(t), it is necessary to pursue this iteration process until Δi, Δs, and Δr become negligibly small. One might be willing to do this once, but for many determinations at different values of a, b, and n, the exercise becomes an enormous piece of computational effort. Either a digital or analog computer can be used to effect the necessary integrations.

The manual procedure would be to construct a table with a = 0.01, b = 0.5, and n = 40.

t	s	i	r	ais	bi	ais - bi
0	39	1	0	0.39	0.5	-0.11
1	38.61	0.89	0.5			

From the original entries (t = 0, s = 39, i = 1, and r = 0), we determine

$$ais = 0.01(1)(39) = 0.39 \quad bi = 0.5(1) = 0.5$$
$$ais - bi = 0.39 - 0.5 = -0.11$$

creating the values of s, i, and r at t = 1,

$$s = 39 - 0.39 = 38.61$$
$$i = 1 + (-0.11) = 0.89$$
$$r = 0 + 0.5 = 0.5$$

which are entered in the table. The process continues until no change occurs. Such a routine procedure is easily programmed on a programmable hand-held calculator. Table 7.1 displays such a program. The results are as follows for a = 0.01, b = 0.5, n = 40, and t in days:

t	s	i	r	t	s	i	r
0	39	1	0	10	36.70	0.27	3.04
1	38.61	0.89	0.50	20	36.14	0.06	3.80
2	38.27	0.79	0.95	40	35.98	0.0031	4.0123
3	37.96	0.70	1.34	60	35.98	0.0001	4.0228
4	37.70	0.61	1.69	150	35.9767	≈ 0	4.0223
5	37.47	0.54	1.99				

Table 7.1. Card Programmable Calculator Program for Epidemic Model

Entry	Program Steps	Display
a	[A] ST01 R/S	a
b	ST02 R/S	b
n	ST03 R/S	n
0	ST00 ST04 ST05 e^x ST09 RCL0 -x-	t_0
	RCL3 -x-	s_0
	RCL4 -x-	i_0
	RCL5 -x- R/S	r_0
i	[B] enter ST04 RCL1 RCL4 × RCL3 × ST06 RCL2 RCL4 × ST07 RCL6 RCL7 - RCL4 + ST04	
	RCL7 RCL5 + ST00 -x-	t
	RCL3 -x-	s
	RCL4 -x-	i
	RCL5 -x-	r
	RCL3 RCL4 + RCL5 + -x-	s+i+r
	RCL4 GTOB	

These results are plotted in Fig. 7.35(a). The number of infecteds is continuously decreasing and the number of recovereds rises to 4.0233. This number includes the original infector so the prediction of the number of cases is three. This number can be obtained also by subtracting the final number of susceptibles 35.9767 from the initial census of susceptibles 39. What fraction of the original population contracted the disease? The answer is 3.02/39 or 0.0775. This result is similar to our experience when people say, "The flu is around again, have you had it?"

Suppose the population is increased to 100 and a single case occurs. The probability of a susceptible contracting the disease is (83.22 - 1)/99 or 0.831. Examine the plot of this case in Fig. 7.35(c). Note that the cases of infecteds peaked at approximately 18 on the tenth day. Although a minority was sick at any one time, more than 83 percent of the population contracted the disease--it was an epidemic. Note that when 1 of a group of 40 contacts the disease, less than 8 percent contract the disease, but when 1 of a group of 100 contacts the disease, more than 83 percent become ill. Apparently some condition exists that allows the number of infecteds to snowball until a large segment of the population contracts the disease. The signal that this can happen is to be found in the algebraic sign of Δi. When Δi is initially negative, the active cases decline from the outbreak date. When Δi is initially positive, the active cases increase before they can subside because of the declining population of susceptibles.

We could say that the population exhibits stability when subjected to a perturbation (appearance of an infected), and it immediately and consistently reduces the population of infecteds. Conversely, we could say that a popula-

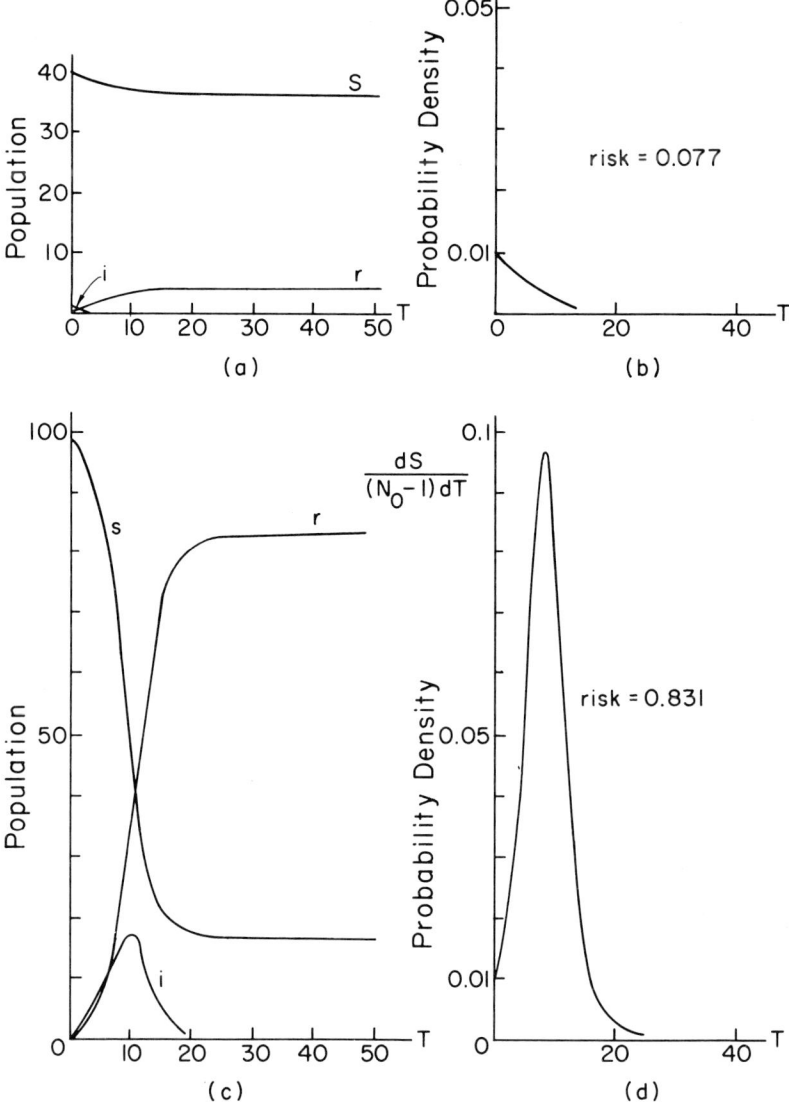

Fig. 7.35. (a) plots of s, i, and r with time when a single infected is injected into 39 susceptibles. (b) Probability density-time plot for (a). Area under the curve is the risk (0.077) of becoming infected if you are one of 39 susceptibles. (c) Plots of s, i, and r with time when a single infected is injected into 99 susceptibles. (d) Probability density-time plot for (c). Area under the curve is the risk (0.831) of becoming infected if you are one of 99 susceptibles.

tion is unstable when it is perturbed and the infected population grows, consuming susceptibles, creating recovereds, and eventually dying out because of loss of supply of susceptibles to sustain the infected population. Now di/dt is zero when i = 0, which occurs when the disease has run its course. Also di/dt can be zero if (as - b) = 0. This condition occurs when the critical

Table 7.2. Probability of Contracting the Disease for Different Populations of Susceptibles

Population of Susceptibles	Probability of Contracting the Disease
1	0.02027
2	.0207
3	.021
4	.0216
9	.0241
19	.0316
29	.0455
39	.0775
49	.171
59	.361
69	.538
79	.699
89	.791
99	.831
199	0.999

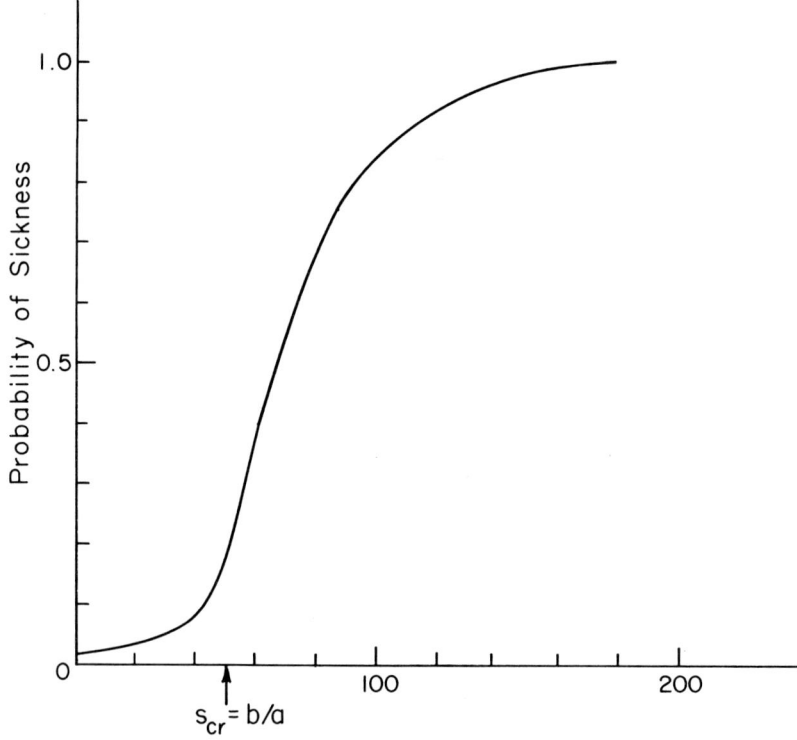

Population of Susceptibles

Fig. 7.36. Probability of sickness as a function of the population of susceptibles where $a = 0.01$ and $b = 0.5$. The critical population of susceptibles is denoted as s_{cr}.

First Principle   247

population of susceptibles is s_{cr} and $s_{cr} = b/a$. In our case where $a = 0.01$ and $b = 0.5$,

$$s_{cr} = 0.5/0.01 = 50$$

and the critical population is $n_{cr} = 50$. The danger of an epidemic is greater when an infected occurs in a large group than when an infected occurs in a small group! What maneuvers are possible to reduce the chance of an epidemic or to minimize its toll if it cannot be prevented? An increase in b corresponds to an improved recovery rate, brought about by improved medical techniques. A decrease in a corresponds to a reduction in intimacy of contact (handshake replaced by a hand wave). The population of susceptibles can be permanently reduced by inoculation, immunization, vaccination, and similar techniques and temporarily reduced by isolation and quarantine.

The chances of a susceptible contracting the disease as a function of the size of the population of susceptibles can be obtained from a study of the model.

You have examined a model. You have seen some of its predictions. Do you believe it? Some of it? None of it? What shall we look for in epidemic data to show conformity of the model to reality or departure from reality? When you can answer some of these questions, you will be progressing toward understanding modeling.

A plot of data in Table 7.2 is provided in Fig. 7.36. The critical population of susceptibles is indicated in the abscissa at $s_{cr} = 50$. Note the rapid increase of hazard with initial population size.

7.11 PROBABILISTIC DEDUCTIVE AND SIMULATIVE MODELS

It should not be surprising that an algebra exists for random variables just as algebras exist for vectors, complex numbers, and real numbers. Such an algebra shares many properties of the familiar real variable algebra; in fact, it contains the real variable algebra as a subset. The random variable algebra exhibits in addition

--uniqueness of sum
--associative law
--identity element
--unary minus operation leading to subtraction
--commutative law

and in multiplication

--uniqueness of product
--associative law
--identity element
--existence of multiplicative inverse leading to division
--commutative law

and the distributive law holds in combinations of addition and multiplication. A body of theorems that rests on the preceding algebraic properties can be developed. This body of theorems is called an algebra of expectation. For a function ϕ of x, the mean and standard deviation are given by

$$\mu_{\phi(x)} = E[\phi(x)] = \int_{-\infty}^{\infty} \phi(x) f(x) \, dx \qquad (7.10)$$

$$\sigma_{\phi(x)} = \{\int_{-\infty}^{\infty} [\phi(x) - \mu_{\phi(x)}]^2 f(x) \, dx\}^{1/2} \tag{7.11}$$

where x is a random variable. For a function of n variates, the following approximations hold for $\phi\{x_n\}$:

$$\mu_\phi = \phi(\mu_{x_1}, \mu_{x_2}, \ldots, \mu_{x_n}) + \frac{1}{2} \sum_{i=1}^{m} \left.\frac{\partial^2 \phi}{\partial x_i^2}\right|_\mu \sigma_{x_i}^2 + \ldots \tag{7.12}$$

$$\sigma_\phi = \left\{\sum_{i=1}^{m} \left[\frac{\partial \phi}{\partial x_i}\right]_\mu^2 \sigma_{x_i}^2 + \frac{1}{2} \sum_{i=1}^{m} \left[\frac{\partial^2 \phi}{\partial x_i^2}\right]_\mu^2 \sigma_{x_i}^4 + \ldots\right\}^{1/2} \tag{7.13}$$

when the distribution of x is either unskewed or lightly skewed. Using Eqs. (7.12) and (7.13) for $\phi(x) = x^n$ for any real n, we obtain

$$\mu_{x^n} = \mu_x^n [1 + (1/2)n(n-1)(\sigma_x/\mu_x)^2 + \ldots]$$

$$\sigma_{x^n} = n\mu_x^{n-1} \sigma_x [1 + (1/4)(n-1)^2 (\sigma_x/\mu_x)^2 + \ldots]$$

Here is a way to answer our previous reciprocal problem. Let $\phi = 1/x$; consequently n = -1 and

$$\mu_{1/x} = (1/\mu_x)[1 + (1/2)(-1)(-1-1)(\sigma_x/\mu_x)^2 + \ldots] = (1/\mu_x)[1 + (\sigma_x/\mu_x)^2 + \ldots]$$

The mean of a reciprocal is always larger than the reciprocal of the mean (by 1 + the coefficient of variation squared) for a symmetric or lightly skewed distribution. Note that we acknowledge the existence of a random variable algebra and then use a result employing real number algebra. This is similar to acknowledging the existence of a vector algebra and then posing the problems in terms of components and employing real number (scalar) algebra. The algebra exists and it is important to mathematicians that all its properties be illuminated. Then the results of the existence of the algebra are placed in real algebraic terms. We do not substitute a random variable into a denominator of a fraction with real number 1 as the numerator and examine the variate that is the quotient as a random variable. We look at selected attributes of the quotient (say the mean) and use a real variable algebraic formula, determining the real variable number that is an excellent approximation of the mean of the quotient. Appendix C provides many such useful expressions for your present and future reference--engineering textbooks are surprisingly mute on this point.

The mathematical model building sequence of steps (isolate, identify, qualify, relate, limit, and solve) that were effective in the deterministic deductive and assertive deductive models are just as useful in probabilistic modeling. The probabilistic deductive mathematical model, examined first, differs from the previous models largely in the solve step, since the model is expressed in an algebra of expectation rather than a real variable algebra.

Figure 7.37 depicts a chemical actuator (commonly called an explosive

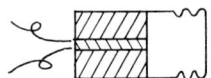

Fig. 7.37. An electrically activated chemical squib (actuator).

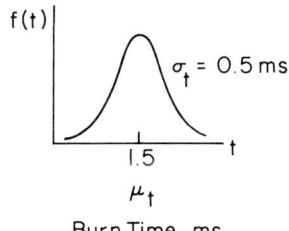

Fig. 7.38. The probability density of actuator burn time.

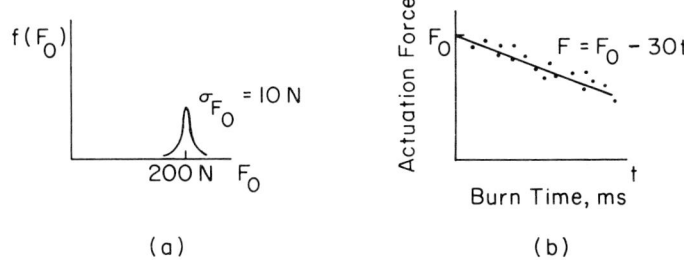

(a)   (b)

Fig. 7.39. The probability density of the largest actuator force, F_0, and the attenuation of the actuation force with increasing burn time.

squib). Such actuators are available in many sizes and are capable of performing work (displacing a force through a distance) of actuation, remotely on electrical command, commonly in adverse environments. The largest force manifests itself when ignition (detonation) is prompt and this force is designated F_0. For the example to follow, $F_0 \sim N(200, 10)$ N, denoting that this largest force is Gaussian (normally) distributed (Fig. 7.38), and exhibits a mean of 200 N and a standard deviation of 10 N. See Fig. 7.39(a). For the usual case of slower burning, the largest force is less than F_0, namely

$$F = F_0 - 30t$$

where the burn time in milliseconds is $t \sim N(1.5, 0.5)$. See Fig. 7.39(b). This indicates that the mean burn time is 1.5 ms with a standard deviation of 0.5 ms and the burn time is normally distributed.

If two squibs are used so that their actuation forces add, what fraction of cables will be severed under circumstances wherein the chisel force requirement (r) to sever is $r \sim N(300, 15)$ N? Figure 7.40(a) depicts the physical scheme envisioned. The system is denoted as the chisel and the equalizer bar. The significant influences of the surroundings are the normal tractions exerted by the squibs and the wire. A sufficiently precise mathematical model of these influences is shown in Fig. 7.40(b) using three parallel force vectors. Newton's second law of motion for a particle (ignoring inertial effects) is $\Sigma F_x = F_1 + F_2 - s = 0$,

$$S = F_1 + F_2$$

Now F_1 and F_2 are independent variates and S is simply their sum. The criterion of success (cables severed) is $S > r$. Both S (due to F_1 and F_2) and r are variates. The inequality $S > r$ can be made an equation by the addition of a "slack variable" m.

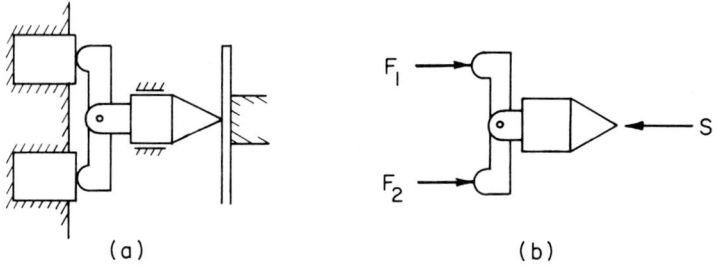

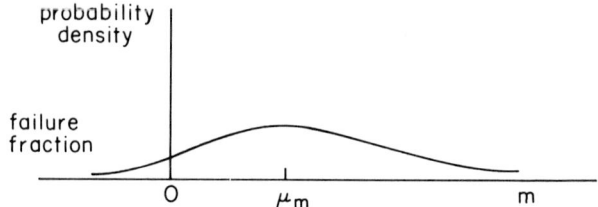

Fig. 7.40. (a) Two actuators, equalizing bar, and cable-cutting chisel. (b) The influence of the surroundings on the bar and chisel as a system. (c) The probability density of the cutting force margin.

$$S > r \quad S = r + m \quad F_1 + F_2 = r + m$$

and therefore

$$m = F_1 + F_2 - r = F_{01} - 30t_1 + F_{02} - 30t_2 - r \quad (7.14)$$

The slack variable m is called the *force margin*, because if it is positive there is sufficient force to sever the cable. If it is negative there is no margin of force and the cable is not severed.

Equation 7.14 is an equation involving random variables. Functionally it is expressible as

$$m = m(F_{01}, t_1, F_{02}, t_2, r)$$

The mean value of m is given by Eq. (7.12), noting that

$$\frac{\partial^2 m}{\partial F_{01}^2} = 0 \quad \frac{\partial^2 m}{\partial t_1^2} = 0 \quad \frac{\partial^2 m}{\partial F_{02}^2} = 0 \quad \frac{\partial^2 m}{\partial t_2^2} = 0 \quad \frac{\partial^2 m}{\partial r^2} = 0$$

Then

$$\mu_m = \mu_{F_{01}} - 30\mu_{t_1} + \mu_{F_{02}} - 30\mu_{t_2} - \mu_r + 0 + \ldots$$

$$= 200 - 30(1.5) + 200 - 30(1.5) - 300 = 10 \text{ N}$$

The standard deviation σ_m is given by Eq. (7.13), noting that

$$\frac{\partial m}{\partial F_{01}} = 1 \quad \frac{\partial m}{\partial t_1} = -30 \quad \frac{\partial m}{\partial F_{02}} = 1 \quad \frac{\partial m}{\partial t_2} = -30 \quad \frac{\partial m}{\partial r} = -1$$

First Principle

$$\frac{\partial^2 m}{\partial F_{01}^2} = 0 \quad \frac{\partial^2 m}{\partial t^2} = 0 \quad \frac{\partial^2 m}{\partial F_{02}^2} = 0 \quad \frac{\partial^2 m}{\partial t_2^2} = 0 \quad \frac{\partial^2 m}{\partial r^2} = 0$$

Then

$$\sigma_m = \left[1^2 \sigma_{F_{01}}^2 + (-30)^2 \sigma_{t_1}^2 + 1^2 \sigma_{F_{02}}^2 + (-30)^2 \sigma_{t_2}^2 + (-1)^2 \sigma_r^2 + 0 + \ldots \right]^{1/2}$$

$$= \left[10^2 + 30^2 (0.5)^2 + 10^2 + 30^2 (0.5)^2 + 15^2 \right]^{1/2} = 29.5 \text{ N}$$

Now the force margin $m \sim N(\mu_m, \sigma_m)$, since sums of Gaussian (normal) variates are distributed normally. Figure 7.40(c) indicates the distribution of m and denotes its mean and the area that is the measure of the failure fraction (probability of failure to sever). The displacement of the origin of the m-distribution from the mean expressed in standard deviations is given by

$$z = (0 - \mu_m)/\sigma_m = (0 - 10)/29.5 = -0.337$$

A table of normalized Gaussian variates would indicate that a 0.6330 probability of severing the cable exists.

Let us add to the foregoing problem that the reliability of a squib activating when electrical potential is applied is 0.97. The fraction of cables cut is then

fraction = (0.6330)(0.97)(0.97) = 0.5956

Note that the deterministic model with two squibs exhibits a force margin of +10 N. The deterministic inference that all cables would be severed is misleading (inconsistent with nature). The probabilistic model with two squibs exhibits a force margin of +10 N also; however, the inference is that cables could be severed in 63 percent of the instances. If probability density distributions on F_0, t, and r are not Gaussian but are known from experiment, the force margin is still +10 N for symmetric or lightly skewed distributions; the inference as to the probability of severing the cable is not 63 percent and is made with appropriate statistics, which may be other than Gaussian. For three equalized squibs acting to cut the cable

$$m = F_1 + F_2 + F_3 - r = F_{01} - 30t_1 + F_{02} - 30t_2 + F_{03} - 30t_3 - r$$

and it follows that

$$\mu_m = \mu_{F_{01}} - 30\mu_{t_1} + \mu_{F_{02}} - 30\mu_{t_2} + \mu_{F_{03}} - 30\mu_{t_3} - \mu_r$$

$$= 200 - 30(1.5) + 200 - 30(1.5) + 200 - 30(1.5) - 300 = 165 \text{ N}$$

$$\sigma_m = \left[1^2 \sigma_{F_{01}}^2 + (-30)^2 \sigma_{t_1}^2 + 1^2 \sigma_{F_{02}}^2 + (-30)^2 \sigma_{t_2}^2 \right.$$
$$\left. + 1^2 \sigma_{F_{03}}^2 + (-30)^2 \sigma_{t_3}^2 + (-1)^2 \sigma_r^2 + 0 + \ldots \right]^{1/2}$$

$$= [10^2 + 30^2 (0.5)^2 + 10^2 + 30^2 (0.5)^2 + 10^2 + 30^2 (0.5)^2 + 15^2 + 0 + \ldots]^{1/2}$$

$$= 35.64 \text{ N}$$

The z-variable corresponding to the displacement of the zero of m from the mean is

$$z = (0 - \mu_m)/\sigma_m = (0 - 165)/34.64 = -4.76$$

which corresponds to a failure to sever of one in a million trials. The reliability is R = 0.999 999.

This model is probabilistic deductive wherein the algebra in the solve step is that of random variables. We are able to assess the reliability easily because the distributions of F, t, and r are Gaussian and the variables appeared in an additive manner. Had $F = F_0 - 20t^2$ with Gaussian distributions of F_0, t, and r, then m would be distorted from Gaussian and z-table information would give an incorrect reliability prediction.

In circumstances wherein the slack variable (say) is of unknown distribution because of the algebraic combinations of the Gaussian F, t, and r, the modeling can be *simulative*.

Let us repeat the two squib problem with a simulative (rather than random variable algebraic) reliability determination. Any large digital computer has a uniform random number generator that can generate a uniform random number in the interval (0, 1). If either or both of two such numbers are greater than 0.97 we have a failure to sever due to misfire(s). Any large digital computer has a Gaussian random number generator. If five Gaussian random numbers are generated of appropriate mean and standard deviation to created instances of F_{01}, t_1, F_{02}, t_2, and r, the slack variable can be calculated directly from Eq. (7.14). If the uniform random numbers are both greater than 0.97 and m is greater than zero, tally one instance of severing the cable; otherwise, tally one failure to sever. The reliability $\hat{R}$ is estimated by computing

$$\hat{R} = (\text{number of severed})/(\text{number of trials})$$

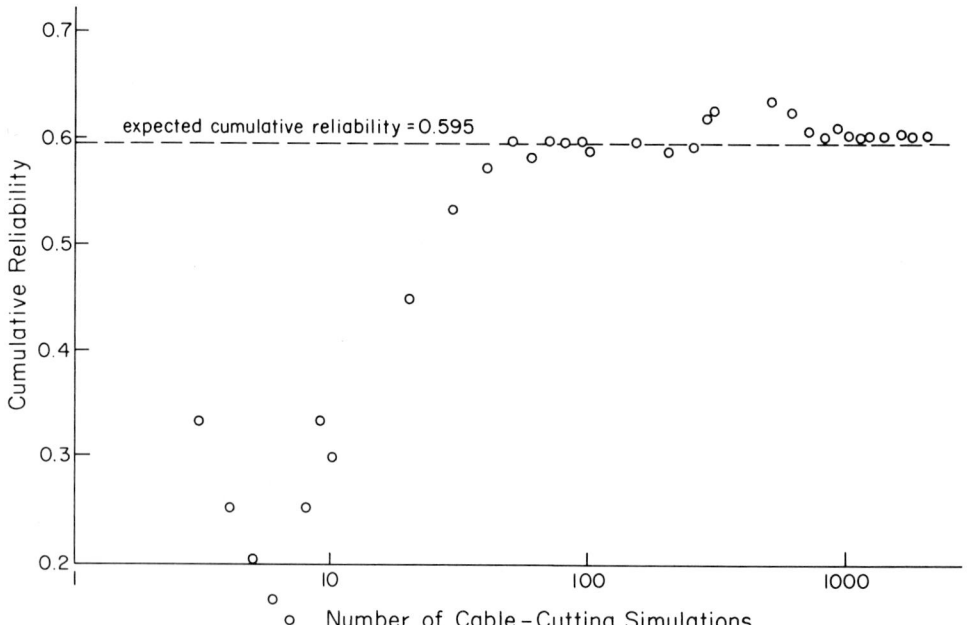

Fig. 7.41. A computer-implemented simulation model of the cable-cutting actuator problem produced this reliability vs. number of simulations plot.

First Principle

Figure 7.41 shows the simulation reliability after various numbers of trials up to 2000 (this required 14 000 random numbers of which 4000 were uniform random and 10 000 were Gaussian; the Gaussians, in turn, required 120 000 uniform random). We expect that as the number of trials approaches infinity the theoretical probability of severing will be approached by this procedure.

The ability to build random numbers of any specified distribution from the basic building block of a uniform random number generator for the interval (0, 1) is important to simulation models. Section 7.12 illuminates the simple yet powerful procedure.

Probabilistic elements enter the simplest of deterministic operations. For example, the life of a pair of Navy socks has a Gaussian distribution exhibiting a mean of 50 days and a standard deviation of 8 days. How long will five pairs last? The mean of all samples of five pairs in concatenation is the sum of the means, namely 250 days, obtained from

$$\mu_l = \mu_{x_1} + \mu_{x_2} + \mu_{x_3} + \mu_{x_4} + \mu_{x_5} = 50 + 50 + 50 + 50 + 50 = 250 \text{ days}$$

The variance in l is related to the variance in x as

$$\sigma_l^2 = \sigma_{x_1}^2 + \sigma_{x_2}^2 + \sigma_{x_3}^2 + \sigma_{x_4}^2 + \sigma_{x_5}^2 = 64 + 64 + 64 + 64 + 64 = 320$$

$$\sigma_l = \sqrt{320} = 17.9 \text{ days}$$

With five pairs of socks, we have a 0.500 chance of observing a sequential life of 250 days. At the 0.95 confidence level,

$$l_{0.95} = \mu_l - z\sigma_l = 250 - 1.645(17.9) = 220.5 \text{ days}$$

using a Gaussian approximation. Something as simple as "the whole equals the sum of its parts" in deterministic context takes on a different meaning in a probabilistic context. In a deterministic model where the life of socks is 50 days, five pairs will last at least 250 days (for certain). In the probabilistic model, five pairs of socks will last at least 250 days with 50 percent assurance or at least 220.5 days with 95 percent assurance.

7.12 GENERATING RANDOM NUMBERS OF SPECIFIED DISTRIBUTIONS

The ability to generate random numbers of uniform distribution on a digital computer (or from a tabulation) opens an entire spectrum of Monte Carlo simulations, permitting the attainment of solutions for which closed form statistical solutions are unavailable. By focusing on the chance element in the microscopic event and tallying statistics on outcomes, probabilities not otherwise assessable can be well estimated.

The mathematics on which this is based is relatively simple and the method so powerful that this early introduction is provided. When y is a function of z, namely $\phi(z)$, the probability density distribution of y is different from that of z and the difference depends uniquely on the nature of $\phi(z)$. Figure 7.42 depicts the pdf (probability density function) of z in the lower right-hand graph. The CDF (cumulative density function) of z is shown above it. The functional relationship between y and z is shown upper right. The CDF of y and the pdf of y are shown to the left. If y is an increasing monotonic function of z, the probability that $y \leq \beta$ is exactly the same as the probability that $z \leq \alpha$. Algebraically this is displayed as

$$\text{Prob}(y \leq \beta) = \text{Prob}(z \leq \alpha)$$

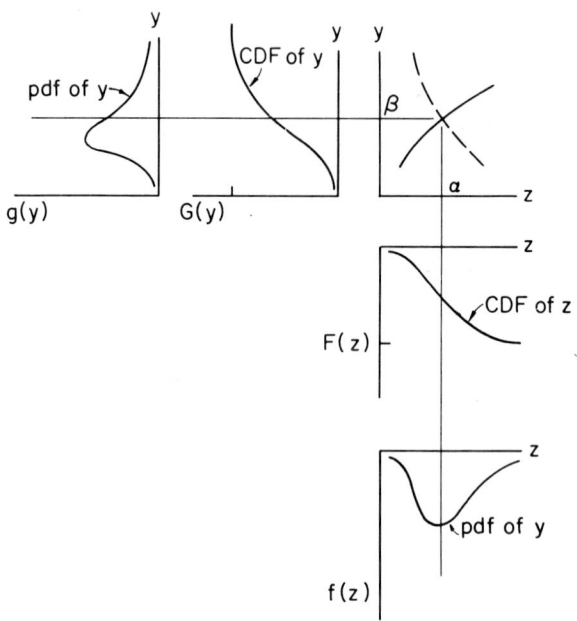

Fig. 7.42. The relation of density functions of y and z when the variables y and z are related functionally.

These probabilities are (by definition) the ordinates to the respective CDF curves. In other words,

$$G(y) = F(z)$$

where $G(y)$ is the CDF of y and F is the CDF of z. Differentiating both sides with respect to y yields

$$\frac{dG(y)}{dy} = \frac{d}{dy} F(z) = \frac{dF(z)}{dz} \frac{dz}{dy}$$

or, solving explicitly for the pdf of y

$$g(y) = f(z) \frac{dz}{dy}$$

This states the simple link between the pdf of y and the pdf of z when $y = \phi(z)$ is a monotonically increasing function. If y is a decreasing monotonic function of z, as depicted in the dashed curve in Fig. 7.42, the probability that $y \leq \beta$ is precisely the same as the probability that $z \leq \alpha$:

$$\text{Prob}(y \leq \beta) = \text{Prob}(z \leq \alpha)$$

The ordinates to the respective CDF curves are involved as $\text{Prob}(y \leq \beta) = G(y)$ and $\text{Prob}(z \leq \alpha) = 1 - F(z)$. Thus

$$G(y) = 1 - F(z)$$

Differentiating with respect to y yields

$$\frac{dG(y)}{dy} = \frac{d}{dy} [1 - F(z)] = - \frac{dF(z)}{dz} \frac{dz}{dy}$$

Solving explicitly for the pdf of y gives

First Principle

$$g(y) = -f(z) \frac{dz}{dy}$$

When y is a monotonically decreasing function of z, the derivative dz/dy is intrinsically negative. The two expressions for g(y) can be combined into one by writing

$$g(y) = f(z) \left|\frac{dz}{dy}\right| \tag{7.15}$$

This simple relationship forms the basis for the generation of random variables of any desired distribution from a uniform random number generator. Let the uniform random distribution in the interval $0 \leq z \leq 1$ be used; then

$$f(z) = 1 \quad (0 \leq z \leq 1)$$
$$= 0 \quad (\text{elsewhere})$$

What follows is a side-by-side development for an increasing and a decreasing monotone, respectively,

$$g(y) = f(z)\frac{dz}{dy} = (1)\frac{dz}{dy} \quad g(y) = -f(z)\frac{dz}{dy} = -\frac{dz}{dy}$$

We can substitute $-dR(y)/dz$ for the density of y where R is the reliability:

$$\frac{-dR(y)}{dy} = \frac{dz}{dy} \quad \frac{dR(y)}{dy} = \frac{dz}{dy}$$

This indicates that

$$-dR(y) = dz \quad dR(y) = dz$$

Integrating between 1 and R(y) and correspondingly between 0 and z yields

$$-\int_1^{R(y)} dR(y) = \int_0^z dz \quad \int_0^{R(y)} dR(y) = \int_0^z dz$$

$$R(y) = 1 - z \quad R(y) = z$$

If we set the explicit equation for R(y) (the survival equation) equal to $1 - z$ in the first instance or z in the second instance, a solution for y in terms of z becomes the required transformation for the generation of random numbers of the y-distribution from the uniform random z. As an example, let a Weibull-distributed random number be required. The Weibull survival equation is

$$R(y) = \exp\left[-\left(\frac{y - y_0}{\theta}\right)^b\right]$$

Using the relation for y as a decreasing monotone of z, we write

$$R(y) = z = \exp\left[-\left(\frac{y - y_0}{\theta}\right)^b\right]$$

Solving explicitly for y yields

$$y = y_0 + \theta[\ln 1/z]^{1/b} \tag{7.16}$$

Equation (7.16) is the transformation to convert a uniform random number z into a Weibull-distributed random number y with a shape factor b, characteristic life θ, and guaranteed life y_0.

The following five numbers are selected from a uniform random number table: $z_1 = 0.2444$, $z_2 = 0.5748$, $z_3 = 0.7761$, $z_4 = 0.6838$, $z_5 = 0.6440$. The required Weibullian random numbers from a distribution with $y_0 = 2$, $\theta = 3$, and $b = 4$ are

$$y_1 = 2 + 3(\ln 1/0.2444)^{1/4} = 5.268$$

$$y_2 = 2 + 3(\ln 1/0.5748)^{1/4} = 4.588$$

$$y_3 = 2 + 3(\ln 1/0.7761)^{1/4} = 4.129$$

$$y_4 = 2 + 3(\ln 1/0.6838)^{1/4} = 4.356$$

$$y_5 = 2 + 3(\ln 1/0.6440)^{1/4} = 4.443$$

In the case of the Gaussian (normal) distribution, an explicit survival equation is unavailable. In fact, numbers of any distribution added together approach a Gaussian distribution more and more closely as the extent of the addition increases (central limit theorem). A dozen uniformly distributed random numbers added together form a sum that is Gaussian distributed to excellent approximation. This is the basis for a large digital computer subroutine GAUSS that calls a uniform random subroutine RANDU a dozen times.

An *approximate* number is one whose value approximates the value of the true number. The number 0.35 is approximate to the number that is sin 20°. It stands alone, without qualification other than the implication that someone finds it useful to some purpose. A *significant* number is one that does not differ from the number it approximates by more than 1/2 in the last recorded digit. Significant numbers can be formed from the true number by the process of truncation known as *rounding*. Significant numbers, despite their widespread use, are not well suited to numerical calculation. Users hope to infer at the end of the calculation that the significant number representing the answer bounds the true answer. Their frustrations in this regard are often labeled "loss of significant figures." Better ways are available to accomplish the intended purpose. The number 0.34 is a significant number representing the number sin 20°. It can be written as 0.34 ± 0.005. An *incomplete* number is an approximation error number in which the error term is omitted. The digits appear to be significant numbers but are not and the result of a calculation may be recorded to any desired number of displayed digits.

The result of a measurement is not a number but a *couple* $(\hat{\mu}_x, \hat{\sigma}_x)$, and the entries $\hat{\mu}_x$ and $\hat{\sigma}_x$ are incomplete numbers representing unbiased estimators of the mean and standard deviation of all possible observations of x. A measurement result is sometimes reported as $\hat{\mu}_x \pm 2\hat{\sigma}_x$ or $\hat{\mu}_x \pm 3\hat{\sigma}_x$. In these incomplete numbers no implication is made concerning accuracy in the digital display. The standard deviation $\hat{\sigma}_x$ qualifies the range in which μ_x is to be found with a stated confidence level. Measurements do not produce numbers but couples $(\hat{\mu}_x, \hat{\sigma}_x)$, and measurement numbers are incomplete numbers that are elements of measurement couples. Measurement couples do not produce range numbers as do significant numbers between which the true value is certain to occur.

People are careless in not declaring the kind of numbers they display; they perform significant number rituals on measurement numbers and do little harm, but they make significant number inferences on elements of measurement couples and run catastrophic risks. The fundamental problem is that approximate and significant numbers are used in arithmetic and incomplete and measurement numbers are used for measurement; these numbers are not always useful outside their disciplines.

First Principle

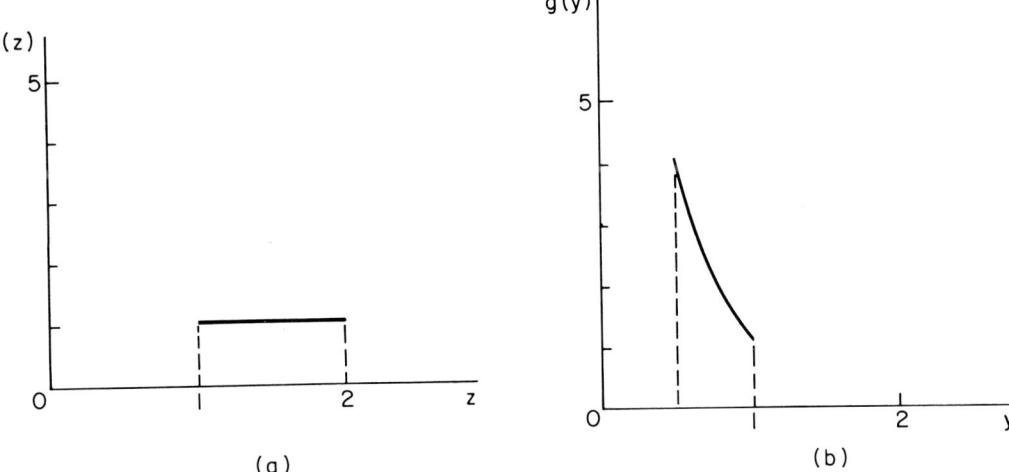

Fig. 7.43. Density function (a) of z, and (b) of y when y = 1/z.

Is the mean of a reciprocal the reciprocal of the mean? Consider a random variable z, Fig. 7.43(a), that is uniformly distributed in the interval $1 \leq z \leq 2$. The reciprocal involves the relationship $y = 1/z$. Does $\mu_y = 1/\mu_z$? The mean of z is, by definition,

$$\mu_z = E(z) = \int_{-\infty}^{\infty} z f(z) \, dz = \int_1^2 z \, dz = 3/2$$

Recall from Eq. (3.5) that $\sigma_z^2 = \mu_{z2} - \mu_z^2$; substituting $E(z^2)$ for μ_{z2}, we have

$$\sigma_z^2 = E(z^2) - \mu_z^2 = \int_{-\infty}^{\infty} z^2 f(z) \, dz - (3/2)^2 = \int_1^2 z^2 \, dz - 9/4 = 1/12$$

or $\sigma_z = 1/(2\sqrt{3})$. To generate the mean and standard deviation of y we need the pdf of y, which we know how to obtain. Noting that y is a decreasing monotone of z, we write

$$g(y) = -f(z) \frac{dz}{dy}$$

Noting that $z = 1/y$ and $dz/dy = -1/y^2$, we write

$$g(y) = (-1)(-1/y^2) = 1/y^2 \quad (1/2 \leq y \leq 1)$$
$$= 3 \quad \text{(elsewhere)}$$

This function is displayed in Fig. 7.43(b). We now have the pdf of y; consequently we may easily find μ_y. First we check to see if the integral of g(y) over the real domain is unity as it must be (check, remember?):

$$\int_{-\infty}^{\infty} g(y) \, dy = \int_{1/2}^{1} dy/y^2 = 1$$

$$G(y) = \int_{-\infty}^{y} g(y) \, dy = \int_{1/2}^{y} dy/y^2 = 2 - 1/y$$

and

$$G(1/2) = 2 - 1/(1/2) = 0 \quad G(1) = 2 - 1/1 = 1$$

as it should be. Now

258   Chapter 7

$$\mu_y = E(y) = \int_{-\infty}^{\infty} y g(y) \, dy = \int_{1/2}^{1} y(1/y^2) \, dy = \ln 2$$

Does $\mu_y = 1/\mu_z$? Clearly the reciprocal of 3/2 is not $\ln 2$. The intuition (savvy) of years of deterministic mathematics is no longer a consistently reliable guide. Until it is replaced by some probabilistic experience, close reliance on definition and deduction is necessary. Even some of your well-learned rules of algebraic manipulation may now be suspect.

To demonstrate how far astray your deterministic experience can lead you, consider the following. There are n people in a classroom. What is the probability that at least two of them have the same birthday, i.e., their birthdays occur on the same day in the same month of the year? If you believe that in a class of 23 the odds are about 51:49 that at least two people have the same birthday your probabilistic sense is good. If you do not feel the problem is an even money bet, your need for a probability and statistics course is well demonstrated.

PROBLEMS

7.1. For the circumstances of Fig. P7.1, show that $(\Sigma p_c - \Sigma p_d)$ vanishes. A particle of mass m at velocity $\dot{x}$ is outside a control surface approaching a stationary particle of mass M within a control surface at time t_1. At time t_2 an inelastic collision has already occurred and the joined particles are moving at speed V. Develop an expression for the velocity V. What is the impulse delivered to the system that is the large particle?

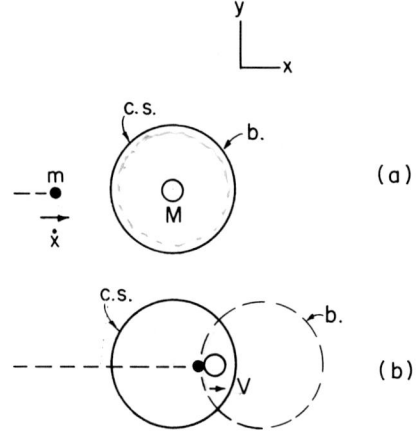

Fig. P7.1

7.2. At time t_1 a particle of mass m_1 within a control surface has a velocity of $\dot{x}_1$ and is approaching a stationary particle of mass M (Fig. P7.2). A particle of mass m_2 with a velocity of $\dot{x}_2$ is approaching the control surface, Fig. 7.12(a). At time t_2 an inelastic collision has already occurred between m_1 and M, which are receding at the velocity of V, (b). Use the accounting interval t_1 to t_2 for the development of an equation predicting V. Use the accounting interval t_1 to t_3 to develop an equation for predicting V.

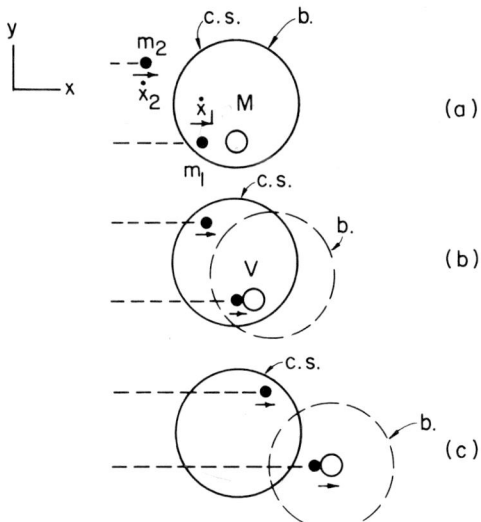

Fig. P7.2

7.3. The conveyor belts of Fig. 7.13 are inclined at angles to the horizontal of θ_1 and θ_2, respectively. What force parallel to belt number 2 is necessary to sustain uniform belt speed? What power is required?

7.4. The belts of Fig. 7.13 are arranged in plan view to be at right angles to each other. If the sand travels east on the first belt it travels south on the second belt. What force parallel to the second belt is necessary to sustain uniform motion? What force transverse to the second belt is necessary to prevent lateral motion? What power is required?

7.5. In the days of steam-powered passenger trains on express runs, it was sometimes arranged that the tender lower a water scoop into a trough (between the rails) that was filled with water. In a mile of track it was possible to refill the tender with water. For a train to maintain its speed while taking on water "on the fly," what additional tractive effort was required of the locomotive?

7.6. A jet of water issues from a nozzle with an exit area of 10 cm^2 and a speed of 10 m/s. The horizontal stream strikes a vertical wall. What force is exerted on the wall by the stream of water?

7.7. A common experimental method for measuring water flow is to exhaust water into a tank on a scale (a weighing tank). If the weights are taken at two instants of time, the difference in weight divided by the time is the average weight rate of flow. A correction is in order because of the falling exhaust stream of water increasing the weight. Investigate the correction and establish a routine procedure for incorporating it in experimental work.

7.8. The jet of Prob. 7.6 is directed against a wall that is moving at a speed u colinear with the jet. What is the force on the wall? If the wall is accelerating, what is the force on the wall?

7.9. A block of wood slides on a horizontal surface. The dynamics of the block are to be investigated. Choose a system and delineate its boundary. Comment on whether your reference frame is inertial or noninertial.

Identify the significant influences of the surroundings.

7.10. A railway train is moving up a curving grade. The dynamics of the train are to be investigated. Choose a system and delineate its boundary. Comment on whether your reference frame is inertial or noninertial. Identify the significant influences of the surroundings.

7.11. A jet aircraft is airborne and its dynamics are to be investigated. Choose a system or control volume and delineate its boundary or control surface. Comment on whether your reference frame is inertial or noninertial. Identify the significant influences of the surroundings.

7.12. An automobile is stopping while traveling in a straight line. Its dynamics are being investigated. Choose a system or a control volume. Comment on whether your reference frames are inertial or noninertial. Identify the significant influences of the surroundings.

7.13. The car of Prob. 7.12 blows its right front tire and the driver brakes hard. Identify the significant influences of the surroundings.

7.14. An internal brake shoe of the automotive type is being analyzed. Choose a system and delineate its boundary. Identify the significant influences of the surroundings.

7.15. The contents of an internal combustion engine cylinder are being analyzed to ascertain the work being done against the piston face. Delineate the system and boundary and indicate the significant influences of the surroundings.

7.16. The dynamics of a simple pendulum are being investigated. Choose a system and delineate its boundary. Comment on whether your reference frame is inertial. Identify the significant influences of the surroundings.

7.17. The dynamics of a person in an elevator are being investigated. Choose a system and delineate its boundary. Comment on whether your reference frame is inertial. Identify the significant influences of the surroundings.

7.18. A straight pipe is being analyzed so that it may be properly supported. Choose a system and delineate its boundary. Identify the significant influences of the surroundings.

7.19. A reducing elbow has an inlet diameter of 25 cm and an outlet diameter of 10 cm. The fluid is turned through an angle of 30°. The inlet pressure is 100 kPa and the outlet pressure is 92 kPa. The flowing fluid is water that enters at 15 m/s. Estimate the reactions R_x and R_y if the weight of the fluid in the control volume is 250 N.

7.20. In the equation

$$p/p_0 = (1 - ax)^{1/(aRT_0)}$$

show that if a approaches zero, p/p_0 approaches $e^{-x/(RT_0)}$.

7.21. In the equation

$$p/p_0 = e^{-x/(RT_0)}$$

show that when x is small, p/p_0 approaches $1 - x/(RT_0)$.

7.22. At 100 000 ft altitude the atmospheric pressure is 0.162 psi. What do our mathematical models predict?

7.23. At 250 000 ft the atmospheric pressure is 0.1139 lbf/ft^2. What do our mathematical models predict?

7.24. Predict the atmospheric pressure at 1000 ft below sea level.

7.25. Check any problems that have been assigned using the methods suggested in the text plus any of your own devising.

First Principle   261

7.26. Exercise the epidemic model for a population of 200. On what day does the epidemic peak (most current cases)?

7.27. A town of 100 susceptibles lies across a railroad track from the adjacent town of 40 susceptibles. Every morning at 8:00 the residents of each town meet on their public squares and each shakes hands with every other person in town including visitors. With each handshake is a chance of transmitting a fatal disease. You are the mayor of the town of 100 and you have just received a public health report stating that someone in the town of 40 died of the fatal disease. You were once an engineer and you understand the model and its implications. You are charged with maintenance of public health and safety. What do you do?

7.28. Build an epidemic model using integer arithmetic and compare results with the continuous integration model of Prob. 7.26.

7.29. What are the mean and standard deviation of $F = kx$ when $\mu_k = 40$ lbf/in., $\sigma_k = 2$ lbf/in., $\mu_x = 3$ in., $\sigma_x = 0.1$ in., and k and x are independent random variables?

7.30. What are the mean and standard deviation of $\sin x$ if $\mu_x = 1.2$ rad and $\sigma_x = 0.1$ rad?

7.31. The random variable $x \sim N(3, 0.1)$ appears in the equation $y = x^{2.5}$. Determine μ_y, σ_y.

7.32. If $s = Mc/I = 6M/(bh^2)$; $\mu_M = 10\,000$ in. lbf; $\sigma_M = 500$ in. lbf; $\mu_b = 1$ in.; $\sigma_b = 0.002$ in.; $\mu_h = 1$ in.; and $\sigma_h = 0.002$ in., determine μ_s, σ_s.

7.33. If $y = e^x$, show that

$$\mu_y = e^{\mu_x}(1 + \sigma_x^2/2) \qquad \sigma_y = e^{\mu_x}\sigma_x(1 + \sigma_x^2/2)^{1/2}$$

7.34. If $y = \ln x$, show that

$$\mu_y = \ln \mu_x - (1/2)(\sigma_x/\mu_x)^2 \qquad \sigma_y = (\sigma_x/\mu_x)[1 + (1/4)(\sigma_x/\mu_x)^2]$$

7.35. If $y = \sin x$, show that

$$\mu_y = \sin \mu_x (1 - \sigma_x^2/2) \qquad \sigma_y = \sigma_x[\cos^2 \mu_x + (\sigma_x^2/2)\sin^2 \mu_x]^{1/2}$$

7.36. If $y = a^x$, show that

$$\mu_y = a^{\mu_x}[1 + (\sigma_x^2/2)\ln a] \qquad \sigma_y = a^{\mu_x}\sigma_x \ln a[1 + (\sigma_x^2/2)\ln^2 a]^{1/2}$$

7.37. Consider a uniform random variable z in the interval $1 \leq z \leq 2$. Show that the mean value of y is

$$\mu_y = [1/(n - 1)](1 - 1/2^{n-1})$$

when $y = 1/z^n$. Show that the pdf of y is

$$g(y) = (1/n)y^{-1/(n-1)} \qquad (1/2)^n < y < 1$$
$$= 0 \qquad \text{(elsewhere)}$$

Particularize your equation for μ_y for $y = 1/z^2$. Compare this result with the approximate form

$$\mu_y = (1/\mu_z^2)[1 + 3(\sigma_z/\mu_z)^2 + \ldots]$$

7.38. Consider a uniform random variable z in the interval $1 \leq z \leq 2$. Show that the mean of the nth power of z is

$$\mu_{z^n} = [-1/(n+1)](1 - 2^{n+1}) = [1/(n+1)](2^{n+1} - 1)$$

For the case where n = 2, show agreement with the expression

$$\mu_{z^2} = \mu_z^2 + \sigma_z^2$$

7.39. The restriction that the creation and destruction of countables takes place only in the system in Eq. (7.1) can be relaxed to allow creation and destruction in the surroundings as well. The equation immediately preceding Eq. (7.1) can be written as

$$\Sigma C_{cd}\bigg|_S = [\Sigma C + \Sigma C]_{\underline{A}\ \underline{B}}\bigg|_{t_2} - [\Sigma C + \Sigma C]_{\underline{A}\ \underline{B}}\bigg|_{t_1} + \Sigma C\bigg|_{\underline{C}\ t_2} - \Sigma C\bigg|_{\underline{A}\ t_2}$$

where $\Sigma C_{cd}\big|_S$ is the net creation of countables in the system during the accounting interval. As before,

$$\Sigma C_f = [\Sigma C + \Sigma C]_{\underline{A}\ \underline{B}}\bigg|_{t_2} \quad \Sigma C_i = [\Sigma C + \Sigma C]_{\underline{A}\ \underline{B}}\bigg|_{t_1}$$

The term representing the census of countables in the domain $\underline{C}$ at the end of the accounting interval can now be replaced with the census of countables crossing the control surface from within to without plus the net census of countables created in the domain $\underline{C}$, or

$$\Sigma C\bigg|_{\underline{C}\ t_2} = \Sigma C_{out} + \Sigma C_{cd}\bigg|_{\underline{C}}$$

Similarly the term representing the census of countables in domain $\underline{A}$ at the end of the accounting interval can now be replaced by the census of countables crossing the control surface from without to within plus the net census of countables created within domain $\underline{A}$, or

$$\Sigma C\bigg|_{\underline{A}\ t_2} = \Sigma C_{in} + \Sigma C_{cd}\bigg|_{\underline{A}}$$

Assemble the terms such that the right side of this new equation is identical with that of Eq. (7.1). Show that the resulting equation can be expressed as

$$[\Sigma C_c - \Sigma C_d]_{c.r.} = [\Sigma C_f - \Sigma C_i]_{c.r.} - [\Sigma C_{in} - \Sigma C_{out}]_{c.s.}$$

where the left side of the above equation is the net creation of countables *in the control region* during the accounting interval. This is Finger-Toe's equation; while less restrictive than Eq. (7.1), it does not link the system and control region viewpoints as explicitly as does Eq. (7.1).

> "O stay," the maiden said, "and rest
> Thy weary head upon this breast!"
> A tear stood in his bright blue eye,
> But still he answered with a sigh,
> "Excelsior!"
>
> **Longfellow**

CHAPTER 8

OPTIMIZATION

8.1 INTRODUCTION

The design imperative is to

> Design (subject to certain problem-solving constraints) a component, system, or process that will perform a specified task (subject to certain solution constraints) optimally.

The last word, *optimally*, means the designer must seek the best (or to good approximation, the best) embodiment of a concept. It does not mean to choose the best of alternative concepts but to seek the best within a concept. If alternative concepts are to be compared, having the best of each allows selection from among optimized alternatives.

Seeking an optimum requires a priori a criterion by which to choose between alternatives within a concept. This criterion is known by various names. Designers speak of merit function and figures of merit. Optimization theorists speak of objective or criterion functions. Whatever the name, this function is essential to the attainment of an optimal solution.

Perhaps an example worked heuristically is a useful vehicle to show that optimization (and its techniques) involves more than meets the casual eye. Consider the following problem. A swimmer at a distance $a = 400$ ft from a beach wishes to proceed to a distance $e = 200$ ft up the beach. At a speed of $v = 1$ ft/s swimming and $V = 2$ ft/s walking, how long does it take the swimmer to move from A to E? What path from A to E (via C) results in the minimum elapsed time? (See Fig. 8.1.) The elapsed time t consists of two intervals: the time to swim from A to C and the time to walk from C to E.

$$t = \frac{(a^2 + c^2)^{1/2}}{v} + \frac{e - c}{V}$$

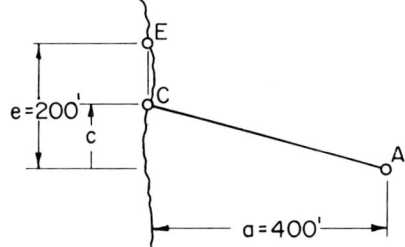

Fig. 8.1. A person swims from A to C and walks from C to E.

If the swimmer chooses to swim directly to the beach and then walk, c = 0 and

$$t = \frac{a}{v} + \frac{e}{V} = \frac{400}{1} + \frac{200}{2} = 500 \text{ s}$$

If the swimmer chooses to swim to a point on the beach where c = 200 (swims directly to E),

$$t = \frac{(a^2 + c^2)^{1/2}}{v} + \frac{e - c}{V} = \frac{(400^2 + 200^2)^{1/2}}{1} + \frac{0}{2} = 447 \text{ s}$$

It seems reasonable to envision a place C on the beach at which the swim/walk combination will result in the least time interval to move from point A to E via C. One is tempted to take the first derivative of the objective function with respect to c, equate to zero, and thereby determine the value of c that results in a minimum time:

$$\frac{dt}{dc} = \frac{c}{v(a^2 + c^2)^{1/2}} - \frac{1}{V} = 0$$

$$c = \frac{av}{(V^2 - v^2)^{1/2}} = \frac{(400)(1)}{(2^2 - 1^2)^{1/2}} = 231 \text{ ft}$$

This procedure tells us the swimmer should move to a point 31 feet above E and the elapsed time will be shorter than any other path. However, since the direct path to E represents a shorter swim and involves no walking, the direct path involves less elapsed time and the path that involves c = 231 ft cannot represent the minimum. Swimming directly to the beach takes 500 s, so the path involving c = 231 cannot represent a maximum either. Something is wrong!

The objective function was constructed by looking at a specific diagram (Fig. 8.1) and we failed to identify an implicit condition. We must require in the objective function that e > c so that walking time cannot be negative. We omitted this restriction and our methodology detected a minimum without significance in the problem. The lesson is that the objective function should be established with great care to assure its fidelity to nature. We can incorporate the restriction in the objective function by writing it as

$$t = \frac{(a^2 + c^2)^{1/2}}{v} + \frac{|e - c|}{V}$$

This may be expressed as two functions as follows. For e > c,

$$t = \frac{(a^2 + c^2)^{1/2}}{v} + \frac{e - c}{V} = t_1 + t_2$$

For c > e,

$$t' = \frac{(a^2 + c^2)^{1/2}}{v} + \frac{c - e}{V} = t_1 + t_2'$$

Table 8.1 can be constructed using a = 400, e = 200, v = 1, and V = 2.

Figure 8.2 indicates the branches and the corrected objective function. Note that the first branch (original objective function) exhibits a least extreme in a location of nonvalidity. Observe that the other branch exhibits a least extreme where c = 0. Neither of these lower extremes is admissible in the problem. The corrected objective function is the heavy line portion of the branches in Fig. 8.2. Note that the least extreme of the corrected objective function involves c = 200 (direct swim). In particular, observe that the

Optimization

Table 8.1. Times Required to Walk or Swim

c	t_1	t_2	t_2'	t	t'
0	400	100	-100	500	300
100	411	50	-50	461	361
200	448	0	0	448	448
300	500	-50	50	450	550
400	560	-100	100	460	660

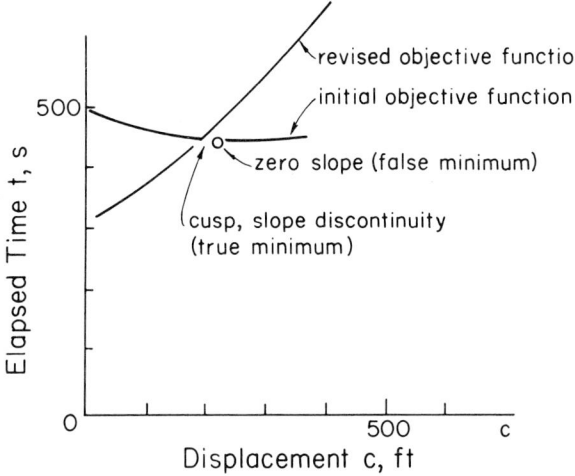

Fig. 8.2. Elapsed time to move from A to E (by swimming to C, then walking to E) as a function of the distance c, showing initial objective function, revised objective function, false minimum (stationary point), and true minimum (a cusp).

least extreme is not a place of zero slope, but a place of slope discontinuity!

Our intuitive method of approaching this problem involved two errors of judgment. We did not incorporate all information in the statement of the criterion function. Also, the intuitive idea that the objective function value dropped from a value of 500 s (c = 0) to a place of horizontal tangency and then rose to a value of 448 s (c = 200) was likewise in error. Because of frustrations and traps (such as encountered in the example) that appear in problems of such great dimensionality that geometric visualization and interpretation is unavailable, the discipline of *optimization* was built. Optimization methodology usually consists of recognizing the *structure* (mathematical nature) of the problem and selecting a method that is efficient for that structure.

8.2 MAXIMA AND MINIMA

Continuous functions in a bounded region exhibit a maximum value and a minimum value of the function. Maximum and minimum values occur either at the boundary of the feasible region or within the feasible region. Weierstrass's theorem gives assurance of their existence. A local extreme is characterized by the property that allowable small perturbations produce no "improvement" in

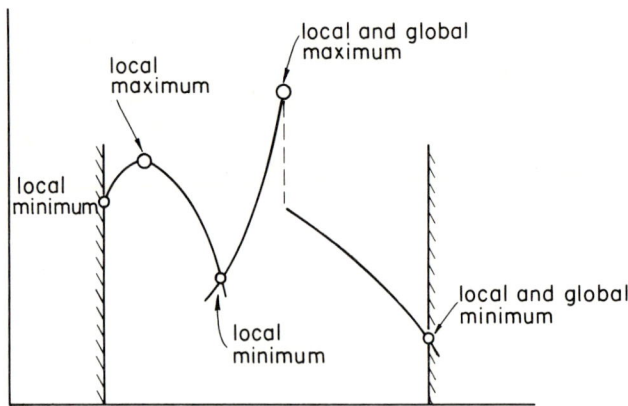

Fig. 8.3. Various locations and functional geometry associated with maximums and minimums of a function of a single independent variable.

the function value, although large perturbations can produce improvement. A global extreme cannot be improved. Interior maxima and minima occur only where the first derivative(s) vanish simultaneously (stationary points) or where derivatives cease to exist at a discontinuity. Maxima and minima can occur at the boundary of the feasible region; the first part of the optimization problem is ascertaining where they can occur. Optima are to be found, as suggested by Fig. 8.3,

--at interior stationary points (continuous first derivatives simultaneously vanish)
--at interior points where discontinuities in the function or the first derivative appear
--at the boundary between feasible and infeasible regions (on a boundary, at intersections of boundaries)

In a function of one independent variable, stationary points exist where the first derivative is zero. If the second derivative is positive we have a minimum, if it is negative we have a maximum, and if it is zero we have a point of inflection. If many successive derivatives are zero with the highest order zero derivative odd, the next subsequent even order derivative determines the character of the stationary point. For example,

$$y = x^2 \qquad \frac{dy}{dx} = 2x \qquad \frac{d^2y}{dx^2} = 2$$

The necessary condition of a stationary point ($dy/dx = 0$) makes $x = 0$ a stationary point, which is a minimum (d^2y/dx^2 is positive). As another example

$$y = x^3 \qquad \frac{dy}{dx} = 3x^2 \qquad \frac{d^2y}{dx^2} = 6x \qquad \frac{d^3y}{dx^3} = 6$$

The necessary condition ($dy/dx = 0$) makes $x = 0$ a stationary point. Since $d^2y/dx^2 = 0$, the stationary point is a point of inflection. Another example is

$$y = x^4 \qquad \frac{dy}{dx} = 4x^3 \qquad \frac{d^2y}{dx^2} = 12x^2 \qquad \frac{d^3y}{dx^3} = 24x \qquad \frac{d^4y}{dx^4} = 24$$

Optimization

The necessary condition (dy/dx = 0) makes x = 0 a stationary point. The fourth derivative is positive (lesser order derivatives being zero); hence the stationary point is a minimum.

To prove that a particular stationary point is a local maximum (minimum), it is necessary to show that in the neighborhood all values of the objective function are inferior (superior) to the value of the objective function at the stationary point. If the Taylor series expansion of the function about the stationary point exists, this procedure can be formalized. (It also becomes involved.)

With two independent variables x_1 and x_2, a stationary point will be a maximum if

$$\frac{\partial^2 y^*}{\partial x_1^2} < 0 \qquad \left(\frac{\partial^2 y^*}{\partial x_1^2}\right)\left(\frac{\partial^2 y^*}{\partial x_2^2}\right) \leq \left(\frac{\partial^2 y^*}{\partial x_1 \partial x_2}\right)^2$$

and a minimum of

$$\frac{\partial^2 y^*}{\partial x_1^2} > 0 \qquad \left(\frac{\partial^2 y^*}{\partial x_1^2}\right)\left(\frac{\partial^2 y^*}{\partial x_2^2}\right) > \left(\frac{\partial^2 y^*}{\partial x_1 \partial x_2}\right)^2$$

where y^* is the objective function evaluated at a stationary point. For example,

$$y = (x_1 - 3)^2 + (x_2 - 4)^2 \qquad \frac{\partial y}{\partial x_1} = 2(x_1 - 3) = 0 \qquad \frac{\partial y}{\partial x_2} = 2(x_2 - 4) = 0$$

Therefore $x_1^* = 3$, $x_2^* = 4$, and consequently

$$\frac{\partial^2 y^*}{\partial x_1^2} = 2 \qquad \frac{\partial^2 y^*}{\partial x_2^2} = 2 \qquad \frac{\partial^2 y^*}{\partial x_1 \partial x_2} = 0$$

$$\left(\frac{\partial^2 y^*}{\partial x_1^2}\right)\left(\frac{\partial^2 y^*}{\partial x_2^2}\right) = (2)(2) = 4 \qquad \left(\frac{\partial^2 y^*}{\partial x_1 \partial x_2}\right)^2 = 0$$

$$\left(\frac{\partial^2 y^*}{\partial x_1^2}\right)\left(\frac{\partial^2 y^*}{\partial x_2^2}\right) > \left(\frac{\partial^2 y^*}{\partial x_1 \partial x_2}\right)^2$$

and the stationary point is a minimum. For n independent variables the determinants D_i are defined:

$$D_i \equiv \begin{vmatrix} \frac{\partial^2 y^*}{\partial x_1 \partial x_1} & \frac{\partial^2 y^*}{\partial x_1 \partial x_2} & \cdots & \frac{\partial^2 y^*}{\partial x_1 \partial x_i} \\ \frac{\partial^2 y^*}{\partial x_2 \partial x_1} & \frac{\partial^2 y^*}{\partial x_2 \partial x_2} & \cdots & \frac{\partial^2 y^*}{\partial x_2 \partial x_i} \\ \vdots & & & \\ \frac{\partial^2 y^*}{\partial x_i \partial x_1} & \frac{\partial^2 y^*}{\partial x_i \partial x_2} & \cdots & \frac{\partial^2 y^*}{\partial x_i \partial x_i} \end{vmatrix}$$

Edelbaum's sufficient condition for a stationary point to be a local minimum is that all D_i be positive:

$$D_i > 0 \quad (i = 1, 2, \ldots, n)$$

Edelbaum's sufficient condition for a stationary point to be a local maximum is that all even order D_i be positive and all odd order D_i be negative:

$$D_i < 0 \quad (i = 1, 3, 5, \ldots)$$

$$D_i > 0 \quad (i = 2, 4, 6, \ldots)$$

As an example, consider

$$y = (x_1 - 1)^2 + (x_2 - 2)^2 + (x_3 - 3)^2$$

$$\frac{\partial y}{\partial x_1} = 2(x_1 - 1) = 0 \quad \frac{\partial y}{\partial x_2} = 2(x_2 - 2) = 0 \quad \frac{\partial y}{\partial x_3} = 2(x_3 - 3) = 0$$

from which the stationary point ($x_1^* = 1$, $x_2^* = 2$, $x_3^* = 3$) is determined. Now

$$\frac{\partial^2 y^*}{\partial x_1 \partial x_1} = 2$$

$$\frac{\partial^2 y^*}{\partial x_1 \partial x_2} = 0$$

$$\frac{\partial^2 y^*}{\partial x_2 \partial x_2} = 2 \qquad \frac{\partial^2 y^*}{\partial x_2 \partial x_3} = 0$$

$$\frac{\partial^2 y^*}{\partial x_1 \partial x_3} = 0$$

$$\frac{\partial^2 y^*}{\partial x_3 \partial x_3} = 2$$

$$D_1 = \left| \frac{\partial^2 y^*}{\partial x_1 \partial x_1} \right| = 2 \qquad D_2 = \begin{vmatrix} \dfrac{\partial^2 y^*}{\partial x_1 \partial x_1} & \dfrac{\partial^2 y^*}{\partial x_1 \partial x_2} \\ \dfrac{\partial^2 y^*}{\partial x_2 \partial x_1} & \dfrac{\partial^2 y^*}{\partial x_2 \partial x_2} \end{vmatrix} = \begin{vmatrix} 2 & 0 \\ 0 & 2 \end{vmatrix} = 4$$

$$D_3 = \begin{vmatrix} \dfrac{\partial^2 y^*}{\partial x_1 \partial x_1} & \dfrac{\partial^2 y^*}{\partial x_1 \partial x_2} & \dfrac{\partial^2 y^*}{\partial x_1 \partial x_3} \\ \dfrac{\partial^2 y^*}{\partial x_2 \partial x_1} & \dfrac{\partial^2 y^*}{\partial x_2 \partial x_2} & \dfrac{\partial^2 y^*}{\partial x_2 \partial x_3} \\ \dfrac{\partial^2 y^*}{\partial x_3 \partial x_1} & \dfrac{\partial^2 y^*}{\partial x_3 \partial x_2} & \dfrac{\partial^2 y^*}{\partial x_3 \partial x_3} \end{vmatrix} = \begin{vmatrix} 2 & 0 & 0 \\ 0 & 2 & 0 \\ 0 & 0 & 2 \end{vmatrix} = 8$$

Since all D_i are positive, the stationary point is a minimum. Note that the determinate D_3 contains D_1 and D_2 as minors.

8.3 METHOD OF LAGRANGE

The method of Comte Joseph Louis Lagrange is to incorporate into the objective function equality constraints in such a manner that the revised objec-

Optimization

tive function contains, as stationary points, the interior stationary points of the original problem and the extremes along the boundary between feasible and infeasible regions. The method *does not*

1. maximize or minimize but establishes a set of points that contains the maximum and minimum of the original problem as a subset
2. necessarily give the stationary points of the Lagrangian the same character (maximum, minimum, saddle) as the corresponding points in the original problem

If the problem is stated as

$$\max y(x_1, x_2) \quad \text{or} \quad \min y(x_1, x_2)$$

subject to $g_1(x_1, x_2) = b_1$

$g_2(x_1, x_2) = b_2$

then the Lagrangian is formed by appending a penalty function to the objective function in the following manner:

$$L = y(x_1, x_2) - \lambda_1[g_1(x_1, x_2) - b_1] - \lambda_2[g_2(x_1, x_2) - b_2] \tag{8.1}$$

where the constants λ_1 and λ_2 are called Lagrangian multipliers. The stationary points of L constitute a set that contains the maximum and the minimum of y. If the constraints include inequalities, they are rendered equalities by the introduction of *slack variables* u_1^2, u_2^2,.... For instance, if the problem were stated

$$\max y(x_1, x_2) \quad \text{or} \quad \min y(x_1, x_2)$$

subject to $g_1(x_1, x_2) \leq b_1$

$g_2(x_1, x_2) \leq b_2$

then the constraints are made into equalities, using slack variables u_1^2 and u_2^2

$$g_1(x_1, x_2) + u_1^2 = b_1$$

$$g_2(x_1, x_2) + u_2^2 = b_2$$

and the Lagrangian is written as

$$L = y(x_1, x_2) - \lambda_1[g_1(x_1, x_2) + u_1^2 - b_1] - \lambda_2[g_2(x_1, x_2) + u_2^2 - b_2] \tag{8.2}$$

The stationary points of the Lagrangian of Eq. (8.1) are established by requiring that

$$\frac{\partial L}{\partial x_1} = 0 \quad \frac{\partial L}{\partial x_2} = 0 \quad \frac{\partial L}{\partial \lambda_1} = 0 \quad \frac{\partial L}{\partial \lambda_2} = 0$$

and the stationary points of the Lagrangian of Eq. (8.2) are established by

$$\frac{\partial L}{\partial x_1} = 0 \quad \frac{\partial L}{\partial x_2} = 0 \quad \frac{\partial L}{\partial \lambda_1} = 0 \quad \frac{\partial L}{\partial \lambda_2} = 0 \quad \frac{\partial L}{\partial u_1} = 0 \quad \frac{\partial L}{\partial u_2} = 0$$

As as example, consider

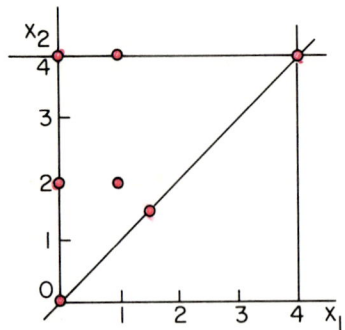

Fig. 8.4. The seven stationary points of the Lagrangian identifying candidates for the extremes of the objective function.

$$\max y = (x_1 - 1)^2 + (x_2 - 2)^2$$

$$\text{subject to } x_2 \leq 4 \quad x_1 \geq 0 \quad x_2 \geq x_1$$

Figure 8.4 depicts the geometry of the problem. In standard form the inequality constraints are expressed as

$$-x_1 \leq 0$$
$$x_2 \leq 4$$
$$x_1 - x_2 \leq 0$$

i.e., equal to or less than a constant or zero. The Lagrangian is

$$L = (x_1 - 1)^2 + (x_2 - 2)^2 - \lambda_1(-x_1 + u_1^2) - \lambda_2(x_2 + u_2^2 - 4) - \lambda_3(x_1 - x_2 + u_3^2)$$

The partial derivatives with respect to x, λ, and u are equated to zero:

$$\frac{\partial L}{\partial x_1} = 2(x_1 - 1) + \lambda_1 - \lambda_3 = 0 \qquad \frac{\partial L}{\partial x_2} = 2(x_2 - 2) - \lambda_2 + \lambda_3 = 0$$

$$\frac{\partial L}{\partial \lambda_1} = -x_1 + u_1^2 = 0 \qquad \frac{\partial L}{\partial \lambda_2} = x_2 + u_2^2 - 4 = 0 \qquad \frac{\partial L}{\partial \lambda_3} = x_1 - x_2 + u_3^2 = 0$$

$$\frac{\partial L}{\partial u_1} = -2u_1\lambda_1 = 0 \qquad \frac{\partial L}{\partial u_2} = -2u_2\lambda_2 = 0 \qquad \frac{\partial L}{\partial u_3} = -2u_3\lambda_3 = 0$$

The partial derivatives with respect to λ_1, λ_2, and λ_3 return the inequality constraints. The partial derivatives with respect to the slack variables u_1, u_2, and u_3 impose conditions called *complementary slackness*. When a λ vanishes, the corresponding point (if optimal) is not affected by that corresponding constraint and the constraint is said to be *slack*. When a slack variable vanishes, the corresponding point is on the corresponding constraint and the constraint is said to be *tight*. The condition of the corresponding λ and slack variable vanishing simultaneously is contradictory, since a constraint may not be tight and slack simultaneously. Thus the condition

$$-u_i \lambda_i = 0$$

imposes the condition $u_i = 0$, $\lambda_i \neq 0$, or $u_i \neq 0$, $\lambda_i = 0$, but not the condition $u_i = 0$, $\lambda_i = 0$. The above partial derivative equations contain three Lagrangian multipliers and three slack variables; consequently eight combinations of zeros are present among the Lagrangian multipliers and slack variables. They are displayed below

λ_1	λ_2	λ_3	u_1^2	u_2^2	u_3^2
0	0	0			
	0	0	0		
0		0		0	
0	0				0
0			0	0	
	0		0		0
		0		0	0
			0	0	0

The number of combinations of n things taken r at a time is $n!/[(n - r)!r!]$, and $0! = 1$. Therefore,

number of combinations of zero λs taken 3 at a time is $3!/[0!3!] = 1$
number of combinations of zero λs taken 2 at a time is $3!/[1!2!] = 3$
number of combinations of zero λs taken 1 at a time is $3!/[2!1!] = 3$
number of combinations of zero λs taken 0 at a time is $3!/[3!0!] = 1$

This results in $1 + 3 + 3 + 1 = 8$ combinations to examine in the search for the stationary points of the Lagrangian. The display above identifies explicitly the eight combinations admissible under the complementary slackness conditions.

The eight partial derivative equations may or may not be a consistent set of equations, given the admissible sets of three zeros and three nonzeros above. A more complete table is displayed below.

x_1	x_2	λ_1	λ_2	λ_3	u_1^2	u_2^2	u_3^2	Lagrangian stationary point	y
1	2	0	0	0	2	2	1	yes	0
0	2	2	0	0	0	2	2	yes	1
1	4	0	4	0	1	0	3	yes	4
3/2	3/2	0	0	1	3/2	5/2	0	yes	1/2
4	4	0	10	6	4	0	0	yes	13
0	0	6	0	4	0	4	0	yes	5
0	4	2	4	0	0	0	4	yes	5
					0	0	0	no	

The set of seven stationary points of the Lagrangian consists of (1, 2), (0, 2), (1, 4), (3/2, 3/2), (4, 4), (0, 0), and (0, 4), circled in Fig. 8.4. The corresponding values of the objective function (*not* the Lagrangian) are 0, 1, 4, 1/2, 13, 5, and 5. Since we are involved in a maximization problem, y*(4, 4) = 13 and y* denotes the optimum. Since u_2^2 and u_3^2 are zero at (4, 4), the second and third constraints are tight. Had we been seeking a minimum, the objective function would have the value of y*(1, 2) = 0. At (1, 2), note that λ_1, λ_2, and λ_3 are zero, all constraints are slack, and the minimum exists at an interior point in the feasible domain.

With either equality or inequality constraints, the partial derivative of the Lagrangian with respect to a lambda is equal to the constraint constant b_i. This fact allows a useful interpretation to be accorded the Lagrangian multipliers. If the constraint constant b_i could be changed, a measure of its influence on the optimal value of the objective function in the neighborhood of the constraint is available, and its measure is the Lagrangian multiplier. In the preceding example, suppose the second constraint, $x_2 \leq 4$, could be changed to $x_2 \leq 4.1$. How much influence would this have on the value of $y^* = 13$? Now

$$y^* = f(b_1, b_2, b_3)$$

$$\Delta y^* = \frac{\partial y^*}{\partial b_1} \Delta b_1 + \frac{\partial y^*}{\partial b_2} \Delta b_2 + \frac{\partial y^*}{\partial b_3} \Delta b_3$$

and at point (4, 4)

$$\frac{\partial y^*}{\partial b_1} = \lambda_1 = 0 \qquad \frac{\partial y^*}{\partial b_2} = \lambda_2 = 10 \qquad \frac{\partial y^*}{\partial b_3} = \lambda_3 = 6$$

$$\Delta b_1 = 0 \qquad \Delta b_2 = 0.1 \qquad \Delta b_3 = 0$$

It follows that

$$\Delta y^* \cong \lambda_1 \Delta b_1 + \lambda_2 \Delta b_2 + \lambda_3 \Delta b_3 = (0)(0) + (10)(0.1) + (6)(0) = 1$$

and the predicted value of $y^* = 13 + 1 = 14$. The exact value of y^* would be attainable from the objective function as

$$y^* = (x_1 - 1)^2 + (x_2 - 2)^2 = (4.1 - 1)^2 + (4.1 - 2)^2 = 14.02$$

Suppose in the third constraint x_2 had to exceed x_1 by 0.1 or more. What would be the change in y^*? If $x_2 > x_1 + 0.1$, then $x_1 - x_2 < -0.1$ and

$$\frac{\partial y^*}{\partial b_1} = \lambda_1 = 0 \qquad \frac{\partial y^*}{\partial b_2} = \lambda_2 = 10 \qquad \frac{\partial y^*}{\partial b_3} = \lambda_3 = 6$$

$$\Delta b_1 = 0 \qquad \Delta b_2 = 0 \qquad \Delta b_3 = -0.1$$

Consequently

$$\Delta y^* = (0)(0) + (10)(0) + (6)(-0.1) = -0.6$$

and the predicted value of $y^* = 13 - 0.6 = 12.4$. The exact value of y^* would be [stationary point of Lagrangian at (3.9, 4)]

$$y^* = (x_1 - 1)^2 + (x_2 - 2)^2 = (3.9 - 1)^2 + (4 - 2)^2 = 12.41$$

Had we been minimizing the function, the optimum $y^*(1, 2) = 0$ would prevail. At this point λ_1, λ_2, and λ_3 are zero; consequently $\partial y^*/\partial b_1 = 0$, $\partial y^*/\partial b_2 = 0$, $\partial y^*/\partial b_3 = 0$, and $\Delta y^* = 0$. A small change in *any* constraint has no effect on the optimal value of y. The examination of the influence of the binding constraint(s) on the optimal value of the objective function is called *sensitivity analysis*. Because of the role of the Lagrangian multipliers in the expression

$$\Delta y^* = \lambda_1 \Delta b_1 + \lambda_2 \Delta b_2 + \lambda_3 \Delta b_3 \tag{8.3}$$

the Lagrangian multipliers are sometimes called *sensitivity coefficients*.

Optimization

Use the method of Lagrange to identify the critical points in the swimmer problem (Fig. 8.1). Stated as an optimization problem, it is

$$\min (a^2 + c^2)^{1/2}/v + |(e - c)/V|$$

subject to $c \geq 0$ and $c \leq 300$ ft, say. We really have two contiguous minimization problems.

$$\min \frac{(a^2 + c^2)^{1/2}}{v} + \frac{e - c}{V} \qquad \min \frac{(a^2 + c^2)^{1/2}}{v} + \frac{c - e}{V}$$

subject to $-c < 0$ subject to $-c < -200$
$c < 200$ $c < 300$

For the left problem the Lagrangian is

$$L = \frac{(a^2 + c^2)^{1/2}}{v} + \frac{e - c}{V} - \lambda_1(-c + u_1^2) - \lambda_2(c + u_2^2 - 200)$$

and for the right problem the Lagrangian is

$$L = \frac{(a^2 + c^2)^{1/2}}{v} + \frac{c - e}{V} - \lambda_3(-c + u_3^2 + 200) - \lambda_4(c + u_4^2 - 300)$$

The derivatives are made to vanish simultaneously to discover the stationary points of the Lagrangians, respectively,

$$\frac{\partial L}{\partial c} = \frac{c}{v(a^2 + c^2)^{1/2}} - \frac{1}{V} + \lambda_1 - \lambda_2 = 0 \qquad \frac{\partial L}{\partial c} = \frac{c}{v(a^2 + c^2)^{1/2}} + \frac{1}{V} + \lambda_3 - \lambda_4 = 0$$

$$\frac{\partial L}{\partial \lambda_1} = -c + u_1^2 = 0 \qquad\qquad\qquad \frac{\partial L}{\partial \lambda_3} = -c + u_3^2 + 200 = 0$$

$$\frac{\partial L}{\partial \lambda_2} = c + u_2^2 - 200 = 0 \qquad\qquad \frac{\partial L}{\partial \lambda_4} = c + u_4^2 - 300 = 0$$

$$\frac{\partial L}{\partial u_1} = -2\lambda_1 u_1 = 0 \qquad\qquad\qquad \frac{\partial L}{\partial u_3} = -2\lambda_3 u_3 = 0$$

$$\frac{\partial L}{\partial u_2} = -2\lambda_2 u_2 = 0 \qquad\qquad\qquad \frac{\partial L}{\partial u_4} = -2\lambda_4 u_4 = 0$$

The two complementary slackness equations within each set lead to four $u\lambda$ combinations. The tables are begun with appropriate zeros entered for u and λ.

c	λ_1	λ_2	u_1^2	u_2^2	Lsp	t		c	λ_3	λ_4	u_3^2	u_4^2	Lsp	t
	0	0			no				0	0			no	
200	0	0.052	200	0	yes	447		300	0	7/8	100	0	yes	550
0	-1/2	0	0	200	yes	500		200	-7/8	0	0	100	yes	447
			0	0	no						0	0	no	

(Lsp is Lagrangian stationary point.) The objective function in both domains is minimum at $c = 200$ with $t^* = 447$ s. Figure 8.5 depicts the two problems. This minimum exists with the real constraints slack ($\lambda_1 = 0$, $\lambda_4 = 0$). The constraints introduced by the discontinuity in slope of the objective function are tight. The introduction of a constraint at a discontinuity of objective

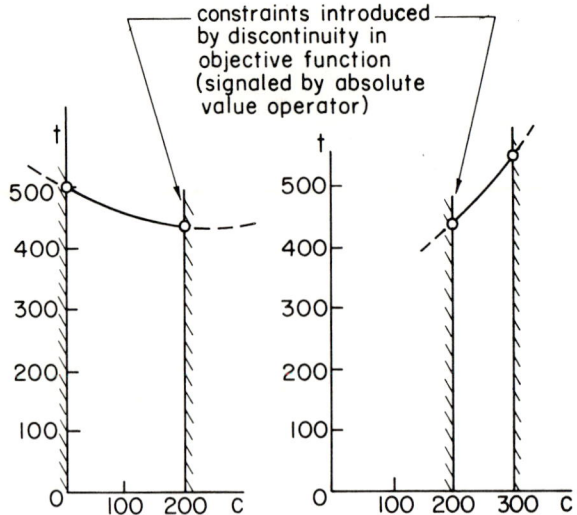

Fig. 8.5. The swimmer problem solved by the method of Lagrange by partitioning into two problems.

function slope and partition into two contiguous optimization problems is a way of forcing Lagrange's method to examine locations of discontinuity of ordinate and/or slope for possible location of an optimum. Its success depends on one's ability to recognize such discontinuities a priori. . In this case the appearance of the absolute value operator in the objective function was symptomatic of the condition of discontinuity in slope.

8.4 STATEMENT OF OPTIMIZATION PROBLEM

The examples of the preceding sections of this chapter emphasize that:

1. The objective function must have a counterpart in reality.
2. All constraints must be explicitly expressed.
3. Procedures must guard against infeasible extremes.
4. When problem dimensionality is too great for geometric interpretation, the method(s) must reduce to algebraic form for implementation.

To meet conditions 3 and 4 the discipline of optimization has developed. In the usual mathematical fashion, problems are classified in terms of structure and methodology. Methods fall into two broad categories:

(a) Indirect methods, which exploit the information contained in the equations representing the objective function and the constraints. Often the problem as posed is converted to another whose solution algorithm is known.
(b) Direct methods, which deal directly with the objective function as a hypersurface. Solution algorithms are called mathematical programs; direct solution methods are called mathematical programming methods.

The function to be optimized is called *objective function*, *criterion function*, *merit function*, *cost function*, *payoff function*, or other names suggested by what is significant in a problem. In optimization literature devoted to problem structure sans application, the names objective function and criterion function are most common.

Optimization

In optimization literature, the problem statement is often of the form

max $y = f(x_1, x_2, \ldots, x_n)$

subject to $g_1(x_1, x_2, \ldots, x_n) \leq b_1$

·

$g_m(x_1, x_2, \ldots, x_n) \leq b_m$ (m inequality constraints)

$h_{m+1}(x_1, x_2, \ldots, x_n) = c_{m+1}$

·

$h_{m+k}(x_1, x_2, \ldots, x_n) = c_{m+k}$ (k equality constraints)

when maximizing an objective function of n design variables subject to m inequality constraints and k equality constraints. The minimization problem is posed with the operator *min* substituted for the operator *max*. For example

max $y = 3/x_1 + 10x_1 x_2$

subject to $x_1 + 3x_2 \leq 6$

is a maximization problem with two design variables x_1 and x_2 and a single linear inequality constraint. The problem

min $y = 4x_1 x_2 x_3 + 100/(x_1 x_2)$

is an unconstrained minimization problem in three design variables. The problem

max $y = x_1 x_2^2 - 6x_2$

subject to $x_1 x_2 = 4$

is a maximization problem with a (nonlinear) equality constraint. Depending on the problem, optimization theory presents some or sometimes no alternative indirect methods for determining the optimal value of the objective function y^* and the corresponding *decision* $(x_1^*, x_2^*, \ldots, x_n^*)$. Some indirect programming methods use Lagrangian multipliers, linear programming, geometric programming, and dynamic programming.

8.5 DIRECT SEARCH METHODS

Consider the problem

max $y = 2x_1 + 3x_2$

subject to $x_1 \leq 4$ $x_2 \leq 3$ $x_1 + x_2 \leq 6$ $-x_1 \leq 0$ $-x_2 \leq 0$

A plot of the feasible domain and contours of y is depicted in Fig. 8.6. It is easy to visualize that the maximum will occur at a constraint on the vertex between two constraints. Let us use the method of Lagrange. Omit the positivity conditions ($-x_1 \leq 0$, $-x_2 \leq 0$). The Lagrangian is

$L = 2x_1 + 3x_2 - \lambda_1(x_1 + u_1^2 - 4) - \lambda_2(x_2 + u_2^2 - 3) - \lambda_3(x_1 + x_2 + u_3^2 - 6)$

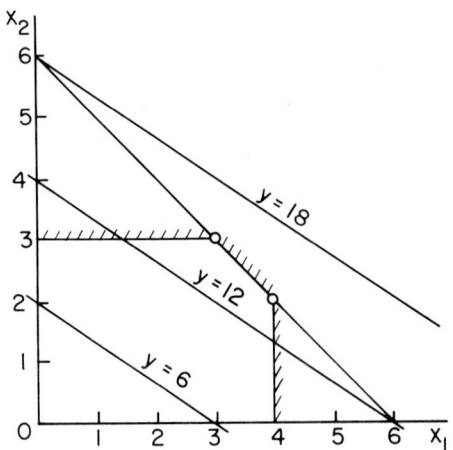

Fig. 8.6. Contours of equal y's on an $x_1 x_2$-plane and delineation of the feasible domain for the first example of Sec. 8.5.

The stationary points of the Lagrangian are established from

$$\frac{\partial L}{\partial x_1} = 2 - \lambda_1 - \lambda_3 = 0 \qquad \frac{\partial L}{\partial x_2} = 3 - \lambda_2 - \lambda_3 = 0$$

$$\frac{\partial L}{\partial \lambda_1} = x_1 + u_1^2 - 4 = 0 \qquad \frac{\partial L}{\partial \lambda_2} = x_2 + u_2^2 - 3 = 0 \qquad \frac{\partial L}{\partial \lambda_3} = x_1 + x_2 + u_3^2 - 6 = 0$$

$$\frac{\partial L}{\partial u_1} = -2u_1 \lambda_1 = 0 \qquad \frac{\partial L}{\partial u_2} = -2u_2 \lambda_2 = 0 \qquad \frac{\partial L}{\partial u_3} = -2u_3 \lambda_3 = 0$$

The three complementary slackness equations lead to eight $u\lambda$ combinations. The table begins with the appropriate zeros entered for u and λ.

x_1	x_2	λ_1	λ_2	λ_3	u_1^2	u_2^2	u_3^2	Lsp	y
3	3	0	1	2	1	0	0	yes	15
4	2	-1	0	3	0	1	0	yes	14
				0	0	0	0	no	
		0	0				0	no	
			0	0		0		no	
		0		0		0		no	
		0	0	0				no	
					0	0	0	no	

The set of stationary points of the Lagrangian contains two points that are circled in Fig. 8.6. Had the positivity conditions been included, the Lagrangian would have been written as

$$L = 2x_1 + 3x_2 - \lambda_1(x_1 + u_1^2 - 4)$$

$$- \lambda_2(x_2 + u_2^2 - 3)$$

$$- \lambda_3(x_1 + x_2 + u_3^2 - 6)$$

$$- \lambda_4(-x_1 + u_4^2)$$

$$- \lambda_5(-x_2 + u_5^2)$$

and the stationary points of the Lagrangian determined from

$$\frac{\partial L}{\partial x_1} = 2 - \lambda_1 - \lambda_3 + \lambda_4 = 0 \qquad \frac{\partial L}{\partial x_2} = 3 - \lambda_2 - \lambda_3 + \lambda_5 = 0$$

$$\frac{\partial L}{\partial \lambda_1} = x_1 + u_1^2 - 4 = 0 \qquad \frac{\partial L}{\partial u_1} = -2u_1\lambda_1 = 0$$

$$\frac{\partial L}{\partial \lambda_2} = x_2 + u_2^2 - 3 = 0 \qquad \frac{\partial L}{\partial u_2} = -2u_2\lambda_2 = 0$$

$$\frac{\partial L}{\partial \lambda_3} = x_1 + x_2 + u_3^2 - 6 = 0 \qquad \frac{\partial L}{\partial u_3} = -2u_3\lambda_3 = 0$$

$$\frac{\partial L}{\partial \lambda_4} = -x_1 + u_4^2 = 0 \qquad \frac{\partial L}{\partial u_4} = -2u_4\lambda_4 = 0$$

$$\frac{\partial L}{\partial \lambda_5} = -x_2 + u_5^2 = 0 \qquad \frac{\partial L}{\partial u_5} = -2u_5\lambda_5 = 0$$

The stationary points of the Lagrangian are found from all combinations of λ taken 5 at a time, 4 at a time, 3 at a time, 2 at a time, 1 at a time, and none at a time. The number of combinations of n things taken r at a time is $n!/[(n-r)!r!]$; consequently

  number of zero λ combinations taken 5 at a time = 5!/[0!5!] = 1
  number of zero λ combinations taken 4 at a time = 5!/[1!4!] = 5
  number of zero λ combinations taken 3 at a time = 5!/[2!3!] = 10
  number of zero λ combinations taken 2 at a time = 5!/[3!2!] = 10
  number of zero λ combinations taken 1 at a time = 5!/[4!1!] = 5
  number of zero λ combinations taken 0 at a time = 5!/[5!0!] = 1

A total of 1 + 5 + 10 + 10 + 5 + 1 = 32 combinations must be identified without omission and investigated to find the stationary points of the Lagrangian.

The basic problem increases in complexity as the number of constraints increases. As a problem becomes larger in the number of design variables and the number of constraints, it becomes algebraically unwieldy. With 32 combinations we have three more stationary points in the set: (0, 0), (0, 3), and (4, 0).

George Dantzig simplified the solution of this type of problem by starting at a feasible vertex and stepping to the next feasible vertex until the maximum or minimum is located. Linear programming was developed by Dantzig in 1947 for solving large problems in which the objective function and the constraints are linear. The subject matter associated with his method is called *linear programming*. Linear programming is studied and used to avoid the complexities introduced by the Lagrangian method as the dimensionality (n + m + k) grows in problems involving linear objective functions and constraints. It is programmable on the digital computer and computation centers have programs available for use in their libraries.

Algorithmic approaches to optimization that rely on evaluations of the merit function without attempting explicit evaluations of derivatives are numerous in mathematical programming. As examples, the elements of the golden section search and the steepest ascent algorithms of mathematical programming are briefly presented in Secs. 8.6, 8.7, and 8.8. We address ourselves to maxmization problems [note that max y = min(-y)].

A *unimodal* function of one independent variable has a single extreme (maximum) in a given interval, i.e., the function is either monotonically increasing throughout the interval, monotonically decreasing throughout the interval, or monotonically increasing and then monotonically decreasing for the remainder of the interval. Unimodal functions are important, since functions may be divided into regions of contiguous unimodal functions. Our stratagem is to identify a region of a unimodal function and reduce the interval by an amount specified a priori, usually a fraction of the original interval. The original interval contains the optimum; the final interval will contain the optimum. In the original interval we are uncertain as to where the optimum is located. In the final interval we are still uncertain where the optimum is, but the interval is sufficiently small that *any* abscissa of the final interval and *any* ordinate of the final interval is a satisfactory approximation to the optimal solution. For this reason the interval, final or initial, is called the *interval of uncertainty*.

The object of our search is an extremum (in particular the largest ordinate, corresponding abscissa, or both) of the formal figure of merit function or the tactical figure of merit function. An a priori knowledge of the ordinates is presumed to be lacking, but a knowledge of the bounds of the domain within which suitable alternatives are to be found is presumed complete. To draw a distinction, consider the problem of discovering, in one-dimensional search, the interval in which the extremum of a unimodal function lies--i.e., narrowing down the interval in which the extremum lies from the initial domain to some fraction of the initial domain.

A unimodal function has a single peak in a given interval, and each successive ordinate is progressively larger than the last until the peak is reached; then each successive ordinate is progressively less than the last. Nonunimodal functions can be divided into a number of contiguous unimodal regions. Examples of unimodal functions appear in Fig. 8.7. We are given the unimodality of the function y of independent variable x. In narrowing the interval in which the extremum may be found with certainty from the original range $a \leq x \leq b$ to something less, a problem arises concerning the expenditure of function evaluations. Suppose four function evaluations may be expended. Should we expend them simultaneously or one at a time?

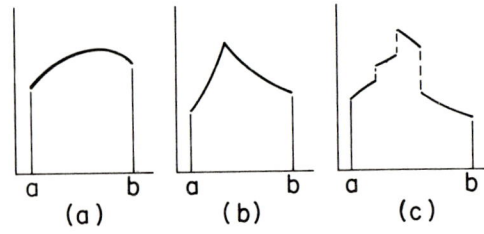

Fig. 8.7. Examples of functions that are unimodel for maximization. (From C. R. Mischke, *An Introduction to Computer-Aided Design*, 1968, Prentice-Hall, Englewood Cliffs, N.J., p. 56. Used with permission.)

Optimization

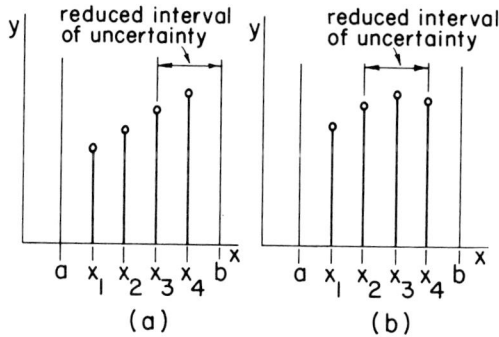

Fig. 8.8. Simultaneous evaluation of the function in four locations and the resulting reduction in the interval of uncertainty. (From Mischke, *Computer-Aided Design*, p. 57. Used with permission.)

The simultaneous approach is something of a shotgun approach, and a four-ordinate simultaneous function evaluation might appear as indicated in Fig. 8.8. In Fig. 8.8(a) the extremum lies in the interval $x_3 \leq x \leq b$ and in Fig. 8.8(b) in the range $x_2 \leq x \leq x_4$. In the absence of knowledge of the relationship between x and y, four ordinates would be placed, equally spaced, in the interval; and the domain of the extremum is narrowed to 0.4 of the interval (a, b).

If we expend our ordinate evaluations first as a pair to determine the trend of the function and then singly in the reduced certain domain of the ex-

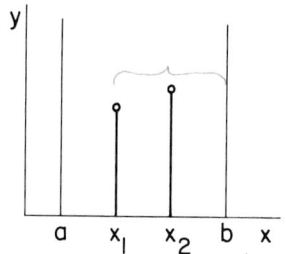

Fig. 8.9. A pair of simultaneous function evaluations in the interval (a, b). From Mischke, *Computer-Aided Design*, p. 57. Used with permission.)

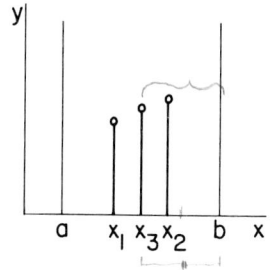

Fig. 8.10. An additional ordinate at x_3 between x_1 and x_2. (From Mischke, *Computer-Aided Design*, p. 57. Used with permission.)

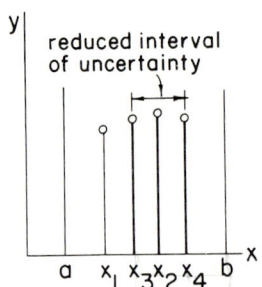

Fig. 8.11. An additional ordinate at x_4 between x_2 and b. (From Mischke, *Computer-Aided Design*, p. 58. Used with permission.)

tremum, we would have knowledge with the first step that the extremum is certainly in the interval $x_1 \leq x \leq b$ as indicated in Fig. 8.9. Now, by placing an ordinate between x_1 and x_2 we obtain the situation depicted in Fig. 8.10 and have a new reduced interval of $x_3 \leq x \leq b$. The placing of our last ordinate evaluation between x_2 and b yields the still further reduced interval $x_3 \leq x \leq x_4$, as shown in Fig. 8.11. If we place initial ordinates at the one-third and two-thirds points of the original interval and successive ordinates bisecting previous spacings, the final region of uncertainty would be one-third of the interval (a, b).

The second method employed exploits the results of prior trials, and ordinate evaluations are expended where they contribute the most needed information. The first method is called a simultaneous search plan and the latter is called a sequential search plan. The second approach is preferable as it expends fewer functional evaluations in reducing the interval of certainty to a specified level. The former plan is used where time is not available to obtain partial results and to act on that information for future evaluations. Since the number of function evaluations is an index to effort expended in narrowing the region of uncertainty, economy of search effort and expeditious search techniques go hand in hand.

If the computer is to execute the search and it can calculate quickly, why worry over some extra function evaluations? If one evaluation of a function takes sizable computer time, then whether 10^2 or 10^6 function evaluations are necessary is significant.

8.6 EXHAUSTIVE SEARCH

Consider a unimodal function y in the interval $a \leq x \leq b$ to be evaluated at n evenly spaced abscissas, as shown in Fig. 8.12 in a simultaneous search. The abscissa spacing is $1/(n + 1)$ of the interval (a, b) and the interval of uncertainty has been reduced to two spaces. The functional reduction of the original interval of uncertainty, F, is $2/(n + 1)$. The following data indicate the cost in function evaluations (n) for fractional interval reduction (F): $n = 1, F = 1; n = 2, F = 2/3; n = 3, F = 1/2; n = 4, F = 2/5;...; n = 199, F = 1/100;...; n, F = 2/(n + 1)$. Clearly, $F = 2/(n + 1)$ and

$$n = \left[\frac{(2 - F)}{F} \right]_+ \qquad (8.4)$$

where the nearest integer $\geq (2 - F)/F$ is enclosed in brackets. For $F = 0.01$,

$$n = [(2 - 0.01)/0.01]_+ = 199$$

n = no. of evenly spaced abscissas.
F = functional reduction of original interval of uncertainty.

Optimization

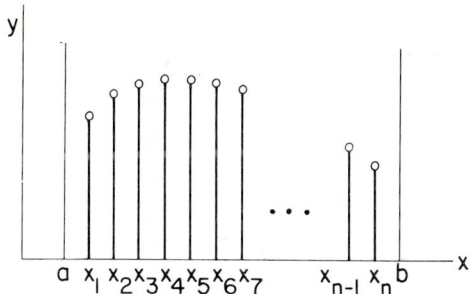

Fig. 8.12. Simultaneous evaluation of n ordinates in interval (a, b), forming the pattern of ordinates for an exhaustive search in the interval. (From Mischke, *Computer-Aided Design*, p. 58. Used with permission.)

As a practical matter in programming such a search routine, it can be desirable to evaluate the function also at each end of the interval. Although not helpful in reducing the interval of uncertainty, it is useful when points for plotting purposes are sought and the use of an exhaustive search routine to spray an interval full of equally spaced ordinates is convenient.

8.7 INTERVAL-HALVING AND GOLDEN SECTION SEARCHES

In the previous section we discovered that in a simultaneous search for a fractional reduction in the interval of uncertainty $F = 0.1$, 19 functional evaluations would be required; for $F = 0.01$, 199 evaluations would be involved; and for $F = 0.001$, 1999 evaluations would be necessary. In view of the rapidly increasing cost to reduce the interval of uncertainty, it might be advantageous to nest the search by spending 19 function evaluations to reduce the interval to 0.1, then spend 19 more in the reduced interval to contract the interval of uncertainty to 0.01, and finally spend 19 more in the twice reduced interval. The resultant expenditure in functional evaluations is $19 + 19 + 19 = 57$ and accomplishes in a combined sequential and simultaneous strategy the result that requires 1999 evaluations in a simultaneous approach. The gain in economical expenditure of function evaluations is impressive.

The choice of F for each simultaneous subsearch was arbitrary. What is the best choice of F for the subsearch? We had

19 evaluations reducing original interval to 0.1
19 evaluations reducing original interval to $(0.1)^2$
19 evaluations reducing original interval to $(0.1)^3$

and in general we would have

n evaluations reducing the original interval to $2/(n + 1)$
n evaluations reducing the original interval to $[2/(n + 1)]^2$
·
n evaluations reducing the original interval to $[2/(n + 1)]^m$

The total expenditure of function evaluations, $N = mn$, and the final reduction in the interval of uncertainty is designated F. Thus

$$F = [2/(n + 1)]^m = [2/(n + 1)]^{N/n}$$

since $m = N/n$. Solution of the above equation for N yields

$$N = \frac{n \ln(1/F)}{\ln[(n + 1)/2]}$$

A minimum N for a specified F is associated with a definite n. Since n is limited to integer values and differentiation of the equation for N with respect to n leads to a transcendental equation, the smallest value of N is found by simple substitution.

$$N\big|_{n=1} = \infty$$

$$N\big|_{n=2} = \frac{2}{\ln 3/2} \ln(1/F) = 4.93 \ln(1/F)$$

$$N\big|_{n=3} = \frac{3}{\ln 4/2} \ln(1/F) = 4.32 \ln(1/F)$$

$$N\big|_{n=4} = \frac{4}{\ln 5/2} \ln(1/F) = 4.37 \ln(1/F)$$

Higher values of n result in continuously increasing values of N. The strategy to use for the least N is the one involving the placing of n = 3 ordinates within the search interval. The optimal choice of f for the fractional reduction sought during each subsearch is $f = 2/(n + 1) = 1/2$. This optimal blending of sequential and simultaneous approaches is called *interval-halving*.

Consider a unimodal function y in the interval $a \leq x \leq b$ that will be evaluated with three equally spaced ordinates in the intervals shown in Fig. 8.13. In all the cases in Fig. 8.13, the interval of uncertainty has been reduced to half the original interval. This allows half the interval to be discarded and attention turned to the remaining interval of uncertainty. Note that in each of the situations of Fig. 8.13, the central ordinate has already been evaluated. Now two more ordinates are evaluated at the one-fourth and three-fourths points of the shortened interval of uncertainty as shown in Fig. 8.14. We observe from Fig. 8.14 that the new region of uncertainty can be again halved and the extremum now lies with in a range of one-fourth of the original interval. Only five functional evaluations have been expended up to this time. The following data indicate the cost in function evaluations (N)

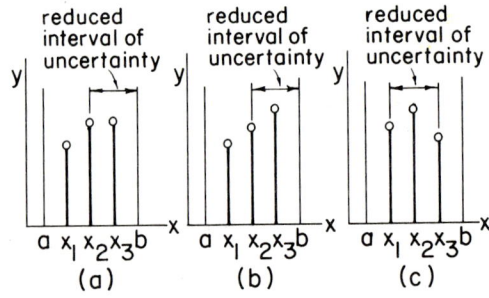

Fig. 8.13. Interval-halving search with (a) two ordinates equal; (b) all ordinates unequal, $x_1 < x_2 < x_3$; (c) all ordinates unequal, $x_3 < x_2 < x_1$, for unimodal function being maximized. Reduced intervals of uncertainty are indicated. (From Mischke, *Computer-Aided Design*, p. 61. Used with permission.)

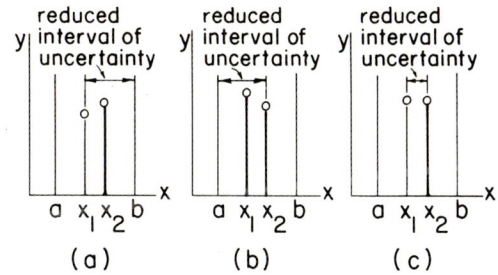

Fig. 8.14. Placing unequally spaced ordinate pairs in an interval (a, b) of a unimodal function being maximized with (a) ordinates unequal, $x_1 < x_2$; (b) ordinates unequal, $x_1 > x_2$; (c) ordinates equal, $x_1 = x_2$, and the corresponding reduction in this area of uncertainty. (From Mischke, *Computer-Aided Design*, p. 65. Used with permission.)

for fractional interval reduction (F): $N = 3$, $F = 1/2$; $N = 5$, $F = 1/4$; $N = 7$, $F = 1/8$; $N = 9$, $F = 1/16$;...; N, $F = 1/2^{(N-1)/2}$. Clearly $F = 1/[2^{(N-1)/2}]$ and

$$N = \left[1 + \frac{2 \ln(1/F)}{\ln 2}\right]_{odd+} \tag{8.5}$$

For $F = 0.01$

$$N = \left[1 + \frac{2 \ln(1/0.01)}{\ln 2}\right]_{odd+} = \{14.4\}_{odd+} = 15$$

If the problem stated is to discover $x^* \pm a$, the fractional reduction in the interval of uncertainty is

$$F = \Delta x'/\Delta x = 2a/\Delta x$$

Then, from the function evaluation budget equation,

$$N = \left[1 + \frac{2 \ln 1/F}{\ln 2}\right]_{odd+}$$

$$N = \left[1 + \frac{2 \ln \Delta x/2a}{\ln 2}\right]_{odd+}$$

$$N = \left[2.88 \ln \frac{\Delta x}{a} - 1\right]_{odd+} \tag{8.6}$$

For example, if the speed of maximum engine torque of an internal combustion engine is desired to be ±25 rpm and the original interval of uncertainty is bounded by 200 and 2200 rpm,

$a = 25$ rpm

$\Delta x = 2200 - 200 = 2000$ rpm

and the function evaluation budget is

$$N = [2.88 \ln(2000/25) - 1]_{odd+} = [11.62]_{odd+} = 13$$

Interval-Halving Example

Maximize $y = x$ $x < 0.5$ $N = [2.88 \ln(0.7/0.03) - 1]_{odd+} = 9$
$y = 1 - x$ $x \geq 0.5$

in the interval $0.3 \leq x \leq 1$, establishing $x^* \pm 0.03$ by interval-halving interval $(0.3, 1)$, $\Delta x = 1 - 0.3 = 0.7$.

$x_1 = 0.3 + (1/4)(0.7) = 0.475$ $y_1 = 0.475$

$x_2 = 0.3 + (2/4)(0.7) = 0.650$ $y_2 = 0.350$

$x_3 = 0.3 + (3/4)(0.7) = 0.825$ $y_3 = 0.175$

New interval $(0.3, 0.650)$, $\Delta x = 0.65 - 0.3 = 0.35$.

$x_4 = 0.3 + (1/4)(0.35) = 0.388$ $y_4 = 0.388$

$x = 0.475 \leftrightarrow y = 0.475$

$x_5 = 0.3 + (3/4)(0.35) = 0.563$ $y_5 = \sout{0.0437}$ 0.437

New interval $(0.388, 0.563)$, $\Delta x = 0.563 - 0.388 = 0.175$.

$x_6 = 0.388 + (1/4)(0.175) = 0.432$ $y_6 = 0.432$

$x = 0.475 \leftrightarrow y = 0.475$

$x_7 = 0.388 + (3/4)(0.175) = 0.519$ $y_7 = 0.481$

New interval $(0.475, 0.563)$, $\Delta x = 0.563 - 0.475 = 0.088$

$x_8 = 0.475 + (1/4)(0.088) = 0.497$ $y_8 = 0.497$

$x = 0.519 \leftrightarrow y = 0.481$

$x_9 = 0.475 + (3/4)(0.088) = 0.541$ $y_9 = 0.459$

Final interval $(0.475, 0.519)$, $\Delta x = 0.044$.

$x^* = 0.497$ $0.475 < x^* < 0.519$

$y^* = 0.497$

Solution is within ± 0.03 of $x^*_{true} = 0.5$.

The interval-halving technique has reduced function evaluations for a final interval of uncertainty of 1/100 the original interval from 199 to 15, a remarkable reduction.

As impressive as was the effort in function evaluation introduced by the interval-halving stratagem, an intuitive feeling lingers that further improvement is possible. In the interval-halving technique, function evaluations are expended two by two. Perhaps we can expend function evaluations one by one, accepting smaller reductions of interval. It will require us to relax the restriction of equally spaced function evaluations. Now our task is to find the proper positioning of the two ordinates so that whichever portion of the interval is eliminated, the remaining ordinate is in the proper position to be one of the two that will survive in the reduced interval, saving a function evaluation at every decision level. This technique leads to a golden section search.

Consider the unimodal function y in the interval $a \leq x \leq b$ evaluated at two intermediate ordinates as shown in Fig. 8.14. In Fig. 8.14(a) the extreme lies in the interval $x_1 \leq x \leq b$. We shall take the first case and the nomenclature of Fig. 8.15 and place a third ordinate in Fig. 8.16. We wish to de-

Optimization

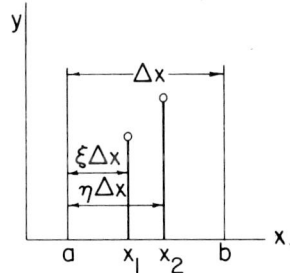

Fig. 8.15. The definitions of ξ and η for golden section search. (From Mischke, *Computer-Aided Design*, p. 65. Used with permission.)

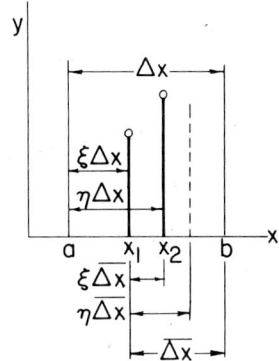

Fig. 8.16. The placing of the third ordinate for case (a) of Fig. 8.14. (From Mischke, *Computer-Aided Design*, p. 65. Used with permission.)

fine ξ and η such that if region $a \le x \le x_1$ is eliminated, the ordinate x_2 is in the proper location--it is the *first* ordinate of the new interval $\overline{\Delta x} = (b - x_1)$--i.e., located at $\xi \overline{\Delta x}$ from x_1, the new left end of the interval of uncertainty (see Fig. 8.17). Similarly, if the region $x_2 \le x \le b$ is eliminated by x_1 being greater than x_2, then x_1 is in the proper location to be the *second* ordinate in the new interval $\overline{\Delta x}' = (x_2 - a)$, i.e., located at $\eta \overline{\Delta x}'$ from a. This readiness for either alternative demands a symmetry about the center of the interval in the location of x_1 and x_2. The geometry of Fig. 8.17 indicates this symmetry. In Fig. 8.17 we can sum the distances that add to Δx and obtain

$$\xi \Delta x + (\eta \Delta x - \xi \Delta x) + \xi \Delta x = \Delta x \quad \text{or} \quad \xi + \eta = 1$$

From the relationship between Δx and $\overline{\Delta x}$ we obtain another requirement from Fig. 8.16.

$$\Delta x = \xi \Delta x + \overline{\Delta x}$$

$$\xi \Delta x + \xi \overline{\Delta x} = \eta \Delta x$$

Solving for $\overline{\Delta x}/\Delta x$ in the two preceding equations and equating, we obtain

$$(1 - \xi) = \overline{\Delta x}/\Delta x = (\eta - \xi)/\xi$$

The outer equality yields $\xi(1 - \xi) = \eta - \xi = (1 - \xi) - \xi$; the resulting quadratic in ξ is

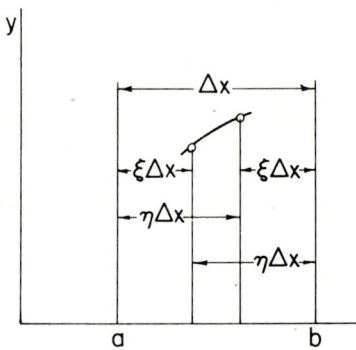

Fig. 8.17. The geometry and symmetry of the placement of ordinates in the golden section search for the maximum of a unimodal function. (From Mischke, *Computer-Aided Design*, p. 66. Used with permission.)

$$\xi^2 - 3\xi + 1 = 0$$

from which $\xi = 3/2 \pm (1/2)\sqrt{5}$. Since $\xi < 1$, the negative sign prevails and

$$\xi = (3 - \sqrt{5})/2 = 0.381\ 966\ 011$$

It follows that $\eta = 0.618\ 033\ 989$ and the fractional reduction of interval per decision is

$$\overline{\Delta x}/\Delta x = (1 - \xi) = \eta = 0.618\ 033\ 989$$

The following data indicate the cost in function evaluations for interval reduction: $N = 2$, $F = \eta$; $N = 3$, $F = \eta^2$; $N = 4$, $F = \eta^3$; ...; N, $F = \eta^{N-1}$. Clearly $F = \eta^{N-1}$ and

$$N = [1 + (\ln F/\ln \eta)]_+ \qquad (8.7)$$

where the brackets enclosing a noninteger indicate that N is assigned the value of the next larger integer. When $F = 0.01$,

$$N = \left[1 + \frac{\ln(0.01)}{\ln(0.618\ 033\ 989)}\right]_+ = [1 + 9.58]_+ = [10.58]_+ = 11$$

This compares favorably with the cost of an exhaustive search to reduce the interval to 1/100 its original size, which is 199 function evaluations.

As an example of application of this golden section strategy, consider the following problem

$$\max y = -x(x - 1)$$
subject to $-x \leq 0.1$
$x \leq 0.6$

Reduction of the interval of uncertainty to $F = 0.1$ is satisfactory. The predicted number of function evaluations is

$$N = [1 + \ln F/\ln \eta]_+ = [1 + \ln 0.1/\ln 0.618]_+ = 6$$

The original interval of uncertainty is $\Delta x = 0.6 - 0.1 = 0.5$. Now

$$x_1 = 0.1 + 0.382(0.5) = 0.2910 \qquad y(0.2910) = 0.2063 \qquad \text{(evaluation 1)}$$

Optimization

$x_2 = 0.1 + 0.618(0.5) = 0.4090 \quad y(0.4090) = 0.2417 \quad$ (evaluation 2)

Since the second ordinate is larger, the left end of the interval below $x = 0.2910$ is eliminated and $\overline{\Delta x} = 0.6 - 0.2910 = 0.3090$. Consequently,

$x_1 = 0.2910 + 0.382(0.3090) = 0.4090 \quad y(0.4090) = 0.2417 \quad$ (old x_2)

$x_2 = 0.2910 + 0.618(0.3090) = 0.4819 \quad y(0.4819) = 0.2497 \quad$ (evaluation 3)

Since the second ordinate is larger than the first, the left end of the interval below $x = 0.4090$ is eliminated and $\overline{\Delta x} = 0.6 - 0.4090 = 0.1910$. Consequently,

$x_1 = 0.4090 + 0.382(0.1910) = 0.4819 \quad y(0.4819) = 0.2497 \quad$ (old x_1)

$x_2 = 0.4090 + 0.618(0.1910) = 0.5271 \quad y(0.5271) = 0.2493 \quad$ (evaluation 4)

The first ordinate is larger, the right side of the interval above $x = 0.5271$ is eliminated, and the new interval is $\overline{\Delta x} = 0.5271 - 0.4090 = 0.1181$. Consequently,

$x_1 = 0.4090 + 0.382(0.1181) = 0.4541 \quad y(0.4541) = 0.2479 \quad$ (evaluation 5)

$x_2 = 0.4090 + 0.618(0.1181) = 0.4819 \quad y(0.4819) = 0.2497 \quad$ (old x_1)

The second ordinate is larger, the left side of the interval below $x = 0.4541$ is eliminated, and the interval of uncertainty is reduced to $\overline{\Delta x} = 0.5271 - 0.4541 = 0.0330$. Consequently,

$x_1 = 0.4541 + 0.382(0.0330) = 0.4819 \quad y(0.4819) = 0.2497 \quad$ (old x_2)

$x_2 = 0.4541 + 0.618(0.0330) = 0.4992 \quad y(0.4992) = 0.249999 \quad$ (evaluation 6)

We are certain that the abscissa x^* lies between $x = 0.4819$ and $x = 0.5271$. The largest of the last two ordinates is $y = 0.249\,999$, which is taken as an approximation to y^* as is its corresponding abscissa to $x^* = 0.4992$. Note that this was accomplished with six function evaluations and the final interval was less than 10 percent of the original interval.

If the problem is stated to discover $x^* \pm a$, in the final interval of uncertainty there are two function evaluations and the longest possible distance from the tallest to the edge of the interval is $\Delta x'$, so we may write

$a = 0.618\,\Delta x'$

$\Delta x' = 1.618a$

$F = \Delta x'/\Delta x = 1.618a/\Delta x$

and from the function evaluation budget equation

$$N = \left[1 + \frac{\ln F}{\ln \eta}\right]_+ = \left[1 + \frac{\ln a/(\eta\,\Delta x)}{\ln \eta}\right]_+ = \left[1 + \frac{\ln a - \ln \eta - \ln \Delta x}{\ln \eta}\right]_+$$

$$= \left[2.08\,\ln \frac{\Delta x}{a}\right]_+ \tag{8.8}$$

From the example of speed of maximum engine torque seeking $x^* \pm 25$ rpm on the interval 200, 2200, we have

$a = 25$ rpm $\Delta x = 2200 - 200 = 2000$ rpm

$N = [2.08 \ln(2000/25]_+ = [9.07]_+ = 10$

This search is more efficient than the 13 function evaluations necessary to implement an interval-halving search strategy.

Golden Section Example

Maximize $y = x$ $x < 0.5$ $N = [2.08 \ln(0.7/0.03)]_+ = 7$

$y = 1 - x$ $x \geq 0.5$

in the interval $0.3 \leq x \leq 1$, establishing $x^* \pm 0.03$ by golden section, interval $(0.3, 1)$, $\Delta x = 1 - 0.3 = 0.7$.

$x_1 = 0.3 + 0.382(0.7) = 0.567$ $y_1 = 0.433$

$x_2 = 0.3 + 0.618(0.7) = 0.733$ $y_2 = 0.267$

New interval $(0.3, 0.733)$, $\Delta x = 0.733 - 0.3 = 0.433$.

$x_3 = 0.3 + 0.382(0.433) = 0.465$ $y_3 = 0.465$

$x_1 = 0.567$ ↔ $y_1 = 0.433$

New interval $(0.3, 0.567)$, $\Delta x = 0.567 - 0.3 = 0.267$.

$x_4 = 0.3 + 0.382(0.267) = 0.402$ $y_4 = 0.402$

$x_3 = 0.465$ ↔ $y_3 = 0.465$

New interval $(0.402, 0.567)$, $\Delta x = 0.567 - 0.402 = 0.165$.

$x_3 = 0.465$ ↔ $y_3 = 0.465$

$x_5 = 0.402 + 0.618(0.165) = 0.504$ $y_5 = 0.496$

New interval $(0.465, 0.567)$, $\Delta x = 0.567 - 0.455 = 0.102$.

$x_5 = 0.504$ ↔ $y_5 = 0.496$

$x_6 = 0.465 + 0.618(0.102) = 0.528$ $y_6 = 0.472$

New interval $(0.465, 0.528)$, $\Delta x = 0.528 - 0.465 = 0.063$.

$x_7 = 0.465 + 0.382(0.063) = 0.489$ $y_7 = 0.489$

$x_5 = 0.504$ ↔ $y_5 = 0.496$

Final interval $(0.489, 0.528)$, $\Delta x = 0.528 - 0.489 = 0.039$.

$x^* = 0.504$ $0.489 \leq x^* \leq 0.528$

$y^* = 0.496$

Solution within ± 0.03 of $x^*_{true} = 0.5$.

Summary of One-Dimensional Search Stratagems and Function Evaluation Budgets

Exhaustive search stratagem:

$N = [(2 - F)/F]_+$ or $N = [(\Delta x/a) - 1]_+$

Optimization

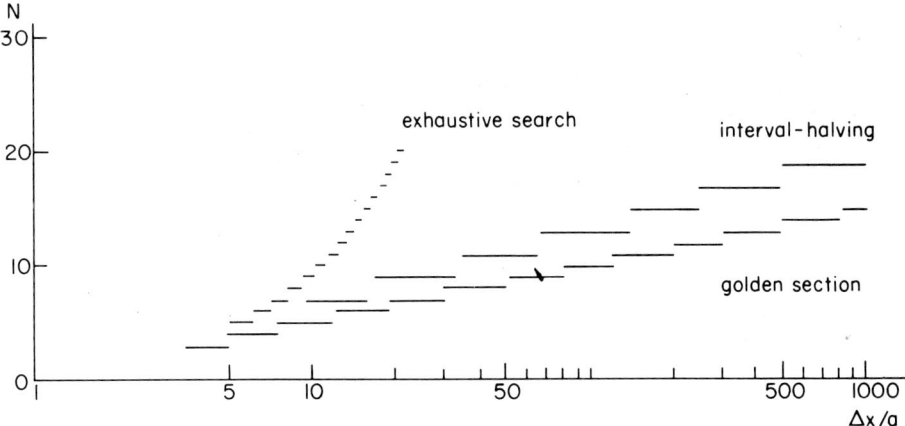

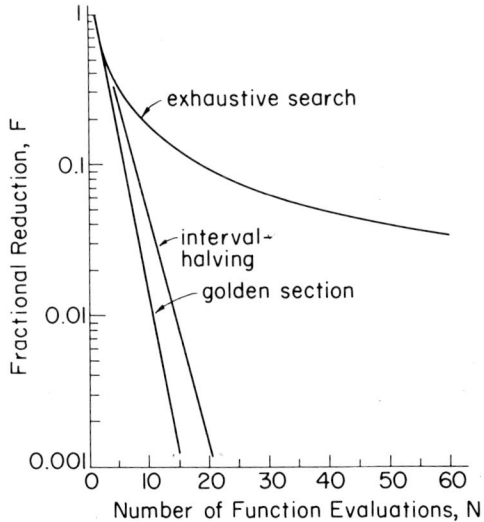

Fig. 8.18. (above) Number of function evaluations, N, for maximization by exhaustive, interval-halving, and golden section searches for $x^* \pm a$. (below) Relative effectiveness of exhaustive, interval-halving, and golden section searches for fractional reduction F in the uncertainty interval. (From Mischke, *Computer-Aided Design*, p. 68. Used with permission.)

Interval-halving search stratagem:

$$N = \left[1 + \frac{2 \ln(1/F)}{\ln 2}\right]_{odd+} \quad \text{or} \quad N = [2.88 \ln(\Delta x/a) - 1]_{odd+}$$

Golden section search stratagem:

$$N = [1 + 2.08 \ln(1/F)]_+ \quad \text{or} \quad N = [2.08 \ln(\Delta x/a)]_+$$

Example. How many experiments are necessary to find the speed of maximum torque of tractor engine in interval 200-2200 rpm declaring $rpm^* \pm 25$?

Exhaustive:

N = $[2000/25 - 1]_+$ = 79

Interval-halving:

N = $[2.88 \ln(2000/25) - 1]_{odd+}$ = 13

Golden section:

N = $[2.08 \ln(2000/25)]_+$ = 10

Figure 8.18(a) and (b) indicate graphically the relative efforts necessary to carry out a one-dimensional search using the methods mentioned so far in this chapter.

8.8 MULTIDIMENSIONAL SEARCHES

When a function of one independent variable is explored for an extremum, a convenient geometric model is a functional locus in a plane with the search interval as a line segment as indicated by Fig. 8.19. When a function of two independent variables is explored, the concept of a surface above a plane is a useful geometric interpretation, as indicated in Fig. 8.20. A function of three independent design variables (a merit hypersurface) can be given a geometric interpretation by using three space variables (x_1, x_2, x_3) and thinking of the magnitude of an ordinate to the hypersurface (Fig. 8.21).

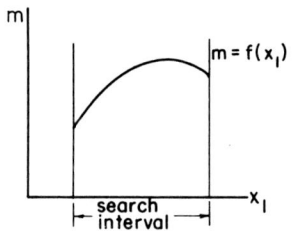

Fig. 8.19. A merit function of a single independent design variable. (From Mischke, *Computer-Aided Design*, p. 69. Used with permission.)

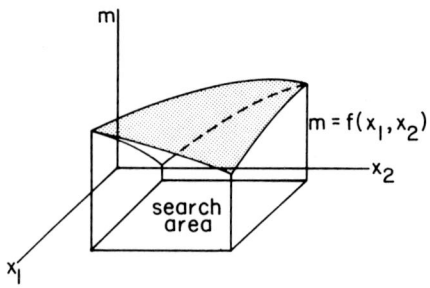

Fig. 8.20. A merit surface representing a function of two independent design variables. (From Mischke, *Computer-Aided Design*, p. 69. Used with permission.)

Optimization

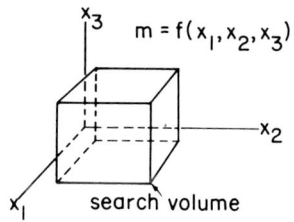

Fig. 8.21. A merit hypersurface defined by three independent design variables. (From Mischke, *Computer-Aided Design*, p. 69. Used with permission.)

When a function of more than three independent variables arises, geometric visualization is impossible. We have the convenience of geometric thinking in simple problems; in more complicated problems algebraic methods must be used despite the lack of geometric insight. An intuitive approach to multidimensional problems is to use algebraic methods (as we must)--develop, prove, and gain experience with them in problems with two independent variables and then extend the method into n-dimensional space.

A function of n variables,

$$y = M(x_1, x_2, \ldots, x_n)$$

can be represented in a hyperspace of (n + 1) orthogonal axes or an (n + 1)-space. The function y is said to define a hypersurface. Thus the merit function

$$M = M(x_1, x_2, x_3)$$

is represented in a 4-space hyperspace and M defines a hypersurface.

Several difficulties are encountered in multidimensional searches:

--Multidimensionality overpowers the effective search techniques available in a 1-space.
--The enormity of n-space is overwhelming.
--A lack of objective comparisons of searching techniques prevents construction of optimal search plans.

Suppose a search region has c cells and we wish to evaluate the merit function in the b best cells. The probability of choosing at random a cell that is in the group of best cells (b in number) is b/c. The probability of missing a cell of the best group in a single try is (1 - b/c). The selection of two cells at random involves a $(1 - b/c)^2$ probability of a double miss. The probability of finding one or more (at least one) cell in n tries is $[1 - (1 - b/c)^n]$. If the fraction b/c is denoted as f, then Prob(f), the probability of finding at least one cell of the best group (b = df), is

$$\text{Prob}(f) = 1 - (1 - f)^n$$

$$n = \left[\frac{\ln(1 - \text{Prob}(f))}{\ln(1 - f)} \right]_+ \tag{8.9}$$

If f = 0.01 and Prob(f) = 0.99, then

$$n = \left[\frac{\ln(1 - 0.99)}{\ln(1 - 0.01)} \right]_+ = [458.21]_+ = 459$$

In other words, it would take 459 tries to have a 0.99 certainty of selecting at random one or more best cells (of a group fc = 0.01c).

Multivariable searches are not unlike a strange form of mountain climbing (to use a three-dimensional example) in which a climbing director, at a radio but out of sight of the mountain, has a robot mountain climber to report local elevations above sea level and move from place to place on command. The climb director has no picture of the mountain and would like a picture but cannot afford the necessary survey. The director settles for less and is always torn between two alternatives:

--Should the robot explore within a locality to get a clear picture of the lay of the land and to gain the most elevation in the next climbing move?
--Should the robot move again in a previously elevation gaining direction, settling for more but smaller gains per function evaluation?

In the beginning the climbing director must explore a randomly chosen region because nothing is known. Then, having some feel for the local topography, moves are ordered in the most promising direction. If elevation is gained, more moves in the same direction are ordered; then a little local exploration is in order to improve the direction of movement. Local exploration is ordered any time a move does not result in a gain in elevation. As the summit is approached, more exploration is required to find the highest point in the case of nearly flat domed mountains.

The method of steepest ascent (gradient projection) is suitable to n-dimensional problems of an unconstrained nature such as

$$\max y = f(x_1, x_2, \ldots, x_n)$$

The algorithm plan is to start at a feasible point, determine which direction is steepest in ascent rate, and move in that direction. If the problem is

$$\max y = f(x_1, x_2)$$

then the first function evaluation would be made at (a, b). This ordinate is indicated in Fig. 8.22. Shifting to Fig. 8.23, we evaluate m_1 at position (a + Δx_1, b) and m_2 at position (a, b + Δx_2) and have three ordinate evaluations in the locality of m_0. These three ordinates are m_0, m_1, and m_2. A plane passing through these three points in the merit surface is approximately tangent to the merit surface, because in the limit as $\Delta x_1 \to 0$ and $\Delta x_2 \to 0$, the plane so defined becomes coincident with the plane tangent to the merit surface at (a, b). Since we cannot let Δx_1 and Δx_2 become zero, we must settle for a good approximation. This approximation to the tangent plane (hereafter

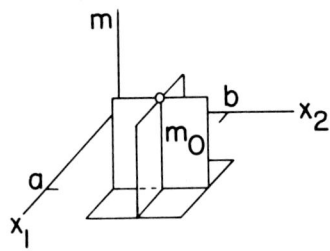

Fig. 8.22. An ordinate to a merit surface erected at (a, b). From Mischke, *Computer-Aided Design*, p. 75. Used with permission.)

Optimization

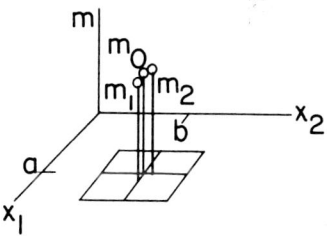

Fig. 8.23. Two additional ordinates to the merit surface in Fig. 8.22 are erected at $a + \Delta x_1$ and $b + \Delta x_2$. (From Mischke, *Computer-Aided Design*, p. 75. Used with permission.)

called the tangent plane) is denoted functionally as $p(x_1, x_2)$ and has the equation

$$p(x_1, x_2) = p_0 + p_1 x_1 + p_2 x_2$$

The constants p_0, p_1, and p_2 may be calculated from the simultaneous solution of the above equation at the three known ordinates m_0, m_1, and m_2:

$$m_0 = p_0 + p_1 a + p_2 b$$
$$m_1 = p_0 + p_1(a + \Delta x_1) + p_2 b$$
$$m_2 = p_0 + p_1 a + p_2(b + \Delta x_2)$$

Subtracting the second equation from the first yields $m_0 - m_1 = -p_1 \Delta x_1$ or

$$p_1 = (m_1 - m_0)/\Delta x_1$$

Subtracting the third equation from the first yields $m_0 - m_2 = -p_2 \Delta x_2$ or

$$p_2 = (m_2 - m_0)/\Delta x_2$$

and it follows that

$$p_0 = m_0 - p_1 a - p_2 b = m_0 - \frac{(m_1 - m_0)a}{\Delta x_1} - \frac{(m_2 - m_0)b}{\Delta x_2}$$

The slope of the tangent plane in the x_1-direction is

$$\frac{\partial p}{\partial x_1} = p_1 = \frac{(m_1 - m_0)}{\Delta x_1}$$

and the slope in the x_2-direction is

$$\frac{\partial p}{\partial x_2} = p_2 = \frac{(m_2 - m_0)}{\Delta x_2}$$

If p_1 and p_2 are positive, the plane is uphill in both the x_1-direction and the x_2-direction. If the climbing director wishes to proceed in the direction of greatest rate of ascent, this direction is determinable from the tangent plane (examine Fig. 8.24). The change Δp, due to movement Δx_1 in the x_1-direction and Δx_2 in the x_2-direction, is given by

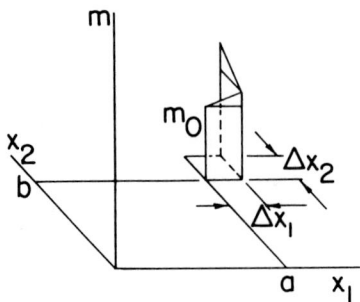

Fig. 8.24. The geometry of the gain in elevation for a move Δx_1 in the x_1-direction and Δx_2 in the x_2-direction from (a, b). (From Mischke, *Computer-Aided Design*, p. 76. Used with permission.)

$$\Delta p = \frac{\partial p}{\partial x_1} \Delta x_1 + \frac{\partial p}{\partial x_2} \Delta x_2 = p_1 \Delta x_1 + p_2 \Delta x_2$$

The slope of the diagonal line from point (a, b) to $(a + \Delta x_1, b + \Delta x_1)$ is

$$S = \frac{\Delta p}{(\Delta x_1^2 + \Delta x_2^2)^{1/2}} = \frac{p_1 \Delta x_1 + p_2 \Delta x_2}{(\Delta x_1^2 + \Delta x_2^2)^{1/2}}$$

The slope is greatest (maximum) for a judiciously selected advance whose components are $\overline{\Delta x_1}$ and $\overline{\Delta x_2}$. Satisfying the conditions $\partial S/\Delta \partial x_1 = 0$ and $\partial S/\Delta \partial x_2 = 0$ leads to step $\overline{\Delta x_2}$ in the x_2-direction, given by

$$\overline{\Delta x_2} = (p_2/p_1)\overline{\Delta x_1}$$

If the searcher decides on an advance of $\overline{\Delta x_1}$ in the x_1-direction (in the case of a plane with no tilt in the x_2-direction, then $p_2 = 0$ and the appropriate x_2-coordinate of the advance is

$$\Delta x_2 = (0/p_1)\Delta x_1 = 0$$

and the steepest ascent is in the x_1-direction.

The previous arguments may be extended to a hyperspace of n abscissas as follows: The hyperplane $p(x_1, x_2, \ldots, x_n)$ that is tangent to the hypersurface (the tangent hyperplane) is denoted functionally as

$$p(x_1, x_2, \ldots, x_n) = p_0 + p_1 x_1 + p_2 x_2 + \ldots + p_n x_n$$

The ordinates at the survey points are

$$m_0 = p_0 + p_1 a + p_2 b + \quad \ldots + p_n z$$
$$m_1 = p_0 + p_1(a + \Delta x_1) + p_2 b + \ldots + p_n z$$
$$m_2 = p_0 + p_1 a + p_2(b + \Delta x_2) + \ldots + p_n z$$
$$\vdots \qquad \vdots$$
$$m_n = p_0 + p_1 a + p_2 b + \quad \ldots p_n(z + \Delta x_n)$$

Subtracting each equation in turn from the first equation above yields

Optimization

$$m_0 - m_1 = -p_1 \Delta x_1$$
$$m_0 - m_2 = -p_2 \Delta x_2$$
$$\cdot$$
$$\cdot$$
$$m_0 - m_n = -p_n \Delta x_n$$

It follows that the slopes of the tangent plane in the ordinal directions are

$$p_1 = (m_1 - m_0)/\Delta x_1$$
$$p_2 = (m_2 - m_0)/\Delta x_2$$
$$\cdot$$
$$\cdot$$
$$p_n = (m_n - m_0)/\Delta x_n$$

The change in the ordinate to the plane p due to small movement from the survey position is given by

$$\Delta p = \frac{\partial p}{\partial x_1} \Delta x_1 + \frac{\partial p}{\partial x_2} \Delta x_2 + \ldots + \frac{\partial p}{\partial x_n} \Delta x_n$$

In terms of the slopes of the tangent plane in ordinal directions, this equation is written as

$$\Delta p = p_1 \Delta x_1 + p_2 \Delta x_2 + \ldots + p_n \Delta x_n$$

The slope of the hyperline drawn from point $(a, b, \ldots, z)$ to the point $(a + \Delta x_1, b + \Delta x_2, \ldots, z + \Delta x_n)$ is

$$S = \frac{\Delta p}{(\Delta x_1^2 + \Delta x_2^2 + \ldots + \Delta x_n^2)^{1/2}} = \frac{p_1 \Delta x_1 + p_2 \Delta x_2 + \ldots + p_n \Delta x_n}{(\Delta x_1^2 + \Delta x_2^2 + \ldots + \Delta x_n^2)^{1/2}}$$

For the slope to be a maximum, the conditions

$$\frac{\partial S}{\partial \Delta x_1} = 0, \frac{\partial S}{\partial \Delta x_2} = 0, \ldots, \frac{\partial S}{\partial \Delta x_n} = 0$$

must be met. These lead to the relations

$$(\Delta x_1^2 + \Delta x_2^2 + \ldots + \Delta x_n^2) p_1 - (p_1 \Delta x_1 + p_2 \Delta x_2 + \ldots + p_n \Delta x_n) \Delta x_1 = 0$$
$$(\Delta x_1^2 + \Delta x_2^2 + \ldots + \Delta x_n^2) p_2 - (p_1 \Delta x_1 + p_2 \Delta x_2 + \ldots + p_n \Delta x_n) \Delta x_2 = 0$$
$$\cdot \qquad\qquad\qquad\qquad\qquad\qquad\qquad\qquad\qquad\qquad\qquad \cdot$$
$$(\Delta x_1^2 + \Delta x_2^2 + \ldots + \Delta x_n^2) p_n - (p_1 \Delta x_1 + p_2 \Delta x_2 + \ldots + p_n \Delta x_n) \Delta x_n = 0$$

Dividing each equation successively into the first yields

$$\overline{\Delta x_2} = (p_2/p_1) \Delta x_1$$
$$\overline{\Delta x_3} = (p_3/p_1) \Delta x_1$$
$$\cdot$$
$$\cdot$$
$$\overline{\Delta x_n} = (p_n/p_1) \Delta x_1$$

The above increments correspond to the steepest ascent.

If the step size in a two-dimensional search (projected in the $x_1 x_2$-plane) is denoted Δ, then

$$\Delta = (\Delta x_1^2 + \Delta x_2^2)^{1/2}$$

If the search is n-dimensional, the step size Δ is given by

$$\Delta = (\Delta x_1^2 + \Delta x_2^2 + \ldots + \Delta x_n^2)^{1/2}$$

$$= \left[\Delta x_1^2 + \left(\frac{p_2}{p_1}\right)^2 \Delta x_1^2 + \left(\frac{p_3}{p_1}\right)^2 \Delta x_1^2 + \ldots + \left(\frac{p_n}{p_1}\right)^2 \Delta x_1^2 \right]^{1/2}$$

$$\overline{\Delta x_1} = \frac{p_1 \Delta}{(p_1^2 + p_2^2 + p_3^2 + \ldots + p_n^2)^{1/2}} \tag{8.10}$$

The increment $\overline{\Delta x_1}$ is correct for effecting the step size Δ along the steepest ascent.

As an example of determining the direction of steepest ascent, consider the function

$$y = x_1^2 + 2x_2 + x_3^3/3$$

at the position (1, 1, 1). We wish to step in the direction of steepest ascent a distance of $\Delta = 0.1$. What are the proper increments Δx_1, Δx_2, and Δx_3 to assure the movement in the direction of steepest ascent? Since we have the function, we can evaluate the slopes of the tangent plane by evaluating derivatives.

$$\frac{\partial y}{\partial x_1} = 2x_1 \qquad \left.\frac{\partial y}{\partial x_1}\right|_{1,1,1} = p_1 = 2$$

$$\frac{\partial y}{\partial x_2} = 2 \qquad \left.\frac{\partial y}{\partial x_2}\right|_{1,1,1} = p_2 = 2$$

$$\frac{\partial y}{\partial x_3} = x_3^2 \qquad \left.\frac{\partial y}{\partial x_3}\right|_{1,1,1} = p_3 = 1$$

Now

$$\overline{\Delta x_1} = \frac{p_1 \Delta}{(p_1^2 + p_2^2 + p_3^2)^{1/2}} = \frac{2(0.1)}{(4 + 4 + 1)^{1/2}} = 0.0667$$

$$\overline{\Delta x_2} = (p_2/p_1)\Delta x_1 = (2/2)(0.0667) = 0.0667$$

$$\overline{\Delta x_3} = (p_3/p_1)\Delta x_1 = (1/2)(0.0667) = 0.0333$$

The direction cosines (if needed) are

$$l = \cos(\Delta, \overline{\Delta x_1}) = \overline{\Delta x_1}/\Delta = 0.0667/0.1 = 0.667$$

$$m = \cos(\Delta, \overline{\Delta x_2}) = \overline{\Delta x_2}/\Delta = 0.0667/0.1 = 0.667$$

Optimization

$$n = \cos(\Delta, \overline{\Delta x_3}) = \overline{\Delta x_3}/\Delta = 0.0333/0.1 = 0.333$$

8.9 FENCE-WALKING, GRAPHICAL INVOLVEMENT

Some problems in optimization are simple enough to completely visualize for manual solution at the desk. Recognition that optima appear

--at interior stationary points
--at discontinuities in function or first derivative
--along boundaries or at intersections of boundaries

and an intuitive physical "feel" for the problem at hand can lead one to expect that the optimum appears along a constraint rather than at an interior stationary point or at a discontinuity. When the optimum is expected along the boundary, it is necessary to manipulate solution steps such that we "walk the fence."

Consider again the problem

$$\max y = (x_1 - 1)^2 + (x_2 - 2)^2$$

subject to $x_1 \geq 0$

$x_2 \leq 4$

$x_2 \geq x_1$

To walk the "west fence" (whose equation is $x_1 = 0$) of the feasible domain in Fig. 8.25, it is necessary to substitute the constraint equation, $x_1 = 0$, into the objective function. The result is

$$y = (x_2 - 2)^2 + 1 \quad (0 \leq x_2 \leq 4)$$

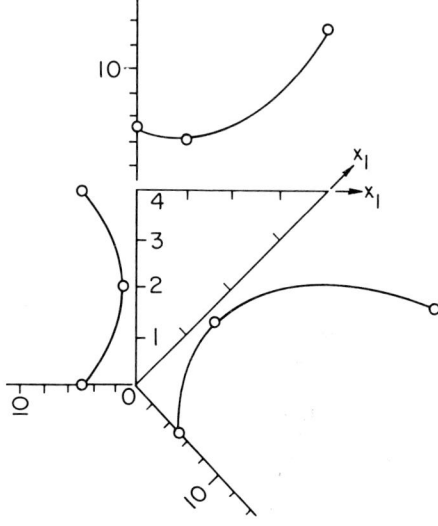

Fig. 8.25. Optimization by "walking the fence" around the feasible domain.

Along the constraint we have a function of a single variable x_2, which is unimodal for minimization and not unimodal for maximization in the interval $0 \le x_2 \le 4$. The stationary point is at $x_2 = 2$ and is characterized as a local minimum, since the second derivative is positive. We evaluate the end points of the interval

$$y(0) = (2 - 0)^2 + 1 = 5 \quad y(4) = (2 - 4)^2 + 1 = 5$$

and display the criterion function along the west constraint in Fig. 8.25. The maximum value of y is 5 and the minimum is 1 along this constraint. To walk the "north fence" whose equation is $x_2 = 4$, it is necessary to substitute $x_2 = 4$ into the criterion function, which results in

$$y = (x_1 - 1)^2 + 4 \quad (0 \le x \le 4)$$

This function exhibits a stationary point (local minimum) at $x_1 = 1$, $y = 4$. Evaluation of y at the ends of the interval results in

$$y(0) = (0 - 1)^2 + 4 = 5 \quad y(4) = (4 - 1)^2 + 4 = 13$$

The maximum of y along the north fence is 13 and the minimum is 1. The function is plotted along the upper constraint in Fig. 8.25. The diagonal constraint is followed by substituting $x_1 = x_2$ in the criterion function, which results in

$$y = (x_1 - 1)^2 + (x_2 - 2)^2 = 2x_1^2 - 6x_1 + 5 \quad (0 \le x \le 4)$$

A stationary local minimum exists at $x_2 = 3/2$, $y = 1/2$. At the ends of the interval

$$y(0) = 2(0)^2 - 6(0) + 5 = 5 \quad y(4) = 2(4)^2 - 6(4) + 5 = 13$$

The maximum value of y is 13 along the diagonal constraint; the minimum is 1/2. We may tabulate our finding, having moved completely around the feasible area along the constraints:

constraint	max	min
west	5	1
north	13	1
diagonal	13	1/2

The function $y = (x_1 - 1)^2 + (x_2 - 2)^2$

exhibits a maximum of 13 at $x_1 = 4$, $x_2 = 4$. The function exhibits an interior stationary minimum of $y = 0$ at $x_1 = 1$, $x_2 = 2$. We have been able to maximize y by moving along the constraints. This maximization was accomplished by back-substituting the equation for the constraint into the criterion function, and noting the valid interval.

When a merit function with n independent variables is linear, i.e.,

$$y = c_1 x_1 + c_2 x_2 + \ldots + c_n x_n$$

there can be no interior stationary point. If there are no discontinuities, the maximum or minimum occurs along a boundary, in this case at an intersection of boundaries. Dantzig's linear programming algorithm for linear merit

Optimization

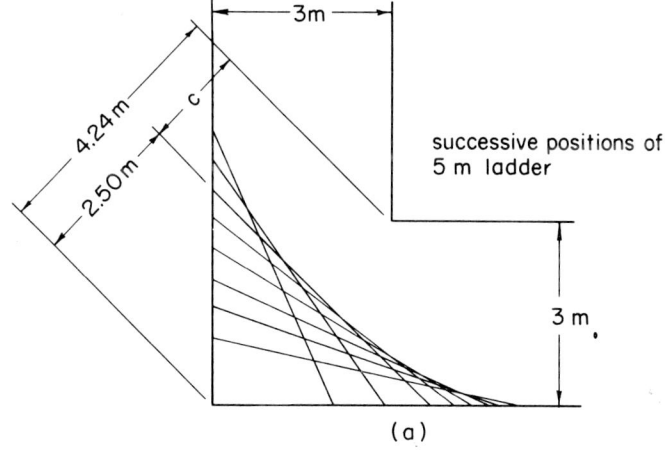

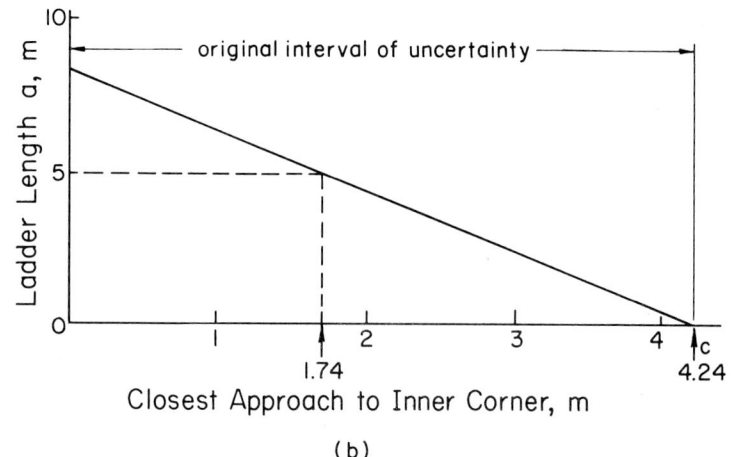

Fig. 8.26. (a) Successive positions of 5 m ladder rounding a corner in a corridor and maintaining contact with the outer wall. (b) Ladder length as a figure of merit as a function of the corner clearance c.

functions is a corner-to-corner improvement scheme because of a priori assurance that the maximum or minimum must occur at constraint intersections.

Optimization problems can include graphical elements and not be entirely algebraic in nature. Graphical techniques are useful when an algebraic approach is elusive, time consuming, or unduly complex. As an example of such a problem, consider the task of ascertaining the length of the longest ladder that can be taken around a right angle in a corridor 3 m wide as in Fig. 8.26(a). The constraint is the inner corner of the corridor turn, and if we visualize the closest point of approach of the passing ladder to the corner as the independent variable of the merit function, the ladder length a is expressed as

$$a = M(c) \quad (0 \leq c \leq 4.24)$$

where c is clearance; 4.24 m is the diagonal distance from inner to outer corners. Figure 8.26(b) delineates the initial interval of uncertainty. Our physical intuition leads us to expect the maximum of the figure of merit, a*, to occur at the left extremity of the interval of uncertainty, where c = 0.

Figure 8.26(a) depicts the corridor drawn to scale. A ladder of arbitrary length, say 5 m, is drawn in successive positions. The lines laid down are all tangent to an envelope curve that no part of the ladder ever crosses. A measurement along the corner-to-corner diagonal shows that the greatest distance of the ladder from the outer corner is 2.50 m; consequently, c = 4.24 − 2.50 = 1.74 m, which is one ordinate to the merit function. Recognizing that a ladder of zero length has a clearance c = 4.24 m gives another point on the merit function. It seems reasonable that a ladder of length 10 m will clear the outer corner by twice the previous 2.50 m, or 5.00 m. The clearance is c = 4.24 − 5.00 = −0.76 m. By simple proportion, from similar triangles in Fig. 8.26(b),

$$5.00/(4.24 - 1.74) = a*/4.24 \quad a* = 8.48 \text{ m}$$

This linearity in the figure of merit leads us to believe that rather than plot ordinate and abscissa intercepts of ladders of various lengths, the ordinate and abscissa intercepts of a ladder of unit length would be useful. Figure 8.27 is a plot of the envelope obtained.

To solve our previous problem directly in Fig. 8.27, we view it as a scale plot of the corridor corner, where the dimensionless variables y/a and

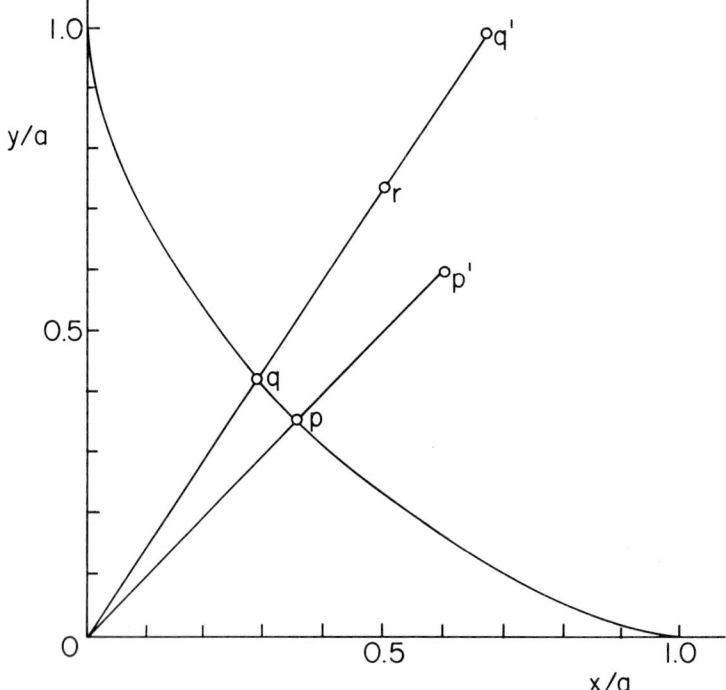

Fig. 8.27. The ladder envelope, suggested by Fig. 8.26(a), normalized by ladder length; consequently true of ladders of all lengths in the neighborhood of a square corner.

Optimization   301

x/a are normalized with the ladder length that results in zero clearance. This places the inner corner at point p. The distance from p to the origin is 4.24/a, which measures 0.5 abscissa units. Therefore

$$4.24/a = 0.5 \quad a = 4.24/0.5 = 8.48 \text{ m}$$

as before. For a 5 m ladder (as the normalization variable requires), the inner corner is plotted 4.24/5 or 0.848 abscissa units from the origin along the diagonal to point p'. The clearance c/a is the distance $\overline{pp'}$ measured in abscissa units:

$$c/a = \overline{pp'} = 0.848 - 0.50 = 0.348$$

$$c = 0.348(a) = 0.348(5) = 1.74 \text{ m}$$

as before.

Consider now the corridor corner with $b_1 = 3$ m and $b_2 = 2$ m [Fig. 8.28(a)]; determine the dimensions of the table of largest area that can be taken around the corridor corner horizontally. Our figure of merit is table area, A = M(a, w). Now a and w are not independent; for a given a there is a specific width that will touch the inner corner. In other words, w is a function of a; consequently A = aw, where w is a function of a. We have back-substituted an equality constraint to reduce the dimensionality of the figure of merit function from two independent variables to one. The relationship between w and a is buried, in part, in Fig. 8.27. Let us find one point on the figure of merit function. Let a = 3 m. The diagonal from outer corner to inner corner of corridor is $(2^2 + 3^2)^{1/2} = 3.61$ m. Along a line of same slope as the diagonal, i.e., 3/2, plot the inner corner 3.61/3 or 1.20 abscissa units along the diagonal on Fig. 8.27 and denote as q'. The clearance c of the left edge of the table is the distance $\overline{qq'}$, which is 1.20 - 0.525 or 0.68 abscissa units, i.e., c/3 = 0.684, from which c = 2.04 m. While this distance 2.04 m locates the closest point of approach of the left edge of the table to the inner corner, it cannot represent the table width, since the table left edge is not perpendicular to the diagonal. To find the lay of the table edge, it is necessary to find the angle made with the abscissa of a line tangent to the envelope at q on Fig. 8.27. Drawing such a tangent and measuring the angle gives 48.9°. From the triangle in Fig. 8.28(a),

$$w = 2.04 \cos(56.3° - 48.9°) = 2.02$$

$$A = aw = 3(2.02) = 6.07 \text{ m}^2$$

For a table length of 4 m, the inner corner is plotted on Fig. 8.27 as 3.61/4 = 0.90 at point r. Then c/4 = 0.90 - 0.525 or

$$c = 1.50 \text{ m}$$

$$w = 1.50 \cos(56.3° - 48.9°) = 1.49 \text{ m}$$

$$A = aw = 4(1.49) = 5.95 \text{ m}^2$$

Values of the merit function for a = 2 to a = 5 are computed and displayed:

a	w	A
2	2.54	5.08
3	2.02	6.07
4	1.49	5.95
5	0.98	4.88

These points are plotted on Fig. 8.28(b). The original interval of uncertain-

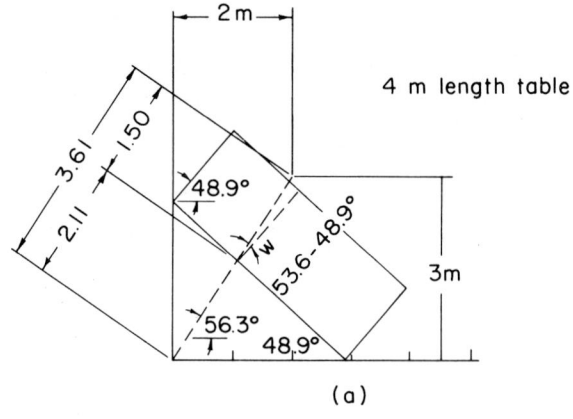

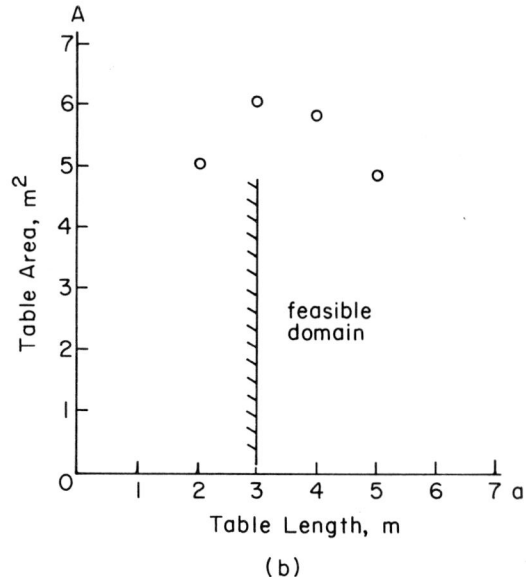

Fig. 8.28. (a) A table of length 4 m being maneuvered around a corridor corner. (b) The figure of merit; table area as function of table length a.

ty extends from a = 0 to a = 6.88 m (longest ladder). With our orientation work thus far we have reduced the interval to 2 ≤ a ≤ 4. A golden section strategy can be used to reduce it further. We have a constraint, expressible as w < 2 m, that can also be expressed approximately as a > 3 m. The lower bound on the interval of uncertainty can be set at 3 m.

Do this problem another way that does not use the information from the ladder problem (Fig. 8.27) and examine agreement or lack of it. Which analysis of the problem do you believe? Why?

PROBLEMS

8.1. The ability of a simply supported rectangular cross-section beam to carry a transverse load is proportional to the width b of the cross section

Optimization

and the (depth d)^2 (Prob. 3.6). Use Edelbaum's criterion to establish if the stationary points are maxima or minima.

8.2. The ability of a simply supported beam of rectangular cross section to resist transverse deflection (sagging) is proportional to the width b and the (depth d)^3 (Prob. 3.7). Use Edelbaum's criterion to establish if the stationary points are maxima or minima.

8.3. A sheet metal tray is to be made from a rectangular sheet of length a and width b, with four squares of size x cut from the sheet corners and the remaining tabs folded up to form the sides of the tray. Determine the length of the side x^* such that the tray volume is a maximum. Use Edelbaum's criterion to establish if the stationary points are maxima or minima.

8.4. Invert Prob. 8.3 by establishing the blank proportions such that a tray holding a litre is optimal for the blank from which it is formed and the blank has minimal area.

8.5. Solve Prob. 2.2, using Edelbaum's criterion to establish if the stationary points are maxima or minima.

8.6. Determine the stationary points of the following surfaces, identifying the character of the stationary points using Edelbaum's criterion.

(a) $x^2 + xy + y^2 - 6x + 2$

(b) $4x + 2y - x^2 + xy - y^2$

(c) $2x^2 - 2xy + y^2 + 5x - 3y$

(d) $x^3 - 3axy + y^3$

(e) $\sin x + \sin y + \sin(x + y)$

(f) $x^2 + xy + y^2 + 3x - 3y + 4$

(g) $x^2 + 3xy + 3y^2 - 6x + 3y - 6$

8.7. A manufacturer produces products A and B, costing $0.50 per kg and $0.60 per kg, respectively. The selling price in cents/kg of product A is x and of B is y. The marketplace will absorb amounts, in kg, of

$$M_A = 250(y - x) \qquad M_B = 32\,000 + 250(x - 2y)$$

To maximize profit, what are the selling prices x^* and y^*?

8.8. Partition the quantity c into three parts such that the product of the parts is a maximum.

8.9. Minimize $x^2 + y^2 + z^2$ subject to $x^2 + y^2 = 1$.

8.10. What point on the plane $5x + 3y + 2z = 19$ is closest to the origin? (This problem is equivalent to minimizing $x^2 + y^2 + z^2$ subject to $5x + 3y + 2z = 19$.)

8.11. Minimize $4x_1^2 + 5x_2^2$ subject to $2x_1 + 3x_2 = 6$.

8.12. Minimize $4x_1^2 + 5x_2^2$ subject to $2x_1 + 3x_2 \geq 6$.

8.13. Minimize $4x_1^2 + 5x_2^2$ subject to $x_1 \geq 2$.

8.14. Using the interval-halving search strategy, determine the optimal dimensions of a rectangular cross-section beam (Prob. 8.1) to be sawed from a right circular cylindrical log of 75 cm diameter within ±1 cm on the width b.

8.15. Using the interval-halving search strategy, determine the optimal dimensions of a rectangular cross-section beam (Prob. 8.2) to be sawed from a right circular cylindrical log of 80 cm diameter within ±2 cm on the width b.

8.16. Using the interval-halving search strategy and the figure of merit Eq. (3.1), establish the wire rope diameter of the optimal single strand hoist within ±1/16 in.

8.17. Using the interval-halving search strategy and your figure of merit from Prob. 8.3, establish the optimal proportions of a tray to be fabricated from a 20 cm by 40 cm sheet within ±5 mm on the depth.

8.18. Using the interval-halving search strategy, establish the optimal heater impedance within 0.1 Ω (Fig. 2.9) given the interval of uncertainty $1 \leq Z_2 \leq 3$ Ω.

8.19. Using the interval-halving search strategy, establish the optimal heater impedance within 0.1 Ω (Fig. 2.9) given the interval of uncertainty $2 \leq Z_2 \leq 3$ Ω.

8.20. Using an interval-halving search strategy, determine the optimal production mix of Prob. 2.2 given the interval of uncertainty $0 \leq x \leq 2$ (hundred) tires with a tolerance of ±5 tires (grade A).

8.21. Using an interval-halving search strategy, determine the optimal production mix of Prob. 2.2 given the interval of uncertainty $2 \leq x \leq 4$ (hundred) tires with a tolerance of ±5 tires (grade A).

8.22. Using a golden section search strategy, determine the optimal dimensions of a rectangular cross-section beam (Prob. 8.1) to be sawed from a right circular cylindrical log of 75 cm diameter within ±1 cm on the width dimension.

8.23. Using a golden section search strategy, determine the optimal dimensions of a rectangular cross-section beam (Prob. 8.2) to be sawed from a right circular cylindrical log of 80 cm diameter within ±2 cm on the width dimension.

8.24. Using a golden section search strategy and the figure of merit Eq. (3.1), establish the wire rope diameter of a two-strand hoist within ±1/16 in. with an interval of uncertainty $0 \leq d \leq 1.25$.

8.25. Using a golden section search strategy and the figure of merit from Prob. 8.3, establish the optimal dimensions of a tray to be fabricated from a 20 cm by 40 cm sheet within ±3 mm on the depth dimension.

8.26. Using a golden section search strategy, establish the optimal heater impedance within ±0.05 Ω (Fig. 2.9) given the interval of uncertainty $1 \leq Z_2 \leq 3$ Ω.

8.27. Using a golden section search strategy, establish the optimal heater impedance within ±0.05 Ω (Fig. 2.9) given the interval of uncertainty $2 \leq Z_2 \leq 4$ Ω.

8.28. Using a golden section search strategy, determine the optimal production mix of Prob. 2.2 given the interval of uncertainty $0 \leq x \leq 2$ (hundred) grade A tires with a tolerance of ±5 grade A tires.

8.29. Using a golden section search strategy, determine the optimal production mix of Prob. 2.2 given the interval of uncertainty $2 \leq x \leq 4$ (hundred) grade A tires with a tolerance of ±5 grade A tires.

8.30. If one were to relax the identical link condition of the example of Fig. 2.10, the figure of merit (M) becomes

$$M = -(\text{weight}) = -(W_1 + W_2) = -(A_1 l_1 + A_2 l_2)\rho$$

and it follows that

$$M = -nPx \frac{\rho}{S}\left[\frac{\cos \beta}{\sin(\alpha + \beta)\cos \alpha} + \frac{\cos \alpha}{\sin(\alpha + \beta)\cos \beta}\right]$$

where α and β are the abscissa angles of links 1 and 2, replacing θ. To establish the optimal geometry, the right-hand bracket must be minimized. For the angles α = 20° and β = 30°, what is the direction of steepest descent? For a step size Δ = 10°, determine Δα and Δβ, locate the new superior position, and evaluate the function at this location.

8.31. A straight corridor 2 m wide makes a turn of 120° in direction. What is the longest ladder that can be moved around the corner in a horizontal position?

8.32. For the conditions of Prob. 8.31, what is the widest table 3 m long that can be maneuvered around the turn?

8.33. An engineer decides to place four equally spaced function evaluations in an interval of a unimodal function before making a reduction in the interval of uncertainty. Derive an expression for the number of function evaluations necessary (a) for a fraction reduction F and (b) for establishing $x^* \pm a$. (c) How does this method compare in numbers of function evaluations with the example at the end of Sec. 8.7?

8.34. Repeat Prob. 8.33 with five equally spaced function evaluations per decision.

8.35. The function $f(x) = x^2 - 15$ has a root in the interval $3.8 \leq x_r \leq 3.9$. Choose figure of merit $M(x) = -|x^2 - 15|$ and find the abscissa corresponding to the maximum ordinate using either interval-halving or golden section search, establishing the root within ±0.001.

8.36. Minimize $(x_1 - 1)^2 + (x_2 - 2)^2$ subject to $2x_1 + x_2 \leq b$ where $b = 2$. Form the Lagrangian, identify the stationary points of the Lagrangian, sketch the contours of y on an $x_1 x_2$-plane and establish the value of y^*. Is the constraint tight or slack? What does Edelbaum's criterion say about the character of the point x_1^*, x_2^*? What is the sensitivity coefficient? If b is increased, would the value of y^* increase or decrease? If the increase in b is Δb, estimate the change in y^*. If $\Delta b = 0.1$, what is Δy^* by sensitivity estimate, and in fact?

8.37. Maximize $(x_1 - 4)^2 + (x_2 - 2)^2$ subject to $x_1^2 + x_2^2 \leq 4$.

8.38. Maximize $x_2^2 - 2x_1 - x_1^2$ subject to $x_1^2 + x_2^2 \leq 1$.

8.39. A total of 300 lineal ft of tubing must be installed in a shell-and-tube heat exchanger to provide the necessary heat transfer area. The cost is

$$\$ = 1500 + 50D^{2.5}L + 40DL$$

where D is the shell diameter, L is the shell length, and the units of \$ are dollars. (a) Minimize \$ subject to $5\pi D^2 L = 300$, $D \geq 0$, $L \leq 20$. (b) Suppose the heat transfer characteristics can be improved so that 290 linear ft of tubing is required. Estimate the new figure of merit $\$^*$ using sensitivity analysis.

8.40. Repeat Prob. 8.39 with the restriction that $L \geq 20$ ft.

8.41. A rectangle has the dimensions a and b. A square of side length a is removed from it, leaving a rectangle of the original proportions b/a, *ad infinitum*. What are the original proportions?

8.42. Show that η of golden section is a number such that $1/\eta = 1 + \eta$.

8.43. Show that the equation of Fig. 8.27 is $(y/a)^{2/3} = 1 - (x/a)^{2/3}$.

8.44. If the corner-to-corner line slope is rise over run, i.e., b_1/b_2, show that the abscissa angle of the ladder at a position just touching the corner (Fig. 8.26) is

$$\theta = \arctan\left[\frac{(b_1/b_2)^{2/3} + 1}{(b_2/b_1)^{2/3} + 1}\right]^{1/2}$$

> One cannot escape the feeling that these mathematical formulae have an independent existence of their own, and an intelligence of their own, that they are wiser than we are, wiser than their discoverers, that we get more out of them than was originally put into them.
>
> H. Hertz

CHAPTER 9

DECISIONS REACHED THROUGH MATHEMATICAL MODEL BUILDING

9.1 INTRODUCTION

We are ready to formulate some decisions employing mathematical modeling in an engineering context. We have labored through eight chapters of ideas and fragments of example

--to build a vocabulary
--to establish a philosophical framework
--to appreciate some tactics
--to examine some strategies
--to entertain some points of view
--to organize what we already know
--to learn a little more
--to identify what we may still need to know

so that we are in a position to appreciate the examples, the fund of knowledge they draw on, and the resources tougher problems could call on--becoming acquainted with resources forms the basis of much of engineering preparation.

When you enroll at your college or university you voluntarily subject yourself to a faculty whose objective is to teach you how to learn, what to learn, to criticize your accomplishments as you learn, and to evaluate your work against the expectation of the engineering profession in industry. Your instructor in this course is largely orienting you to what engineering involves and delineating the kinds of studies that lie ahead. The *pièce de résistance* in your undergraduate education is the final design course where you will have an opportunity to show your mettle in a creative engineering design context.

By coming this far you have done your part in allowing your instructor to show you "what it is all about" in your field of interest. Giving your teacher the opportunity to use mathematical models allows better communication with you than any other method, in my opinion.

There is a fringe benefit too. When the course is completed you will have many ideas related to mathematical model building in your head, as is usually assumed by subsequent instructors; then you will not have to learn the mathematical model building and the subject matter concurrently, with the instructor standing fairly mute on the modeling process.

The examples chosen are technically simple so that you can watch the modeling and the engineering and not be seriously distracted by new (and for present purposes, irrelevant) technical detail. There are no right answers--only right questions. Figure 9.1 outlines steps and questions to consider in solving problems.

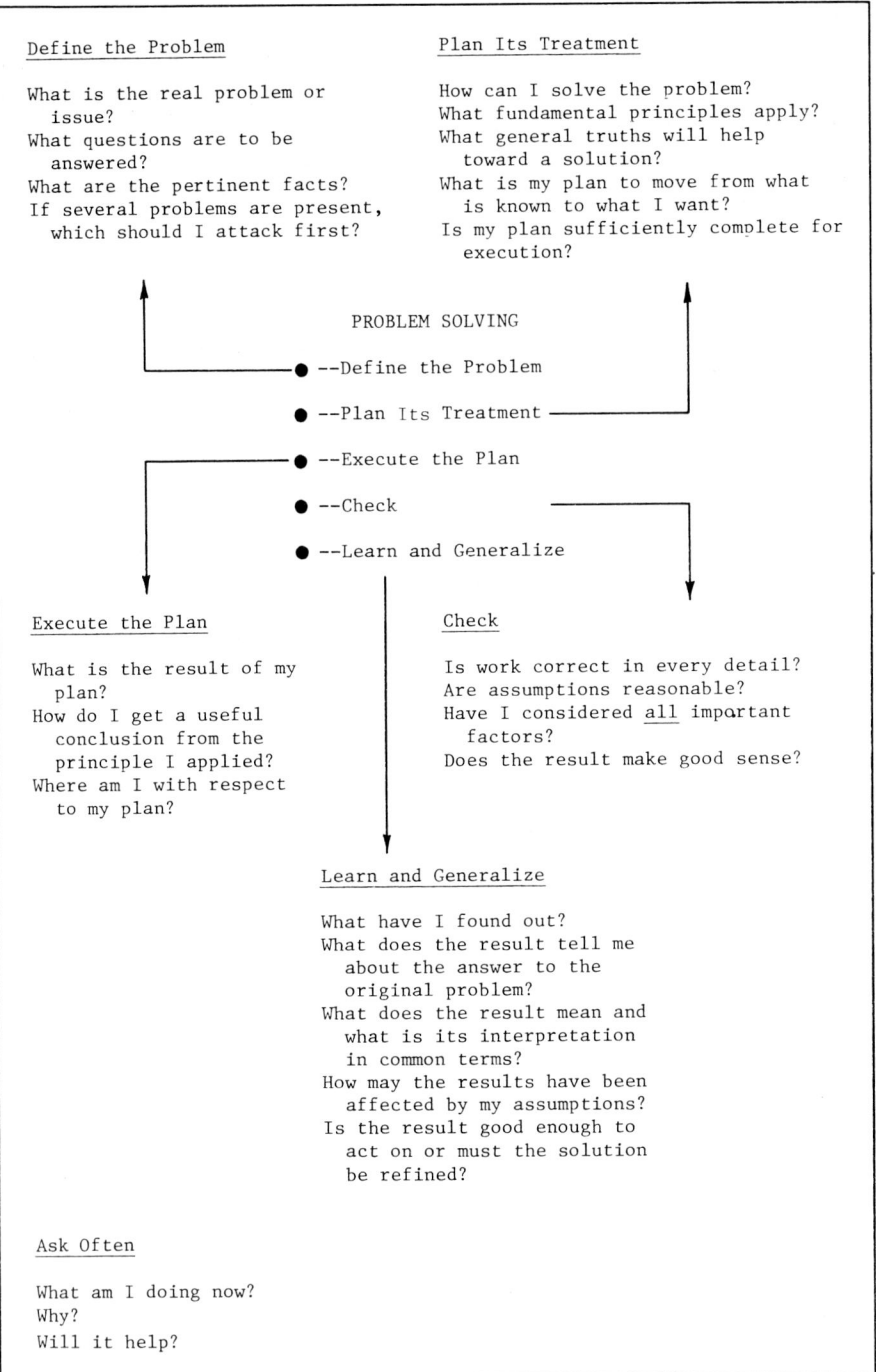

Fig. 9.1. Some pertinent questions to ask yourself while you are solving problems.

Decisions through Model Building   309

9.2 DECISION ON OVERLOADING OF A TRANSFORMER
The phone rings. It's the test technician from the experimental laboratory who has a rush job and no time to order and wait for the delivery of a transformer. The technician has located a transformer rated at 3000 W on the 11.5 V secondary. Twice this power is needed at this voltage but only for about 5 to 10 min. "Can I overload this transformer for 5 min and not damage it?" The name plate data are 3 kVA 115/11.5 V 60 Hz; manufacturer name and number are known. You say that you'll chase down some catalogs for more information, but in the meantime you ask for some additional information. You require the weight of the transformer and the copper loss in watts at rated load at steady state. The outside iron temperature might be useful, so you ask for that as well.

While the technician is wiring an experimental rig to determine the requested information, you have located a manufacturer's catalog. You find that it is an air-cooled transformer and the geometry is such that the iron almost completely surrounds the copper winding. Therefore, nearly all the heat generated in the copper passes through the iron on its way to the ambient air. The catalog gives the allowable copper temperature rise as 63.5°C, core weight 60 lbf, and total weight 95 lbf. The steady-state losses are given as 121 W at rated load.

The iron losses are due to hysteresis and eddy currents. The hysteresis loss is proportional to the (magnetic flux density)$^{1.6}$ and the eddy current losses are proportional to the (flux density)^2. Since a transformer at rated load runs at or near saturation, the doubling of the current load on the transformer will not materially increase the flux density and the iron losses can be considered constant. When the copper losses are reported, they can be subtracted from the 121 W given by the manufacturer to estimate the iron losses. We will follow the engineer's problem-solving process.

How shall I approach the problem? Will my lack of knowledge of geometry prohibit a rational decision? Out come the paper and pencil and the engineer begins writing. Interruption: The lab technician reports the rated load copper losses as 64 W cold and 77 W at rated load. The transformer weighs 94.7 lbf. The outside iron temperature rose 40°C during the test. The following data and thought processes are recorded.

    Transformer: 3 kVA 115/11.5 V 60 Hz
             Core weight: 60 lbf
             Total weight: 95 lbf
             Copper weight: about 30 lbf
             Insulation, varnish, paint: about 5 lbf
             Copper loss: cold, 64 W; hot, 77 W
             Iron loss (constant): 121 - 77 = 44 W
             Copper temperature rise: 63.5°C (114.2°F)
             Iron temperature rise: 40°C (72°F)

How long is it safe to use twice rated current? How does a transformer fail? Insulation failure. How is a transformer damaged? Insulation damage. Insulation life doubles for about every 8°C temperature drop. No damage will be incurred if the copper temperature (hottest in transformer complex) is always less than ambient plus 114.2°F.

Let T_c = copper temperature °F and T_i = iron temperature °F. Are copper and iron temperature the same throughout? The conductivity of copper is ex-

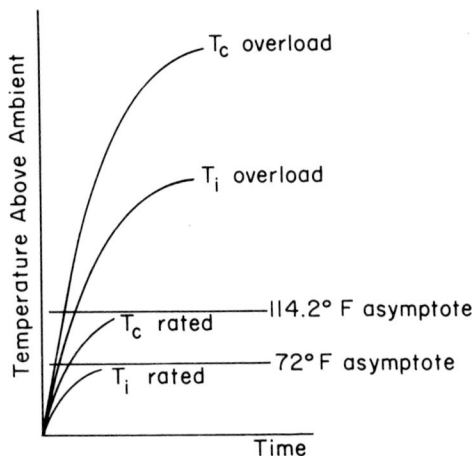

Fig. 9.2. Temperature-time relations for the transformer during rated load and overload.

cellent compared to its insulation. The same is true of iron compared to air but to a lesser degree--there might be a degree or two difference. Consider copper and iron isothermal bodies. A plot of temperature versus time is sketched in Fig. 9.2 to indicate anticipated relationships.

Copper. Take the metallic copper as a system of unknown geometry. The blob in Fig. 9.3 represents the boundary of the system. The boundary is surrounded by insulation. A significant influence of the surroundings on the system is a work effect due to electrical work performed on the system by the surroundings. Another significant influence of the surroundings on the system is a heat effect due to the presence of the surrounding insulation and iron. The heat effect is conductive and/or convective depending on whether insulation touches iron or touches air that touches iron. Consequently the heat transfer between the iron and copper is a mixture of conduction and convection. Both modes are proportional to $(T_c - T_i)$. The heat effect is given by $-\phi(T_c - T_i)$ and the work effect by $-3.413 W_c$ (see Fig. 9.3), where ϕ is a constant, W_c is copper loss (watts) and $3.413 \times$ Btu/hr = W [or 3.413 = Btu/(W·hr)]. The first law of thermodynamics for a system expressed as rates is

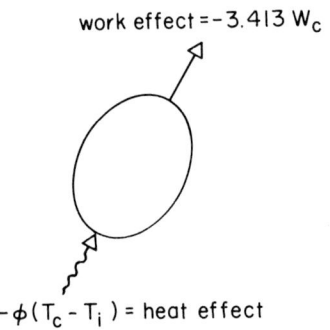

Fig. 9.3. Qualification of the significant influences of the surroundings on the copper of the transformer, which is the system.

Decisions through Model Building

$$\dot{Q} - \dot{W} = \dot{E}$$

where $\dot{E}$ is given by $m_c c_c \dot{T}_c$ and

m_c = copper mass = 30 lbm

c_c = copper specific heat capacity = 0.09 Btu/(lbm·°F)

$\dot{T}_c$ = time rate of change of T_c

The first law becomes

$$-\phi(T_c - T_i) + 3.413 W_c = m_c c_c \dot{T}_c$$

At steady rated load

T_c = 114.2°F (datum is ambient air temperature)

T_i = 72°F (datum is ambient air temperature)

W_c = 77 W

m_c = 30 lbm

c_c = 0.09 Btu/(lbm·°F)

$\dot{T}_c = 0$

and ϕ can be determined

$$-\phi(114.2 - 72) + 3.413(77) = 0$$

$$\phi = \frac{3.413(77)}{114.2 - 72} = 6.23 \text{ Btu/(hr·°F)}$$

The first law statement for the copper at any time becomes

$$-6.23(T_c - T_i) + 3.413 W_c = (30)(0.09)\dot{T}_c$$

From the above equation the rate of copper temperature rise is known any time T_c and T_i are known. They are known initially; consequently,

$$\dot{T}_c = 3.413(64)/[30(0.09)] = 80.9°F/hr$$

At *twice* rated load this rate will be quadrupled. From a cold start at twice rated load,

$$\dot{T}_c = 4(80.9) = 323.6°F/hr$$

This initial rate of copper temperature rise may be used to estimate the allowable time of overload as suggested in Fig. 9.4. The locus of T_c is not known at this point, but its trend is indicated in the graph. (The trend indicates t_A would be the time to stop the transformers.) Its initial slope is known. The 114.2°F allowable copper temperature limit is shown. The intersection of the straight line (initial slope of T_c locus) with the 114.2°F line t_B gives a time at which copper temperature is known to be less than 114.2°F above ambient. Thus

$$t = 114.2/323.6 = 0.353 \text{ hr} = 21.2 \text{ min}$$

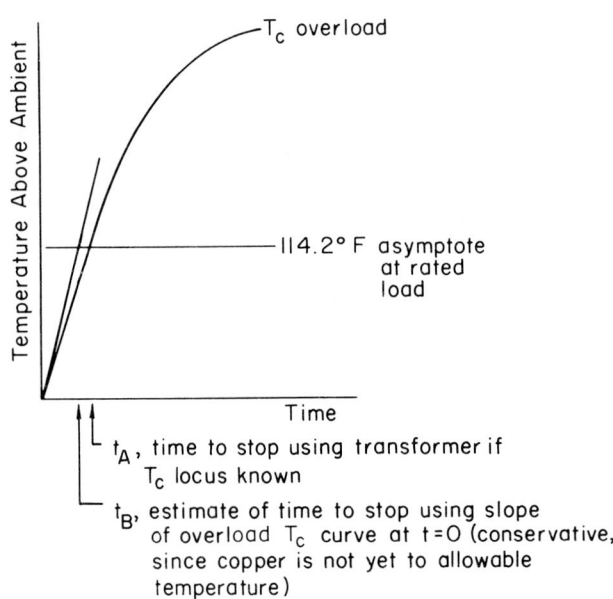

Fig. 9.4. Temperature-time relation of copper during overload; slope to curve at $T_c = 0$.

Summary. The model gives a conservative approximation to the allowable time.

Decision. Recommend overload use for a period not to exceed 10 min.

Resolution. I did not need the iron loss to solve this problem. I was tempted to ask for it to be obtained experimentally but it would have been wasteful of time and money. Resolved to solve problem symbolically before requesting experimental data.

Safeguards. To be sure the transformer is not abused, I will require an output ammeter be installed in circuit and that I be present during use to ensure cold start and timing.

We leave the engineer to call the technician and dictate a memo.

9.3 DECISION ON ROUTING OF A PIPELINE

A pipeline is to be built from Tallahassee, Florida, to Charlotte, North Carolina. Pumping station sites are available in Tallahassee, Tifton, Montgomery, Savannah, Atlanta, Augusta, Columbia, and Ashville. Figure 9.5 displays the topography of the problem. (The available pumping sites were chosen so that the solution is identifiable by inspection.) A computer program has been written that, given data on any two sites and the route between, will determine the parameters and describe the minimum cost configuration between the sites (pipe diameter, wall thickness, pumping station capacity) for the specified throughput. The entries in Table 9.1 represent feasible site-to-site optimal legs with the entry being the cost in millions of dollars. For example, the optimal leg from Tallahassee to Tifton costs $8.8 million for all pipe, installation, and maintenance costs for the next 20 years at the specified throughput. Your assignment: Devise a scheme that will enable the least cost route from Tallahassee to Charlotte to be identified. You write:

Decisions through Model Building 313

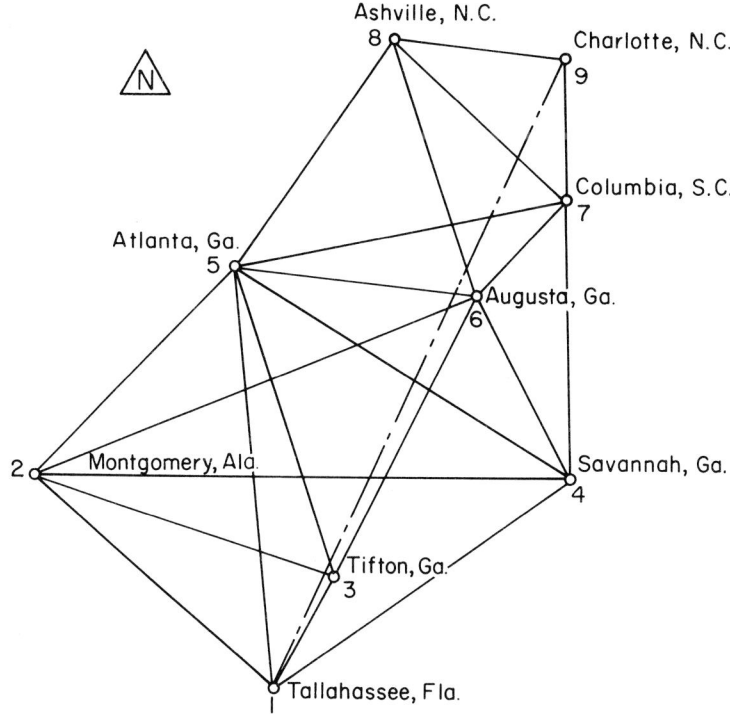

Fig. 9.5. Topographical map of city-to-city possible connections; cities are numbered in direction of progress.

Table 9.1. Merit Function--Computer Estimated Costs of Optimal Legs of Pipeline (circled entries involve back pumping)

		Mont-gomery	Tifton	Savan-nah	Atlanta	Augusta	Colum-bia	Ash-ville	Char-lotte
				(mil $)					
1.	Tallahassee	20.6	8.8	24.7	25.7				
2.	Montgomery		19.8	30.4	17.2	30.1			
3.	Tifton	(19.8)			18.7	19.5			
4.	Savannah	(30.4)			25.4	12.5	14.0		
5.	Atlanta	(17.2)	(18.7)	(25.4)		16.2	21.5	20.5	
6.	Augusta	(30.1)	(19.5)	(12.5)	(16.2)		7.2	17.5	
7.	Columbia			(14.0)	(21.5)	(7.2)		15.7	9.5
8.	Ashville				(20.5)	(17.5)	(15.7)		11.5

My assignment: Design a method that will allow the optimal route to be identified from a tabular figure of merit function.

What are the hidden problem-solving constraints? Computer compatibility for the method of solution is presumed. Exhaustive searches are not feasible for a general method that must be able to solve a problem with 50 pumping sites. I will use intuition to help rule out infeasible alternatives. For example, I would not want to "pump backwards," i.e., with a component of displacement site-to-site directed from Charlotte to Tallahassee. The first step is to secure a topographical map of the area so the connectivity can be displayed (Fig. 9.5). The second step is to draw a line from Tallahassee to Charlotte (or superpose a Cartesian grid on the map such that one grid line contains Tallahassee and Charlotte. Sites can now be numbered in increasing order from Tallahassee to Charlotte as we make progress across the grid. Would a rule of no back pumping make sense? Would we always want to pump from a smaller site number to a larger site number? In a dissipative process such as a fluid pumping, it does make sense, so eliminate all back pumping from the figure of merit table (the circled entries in Table 9.1). If there is any doubt concerning back pumping elimination, leave the entry in. Performing this ritual for the problem at hand confirms the assignment of site numbers (Fig. 9.5). These assignments agree with the order of the entries in Table 9.1. Has someone solved this problem before and am I being tested? Perhaps, but I still need the answer. How can I break a complex problem into simple steps? What problem can I solve right now? The optimal route from site 8 to site 9 costs $11.5 million, since there is no other route from site 8 to site 9. There's the clue! If I am at site 8, the optimal route to 9 is

$$(89) = (89)_0 = 11.5$$

where (89) is the cost of leg site 8 to site 9 and $(89)_0$ is the optimal cost of path from site 8 to site 9.

I have identified the step necessary to success, i.e., getting the fluid from site 8 (if 8 is in the final route) to the terminal at 9. If I find myself at site 8 the answer for optimal completion is (89), the only choice. Second thought: Wherever I am, the remaining path to Charlotte must be an optimal path. Supposing I am at site 7. I can pump to site 8 and complete with the optimal route from 8, which is now known $(89)_0$, or I may go directly from site 7 to site 9. The distinction is a simple numerical comparison between two alternatives and choosing the less costly. If at site 7, possible routes and costs are

$$(78) + (89)_0 = 15.7 + 11.5 = 27.2$$

$$(79) = 9.5 = (79)_0$$

The least cost route is site 7 to site 9 directly *if* I am at site 7. At site 6 I have three completion routes, two of which are feasible. If at site 6, possible routes and costs are

$$(67) + (79)_0 = 7.2 + 9.5 = 16.7 = (679)_0$$

$$(68) + (89)_0 = 17.5 + 11.5 = 29.0$$

(69) is infeasible

The least cost route from site 6 is $(679)_0$, costing $16.7 million. If at site 5, possible routes and costs are

$$(56) + (679)_0 = 16.2 + 16.7 = 32.9$$

$$(57) + (79)_0 = 21.5 + 9.5 = 31.0 = (579)_0$$

$$(58) + (89)_0 = 20.5 + 11.5 = 32.0$$

(59) is infeasible

Decisions through Model Building

If at site 4, possible routes and costs are

$(45) + (579)_0 = 25.4 + 31.0 = 56.4$

$(46) + (679)_0 = 12.5 + 16.7 = 29.2$

$(47) + (79)_0 = 14.0 + 9.5 = 23.5 = (479)_0$

(48) and (49) are infeasible

If at site 3, possible routes and costs are

(34) is infeasible

$(35) + (579)_0 = 18.7 + 31.0 = 49.7$

$(36) + (679)_0 = 19.5 + 16.7 = 36.2 = (3679)_0$

(37), (38), and (39) are infeasible

If at site 2, possible routes and costs are

$(23) + (3679)_0 = 19.8 + 36.2 = 56.0$

$(24) + (479)_0 = 30.4 + 23.5 = 53.9$

$(25) + (579)_0 = 17.2 + 31.0 = 48.2$

$(26) + (679)_0 = 30.1 + 16.7 = 46.8 = (2679)_0$

(27), (28), and (29) are infeasible

If at site 1, possible routes and costs are

$(12) + (2579)_0 = 20.6 + 46.8 = 67.4$

$(13) + (3679)_0 = 8.8 + 36.2 = 45.0 = (13679)_0$

$(14) + (479)_0 = 24.7 + 23.5 = 48.2$

$(15) + (579)_0 = 25.7 + 31.0 = 56.7$; remaining routes from 1 are infeasible

If I am at site 1 the optimal route is $(13679)_0$ at a cost of $45 million; the legs are uncovered the following way. At the site 1 calculation, the optimal route is $(19)_0$, which involves leg (13) plus a $(39)_0$ completion. Going to the site 3 calculation, the $(39)_0$ completion is composed of leg (36) plus $(69)_0$ completion, so the route so far is $(136)_0$. Going to the site 6 calculation block, $(69)_0$ is composed of leg (67) plus a $(79)_0$ completion, making the route so far $(1367)_0$. Going to site 7, we find $(79)_0$ is composed of leg (79). Thus the optimal route is $(13679)_0$ (Tallahassee, Tifton, Augusta, Columbia, Charlotte) if we are at site 1 and wish to proceed optimally to site 9. We are at site 1 and we do want to go to site 9; consequently we have the required solution. The method involves solving the problem backwards; i.e., asking the question, "If I am at the next-to-last site, what is the best way to the terminal?" then marching successively backward. The comparisons become more numerous as we move backward. At site 1 there are eight possible comparisons. The sparseness of the figure of merit matrix allowed only four comparisons to exhaust the possibilities. Figure 9.6(a) depicts the optimal route.

The second most economical route from site 1, (1479), costs $48.2 million

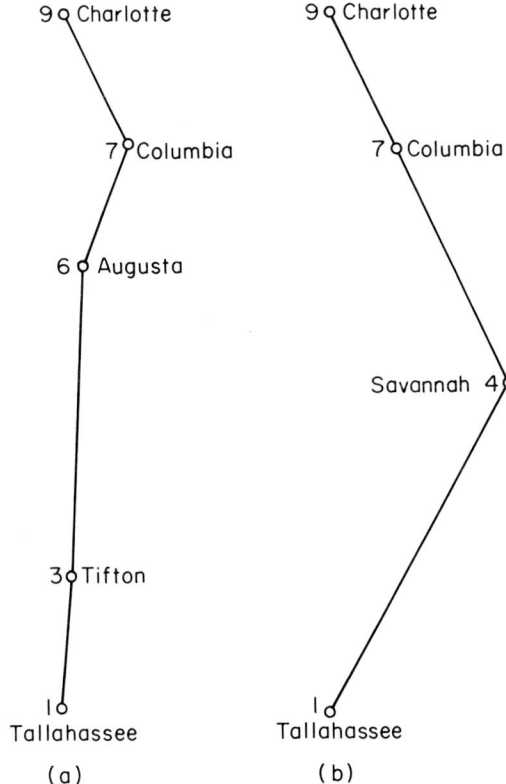

Fig. 9.6. (a) The optimal route from Tallahassee to Charlotte, costing $45 million. (b) The next best route costing $3.2 million more.

and has three pumping sites rather than the four required of the best solution. The economic cost of choosing three pumping sites rather than four is $48.2 million − $45.0 million = $3.2 million, an incremental cost. Figure 9.6(b) depicts this route.

We leave our engineer to write the report. This searching algorithm was coded by C. A. Timko (Prob. 33 of *Computers in Engineering Design Education*, vol. 6, College of Engineering, University of Michigan, Ann Arbor, Apr. 1, 1966). C. Schulz coded a similar approach for IOWA CADET.

9.4 DECISION ON PARAMETERS FOR A CIRCULAR DISK CAM

At times a translating displacement must be uniquely related to a rotational displacement. A mechanism to bring this about is known as a cam with a translating follower. This device is ordinarily expensive because the contouring of the cam surface for functional fidelity is a difficult and costly process. If the cam surface is a right circular cylinder, it can be created with a periphery ground to high precision at modest cost. When the function to be generated can be approximated to the "natural" functional relationship between a circular disk cam and a flat-faced translating follower, the opportunity for an inexpensive cam appears. An optimization process can establish

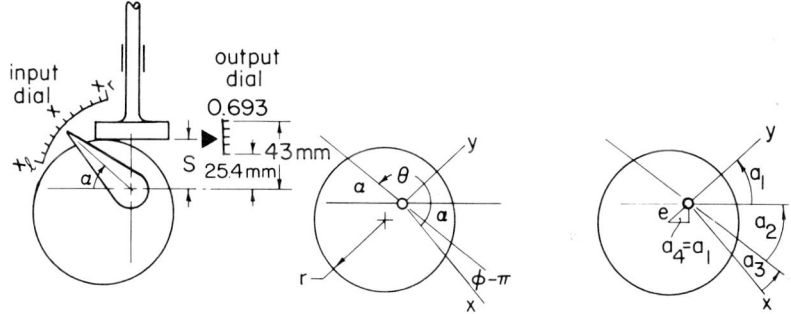

Fig. 9.7. Flat-faced follower circular disk cam as a function generator and the parameter relations.

design variables so that the function generator has the least error of all possible eccentric circular disk cams with flat-faced translating follower. Our problem at the moment is to relate the positions in the follower motion (where function generation is to be exact) to the parameters subject to engineering and manufacturing control such as diameter, eccentricity, and phase angle.

Our first move is to discover the natural functional relationship between the disk angular position and the follower displacement--a geometric problem. The cam and follower are depicted in Fig. 9.7. The clockwise angle between the left drawn horizontal line from cam pivot and the dial cursor is α, the dial angle. The distance of the cam pivot from the disk center is denoted e. The abscissa angle from the x-axis counterclockwise to the dial cursor is phase angle ϕ. The xyz-coordinate system is fixed in the disk with the y-axis containing the disk center and pivot and the x-axis containing the cam pivot. The distance s is the distance between the cam pivot and the face of the follower. The angle $a_3 = \phi - \pi$, the angle $a_2 = \alpha$, and the angle a_1 is complementary to $a_2 + a_3$ or

$$a_1 = \pi/2 - (a_2 + a_3) = \pi/2 - (\alpha + \phi - \pi) = 3\pi/2 - (\alpha + \phi)$$

Angle $a_4 = a_1 = 3\pi/2 - (\alpha + \phi)$. The distance is

$$S = r - e \sin a_4 = r - e \sin [3\pi/2 - (\alpha + \phi)]$$

$$S = r + e \cos(\alpha + \phi) \qquad (9.1)$$

The relationship between the follower position and the dial angle α is cosinusoidal as displayed in Eq. (9.1). The three variables under the designer's control are cam radius r, cam eccentricity e, and cursor abscissa angle ϕ. Since these variables are independent, it is possible to specify three input dial angles α_1, α_2, and α_3 at which the follower position is arbitrary and thereby have r, e, and ϕ determined. If the cam is to develop approximately the functional relation between input dial angle α and follower position S of

$$S = f(\alpha)$$

it follows that the cam will be able to generate $f(\alpha)$ exactly at three input dial angles α_1, α_2, and α_3. Thus

$$S_1 = f(\alpha_1) \quad S_2 = f(\alpha_2) \quad S_3 = f(\alpha_3)$$

and it is possible to write Eq. (9.1) three times as

$$S_1 = r + e\cos(\alpha_1 + \phi)$$

$$S_2 = r + e\cos(\alpha_2 + \phi)$$

$$S_3 = r + e\cos(\alpha_3 + \phi)$$

This set of three transcendental equations may be solved simultaneously for r, e, and ϕ. Subtracting the second equation from the first and the third equation from the first, we obtain

$$S_1 - S_2 = e[\cos(\alpha_1 + \phi) - \cos(\alpha_2 + \phi)]$$

$$S_1 - S_3 = e[\cos(\alpha_1 + \phi) - \cos(\alpha_3 + \phi)]$$

Expanding the trigonometric sums yields

$$S_1 - S_2 = e[(\cos\alpha_1)(\cos\phi) - (\sin\alpha_1)(\sin\phi) - (\cos\alpha_2)(\cos\phi) + (\sin\alpha_2)(\sin\phi)]$$

$$S_1 - S_3 = e[(\cos\alpha_1)(\cos\phi) - (\sin\alpha_1)(\sin\phi) - (\cos\alpha_3)(\cos\phi) + (\sin\alpha_3)(\sin\phi)]$$

Factoring the right-hand sides,

$$S_1 - S_2 = e[(\cos\alpha_1 - \cos\alpha_2)\cos\phi - (\sin\alpha_1 - \sin\alpha_2)\sin\phi]$$

$$S_1 - S_3 = e[(\cos\alpha_1 - \cos\alpha_3)\cos\phi - (\sin\alpha_1 - \sin\alpha_3)\sin\phi]$$

We will simplify by grouping given data as follows:

$$k_1 = \cos\alpha_1 - \cos\alpha_2 \qquad k_4 = \sin\alpha_1 - \sin\alpha_3$$

$$k_2 = \cos\alpha_1 - \cos\alpha_3 \qquad k_5 = S_1 - S_2$$

$$k_3 = \sin\alpha_1 - \sin\alpha_2 \qquad k_6 = S_1 - S_3$$

Substitution yields

$$k_5 = e[k_1 \cos\phi - k_3 \sin\phi] \tag{9.2}$$

$$k_6 = e[k_2 \cos\phi - k_4 \sin\phi] \tag{9.3}$$

Division of Eq. (9.2) by Eq. (9.3) eliminates the variable e:

$$\frac{k_5}{k_6} = \frac{k_1 \cos\phi - k_3 \sin\phi}{k_2 \cos\phi - k_4 \sin\phi}$$

Division of the numerator and denominator of the right-hand side of the above equation by $\cos\phi$ yields

$$\frac{k_5}{k_6} = \frac{k_1 - k_3 \tan\phi}{k_2 - k_4 \tan\phi}$$

The solution for ϕ is given by

$$\phi = \arctan \frac{k_1 k_6 - k_2 k_5}{k_3 k_6 - k_4 k_5} \tag{9.4}$$

Solving Eq. (9.2) for e gives

Decisions through Model Building

$$e = k_5/[k_1 \cos \phi - k_3 \sin \phi] \qquad (9.5)$$

and Eq. (9.1) yields r as

$$r = S_i - e \cos(\alpha_i + \phi) \qquad (i = 1, 2, 3)$$

Equations (9.1), (9.5), and (9.4) determine r, e, and ϕ from α_1, α_2, α_3, and S_1, S_2, S_3. The quadrant of ϕ may be determined from Eqs. (9.2) and (9.3). Solve these equations simultaneously for e sin ϕ and e cos ϕ, using Cramer's rule, and obtain

$$e \sin \phi = \frac{k_1 k_6 - k_2 k_5}{k_2 k_3 - k_1 k_4}$$

$$e \cos \phi = \frac{k_3 k_6 - k_4 k_5}{k_2 k_3 - k_1 k_4}$$

Since the eccentricity cannot be negative, the sign of sin ϕ and the sign of cos ϕ are established by the quotients displayed above. Figure 9.8 indicates the logic flow diagram by which the positive acute angle corresponding to ϕ can be converted to ϕ. Since computer arc tangent subprograms usually return a signed acute angle, it is convenient to define

$$A = \left| \arctan \frac{k_1 k_6 - k_2 k_5}{k_3 k_6 - k_4 k_5} \right|$$

and to define ϕ as illustrated in Fig. 9.8.

Suppose an eccentric circular disk cam needs to be designed to analogically generate the function $y = \ln x$. The variable x will be represented analogically by cam rotation as expressed in dial angle α in the range 45°-135°. The variable y will be represented analogically by the displacement of the follower, with the range from $\ln 1$ to $\ln 2$ represented by a displacement of 17.61 mm and the position $y = 0$ being 25.4 mm from the cam pivot. Figure 9.9(a) depicts the geometry of the situation.

The relationship between the dial angle and the input dial scale engraving of x is linear and can be found from the definition of scale factor:

$$\overline{PQ} = \alpha = s_x(x - \xi)$$

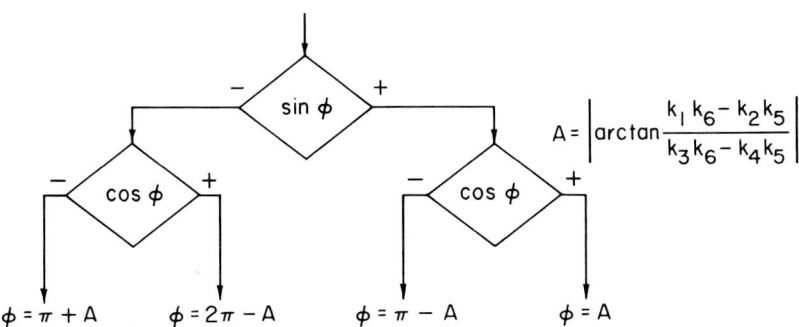

Fig. 9.8. Logic flow diagram for converting the positive acute angle corresponding to ϕ into angle ϕ.

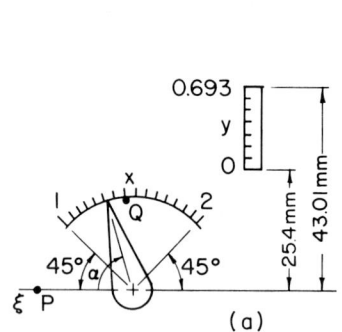
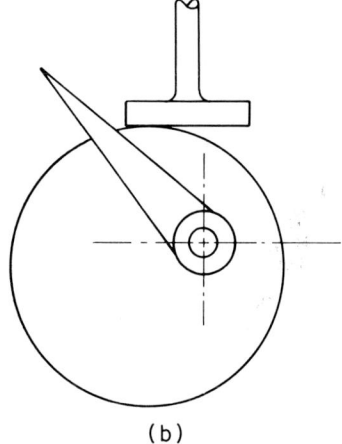

Fig. 9.9. (a) Scale relations for flat-faced function generating circular disk cam. (b) The geometry of the solution.

From the geometry of Fig. 9.9(a) the coordinate of P is ξ, which is numerically 0.5; thus

$$\alpha = s_x(x - 0.5)$$

At $x = 1$ the dial angle is $\alpha = 45°$, therefore

$$\alpha\bigg|_{x=1} = 45° = s_x(1 - 0.5) = 0.5 s_x$$

from which $s_x = 90°$. Consequently

$$\alpha = 90x - 45 \tag{9.6}$$

$$x = 1 + (\alpha - 45)/90 \tag{9.7}$$

The relationship between the cam displacement S and the output scale engraving is linear and can be found from the definition of the scale factor:

$$\overline{PQ} = S = 25.4 + Y = 25.4 + s_y y$$

The distance to $y = 0.693$ is given in Fig. 9.9(a) as 43.01 mm; therefore

$$S\bigg|_{y=0.693} = 25.4 + s_y(0.693) = 43.01$$

from which $s_y = 25.4$ mm. Consequently

$$S = 25.4(1 + y) \tag{9.8}$$

$$y = 0.0394S - 1 \tag{9.9}$$

Equations 9.6-9.9 and Eqs. (9.1), (9.4), and (9.5) become the basis for the synthesis of an eccentric circular disk cam function generator to meet the specified conditions.

The precision points will be placed at dial angles $\alpha_1 = 45°$, $\alpha_2 = 90°$, and $\alpha_3 = 135°$, a satisfactory although not optimal decision. Now

$$x_1 = 1 + (\alpha_1 - 45)/90 = 1 + (45 - 45)/90 = 1$$

$$x_2 = 1 + (\alpha_2 - 45)/90 = 1 + (90 - 45)/90 = 1.5$$

Decisions through Model Building

$$x_3 = 1 + (\alpha_3 - 45)/90 = 1 + (135 - 45)/90 = 2$$

$$y_1 = \ln x_1 = \ln(1) = 0$$

$$y_2 = \ln x_2 = \ln(1.5) = 0.405$$

$$y_3 = \ln x_3 = \ln(2) = 0.693$$

$$S_1 = 25.4 + 25.4y_1 = 25.4 + 25.4(0) = 25.4 \text{ mm}$$

$$S_2 = 25.4 + 25.4y_2 = 25.4 + 25.4(0.405) = 35.69 \text{ mm}$$

$$S_3 = 25.4 + 25.4y_3 = 25.4 + 25.4(0.693) = 43.01 \text{ mm}$$

It follows that

$$k_1 = \cos \alpha_1 - \cos \alpha_2 = \cos 45° - \cos 90° = 0.707$$

$$k_2 = \cos \alpha_1 - \cos \alpha_3 = \cos 45° - \cos 135° = 1.414$$

$$k_3 = \sin \alpha_1 - \sin \alpha_2 = \sin 45° - \sin 90° = -0.293$$

$$k_4 = \sin \alpha_1 - \sin \alpha_3 = \sin 45° - \sin 135° = 0$$

$$k_5 = S_1 - S_2 = 25.4 - 35.69 = -10.29$$

$$k_6 = S_1 - S_3 = 25.4 - 43.01 = -17.61$$

$$(k_1 k_6 - k_2 k_5) = [0.707(-17.61) - 1.414(-10.29)] = 2.10$$

$$(k_3 k_6 - k_4 k_5) = [(-0.293)(-17.61) - 0] = 5.16$$

$$(k_2 k_3 - k_1 k_4) = [1.414(-0.293) - 0] = -0.414$$

$$e \sin \phi = \frac{k_1 k_6 - k_2 k_5}{k_2 k_3 - k_1 k_4} = \frac{2.10}{-0.414} = -5.07$$

$$e \cos \phi = \frac{k_3 k_6 - k_4 k_5}{k_2 k_3 - k_1 k_4} = \frac{5.16}{-0.414} = -12.46$$

The last two equations identify ϕ as the third quadrant angle $\phi = 202.135°$. The eccentricity is

$$e = \frac{k_5}{k_1 \cos \phi - k_3 \sin \phi} = \frac{-10.29}{0.707(\cos 202.1°) + 0.293(\sin 202.1°)} = 13.45 \text{ mm}$$

The cam radius is

$$r = S_1 - e \cos(\alpha_1 + \phi) = 25.4 - 13.45 \cos(45° + 202.1°) = 30.63 \text{ mm}$$

The cam, depicted in Fig. 9.9(b), is specified by a diameter of 61.26 mm, an eccentricity of 13.45 mm, and an input dial cursor abscissa angle of 202.1°. The nature of the error of this device at nonprecision points is determinable by digital computation. The movement of the precision points is possible and a cam may be synthesized with least possible error. Note that the ability to establish r, e, and φ from a specification of α_1, α_2, and α_3 is an essential part of the synthesis process.

This problem was reducible to mathematical modeling entirely from consideration of the geometry and trigonometry of the mechanism part.

9.5 DECISION ON PAN ANGLE FOR A ROCK CRUSHER

Let us say that you have been assigned as the junior member of a project engineering team (on a rotating training assignment).[1] The project engineer is responsible for the design of some rock-crushing equipment. One of the problems concerning the conical pan and roller is whether the apex angle of the crushing surface is influenced principally by material feed considerations or by dynamic considerations. You are assigned the task of determining in what way the dynamic parameters are related.

The project engineer envisions a crusher that will resemble Fig. 9.10 but it will have three or four rollers. The feed to the pan is a granulated,

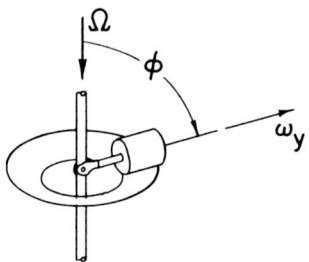

Fig. 9.10. One roller of a conical pan rock crusher. (From C. R. Mischke, 1963, *Elements of Mechanical Analysis*, Addison-Wesley, Reading, Mass., p. 352. (Used with permission.)

brittle material. A search of your immediately available resources gives no direct indication of how crushing is accomplished. From the engineer's verbal description, it appears that the crushing is accomplished mostly by the normal force between the roller and the pan: An oversized stone lifts the roller until it bears the entire "weight" of the roller itself. At this juncture, the stone either fractures or tumbles clear for another try under a subsequent roller. Plow-shaped fixtures on the roller arm return to the roller the material that slips out of its path.

The first decision, then, is that a larger normal force between the roller and the pan makes a more effective crusher. On what does this normal force depend? How large can it be made?

Consider Fig. 9.11. The reference xyz is fixed in the roller with the center of mass as the origin. The inertial reference XYZ is located as shown in the figure. The reference xyz has an angular velocity ω_y (about the Y-axis) that is related to the roller diameter r and to the radius of the roller path. The circumference of this path is $2\pi l \sin \phi$, and the circumference of

---

1. Adapted from Mischke, *Elements of Mechanical Analysis*, Addison-Wesley, Reading, Mass., pp. 351-52, 355. Used with permission.

Decisions through Model Building

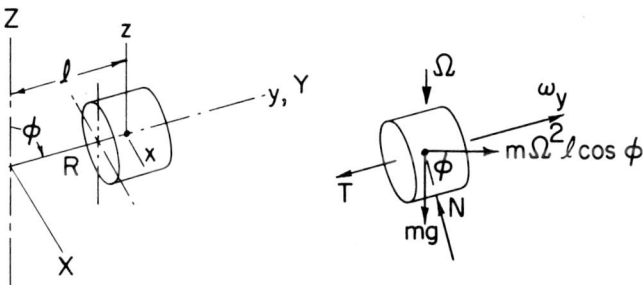

Fig. 9.11. Geometry and free body of a roller. (From Mischke, *Mechanical Analysis*, p. 354. Used with permission.)

the roller is $2\pi r$. The relation between ω_y and Ω is

$$\omega_y/\Omega = l \sin \phi / r$$

In coordinate system XYZ (which is inertial), for the roller as a system,

$$\overline{\Sigma F} = m\overline{a}$$

and in a direction through the origin of xyz and perpendicular to the Z-axis, the body force (centrifugal force due to rotation) is $mr^2 l(\sin \phi)$. The significant influence of the support rock on the roller is modeled as a vector force $\overline{N}$. The weight of the roller is mg in the Z-direction. Taking the component in direction of N,

$$\Sigma F = ma = N - mg(\sin \phi) = -m\Omega^2 l (\sin \phi)(\cos \phi)$$

or dimensionlessly

$$N/(mg) = [\sin \phi + \frac{1}{2}(\Omega^2 l/g)\sin 2\phi]$$

The π-term $N/(mg)$ is first an increasing and then a decreasing function of pan angle ϕ. As a unimodal function with an interior stationary point as a maximum, the maximum $N/(mg)$ can be determined by differentiating $N/(mg)$ with respect to ϕ and equating to zero:

$$\frac{d}{d\phi}\left(\frac{N}{mg}\right) = (\cos \phi + \frac{\Omega^2 l}{g} \cos 2\phi) = 0$$

From the identity $\cos 2\phi = 2\cos^2 \phi - 1$, we write

$$\cos \phi + \frac{\Omega^2 l}{g}(2\cos^2 \phi - 1) = 0$$

$$\cos^2 \phi + \frac{g}{2\Omega^2 l}(\cos \phi) - \frac{1}{2} = 0$$

This is a quadratic in $\cos \phi$ with the roots

$$\cos \phi = \frac{1}{4}\left\{-\frac{g}{\Omega^2 l} \pm \left[\left(\frac{g}{\Omega^2 l}\right)^2 + 8\right]^{1/2}\right\}$$

It is possible for Ω^2 to be zero; consequently $g/(\Omega^2 l) \to \infty$. Since $\cos \phi$ cannot exceed unity, the plus sign prevails. Setting $g/(\Omega^2 l) = \beta$, we obtain

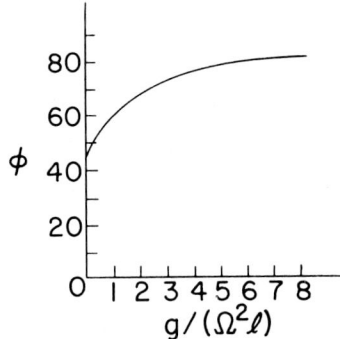

Fig. 9.12. The relation between ϕ and $g/(\Omega^2 l)$. (From Mischke, *Mechanical Analysis*, p. 354. Used with permission.)

$$\cos \phi = (1/4)(-\beta + \sqrt{\beta^2 + 8})$$

As indicated in Fig. 9.12, the optimal angle ϕ for maximum N is found from knowledge of β. Figure 9.13 depicts the equation

$$\frac{N}{mg} = \left(\sin \phi + \frac{\sin 2\phi}{2\beta}\right)$$

This equation is parameterized by β. Clearly, the larger maxima of the normal force increase with increasing angular velocity Ω. In the application at hand, Ω will be small because of the very nature of the crushing process. Therefore, the actual value of $g/(\Omega^2 l)$ will approach infinity and $N/(mg)$ will be given by the lowest curve of the family of Fig. 9.13. This curve is a simple sine curve and the maximum is very broad. In other words, from 80° to 90° there is only a very small change in $N/(mg)$.

In the light of the foregoing analysis, your report to the project engineer should say in substance that the principal parameter in crushing is the weight of the roller, mg, and at probable speeds and geometries the angle ϕ

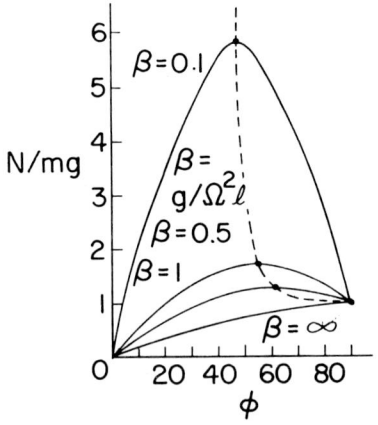

Fig. 9.13. The dimensionless normal force plotted against the conical pan half-apex angle ϕ parameterized by the acceleration ratio. (From Mischke, *Mechanical Analysis*, p. 355. Used with permission.)

Decisions through Model Building

can be determined by parameters other than dynamic ones, since the dynamic parameters have little influence. The slope of the conical crushing pan will be dictated more by the flow and circulation of the granulated material than by dynamic considerations.

9.6 DECISION ON PROPORTIONS OF A CAN

The diameter and height of right circular cylindrical metal food cans are functions of the volume and the pressure of contents, carton stacking, and packing constraints, etc. Our task is to find the proportions of a litre can fabricated of 0.5 mm sheet steel of three parts, as depicted in Fig. 9.14. The can is assembled by dishing and crimping the ends over a right circular cylindrical sleeve. Sheet metal must be provided for the purpose; the allowances are

r_1 = dishing allowance on container ends, mm

r_2 = crimping allowance on container ends, mm

r_3 = seam allowance for forming cylindrical sleeve, mm

h_1 = crimping allowance for attaching ends, mm

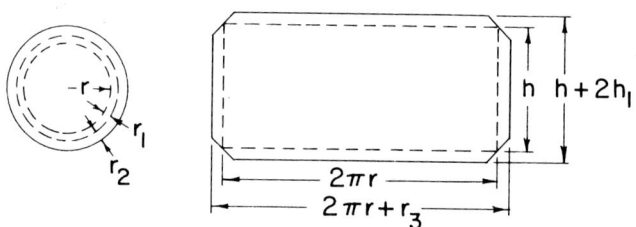

Fig. 9.14. The geometry of parts of a right circular cylindrical sheet metal can. (From C. R. Mischke, 1968, *An Introduction to Computer-Aided Design*, Prentice-Hall, Englewood Cliffs, N.J., p. 97. Used with permission.)

Many combinations of diameter and height will result in a litre interior volume. We choose to seek the combination that results in the least sheet metal requirement. We are presuming for the sake of simplicity that punch press scrap is recyclable at no cost penalty. We estimate the approximate proportions first with the approximation of negligible allowances.

The figure of merit is chosen as the ratio of volume of can interior to the area of sheet metal required to fabricate the can. The volume of a can is approximated by

$$V = \pi r^2 h$$

The sheet metal area of two disks and one rectangle (neglecting allowances) is

$$A = 2\pi r^2 + 2\pi r h$$

and the figure of merit is

$$M = \frac{V}{A} = \frac{\pi r^2 h}{2\pi r^2 + 2\pi rh}$$

The variables r and h are not independent, since the volume of the can must be 1 litre, or 1000 cm³. Thus we have identified a functional constraint

$$\pi r^2 h = 1000 \text{ cm}^3 = 10^6 \text{ mm}^3$$

By back-substituting this equality constraint into the figure of merit we can reduce the dimensionality of the problem to one independent variable. We choose r to be our design variable and eliminate h:

$$M(r) = \frac{\pi r^2 h}{2\pi r^2 + 2\pi rh} = \frac{10^6}{2\pi r^2 + 2(10^6/r)}$$

From the nature of the figure of merit function we observe that the function vanishes when $r \to 0$ and $r \to \infty$. We conclude that the function is unimodal for maximization in the interval $0 < r < \infty$ and the extreme of merit is at an interior stationary point. Therefore calculus would be useful. Setting the first derivative equal to zero, we have $dM/dr = 0$, which requires $4\pi r - 2(10^6/r^2) = 0$ or

$$r = \left[\frac{2(10^6)}{(4\pi)}\right]^{1/3} = 54.2 \text{ mm} \quad h = \frac{10^6}{\pi r^2} = \frac{10^6}{\pi(54.2)^2} = 108.4 \text{ mm}$$

leading us to suspect that the optimal proportion of such a can is diameter equals height.

For the inclusion of allowances we proceed numerically:

$$h = \frac{10^6}{\pi r^2} \qquad M = \frac{\pi r^2 h}{2\pi(r + r_1 + r_2)^2 + (2\pi r + r_3)(h + 2h)}$$

and use an interval-halving or golden section search strategy to determine the optimal proportions.

If the allowances are

$$r_1 = 2 \text{ mm} \quad r_2 = 3 \text{ mm} \quad r_3 = 4 \text{ mm} \quad h = 2 \text{ mm}$$

then

$$M = \frac{10^6}{2\pi(r + 5)^2 + (2\pi r + 4)(h + 4)}$$

Let us choose an interval of uncertainty $25 < r < 75$, comfortably bracketing the $r = 54.2$ mm estimate previously determined (see Fig. 9.15).

Interval of uncertainty (25, 75):

$$x_1 = 25 + 0.382(75 - 25) = 44.1 \qquad h_1 = \frac{10^6}{\pi r_1^2} = 163.7 \qquad M_1 = 16.06$$

$$x_2 = 25 + 0.618(75 - 25) = 55.9 \qquad h_2 = 101.9 \qquad M_2 = 16.41$$

The reduced interval of uncertainty is (44.1, 75):

Decisions through Model Building

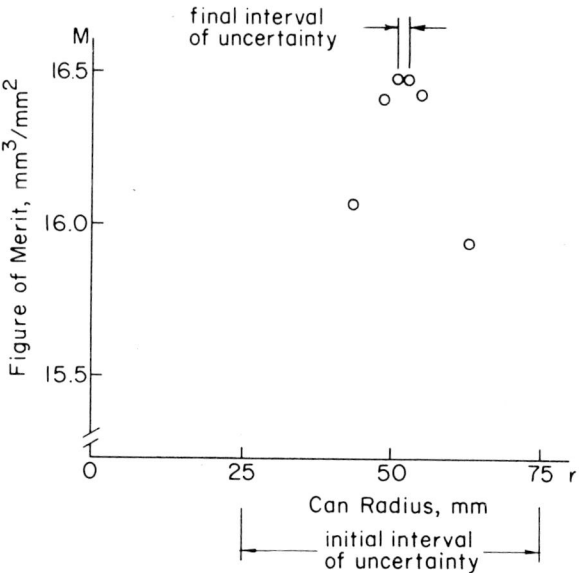

Fig. 9.15. Figure of merit data points of can problem as a function of the can radius.

$x = 44.1 + 0.382(75 - 44.1) = 55.9 \quad h_2 = 101.9 \quad M_2 = 16.41$

$x_3 = 44.1 + 0.618(75 - 44.1) = 63.2 \quad h_3 = 79.7 \quad M_3 = 15.92$

The reduced interval of uncertainty is (44.1, 63.2):

$x_4 = 44.1 + 0.382(63.2 - 44.1) = 51.4 \quad h_4 = 120.5 \quad M_4 = 16.48$

$x = 44.1 + 0.618(63.2 - 44.1) = 55.9 \quad h = 101.9 \quad M = 16.41$

The reduced interval of uncertainty is (44.1, 55.9):

$x_5 = 44.1 + 0.382(55.9 - 44.1) = 48.6 \quad h_5 = 134.8 \quad M_5 = 16.40$

$x = 44.1 + 0.618(55.9 - 44.1) = 51.4 \quad h = 120.5 \quad M = 16.48$

The reduced interval of uncertainty is (48.6, 55.9):

$x = 48.6 + 0.382(55.9 - 48.6) = 51.4 \quad h = 120.5 \quad M = 16.48$

$x_6 = 48.6 + 0.682(55.9 - 48.6) = 53.6 \quad h_6 = 110.8 \quad M_6 = 16.48$

We have been carrying our calculations of radius and height to one decimal place and our calculation of figure of merit to two decimal places. Now we are unable to distinguish between M and M_6. We may increase the number of significant digits carried in our calculations, which becomes a bit more cumbersome; or we may program our pocket calculator such that we store all important numbers, reading the output dial only to monitor the solution. Another alternative is to program this problem on a digital computer. This procedure is simple if a unimodal one-dimensional search strategy is available (as it is in IOWA CADET).

We will allow readers to reduce the interval of uncertainty to 1 mm, carrying more significant digits. Before proceeding, venture an educated prediction as to whether the optimal can proportions are a height longer than the diameter or a height shorter than the diameter.

If a computer program is used, one can observe the sensitivity of the optimal proportions to r_1, r_2, r_3, and h_1.

9.7 PREDICTING BALANCING SPEEDS OF ELECTRIC RAILWAY EQUIPMENT

The Chicago, North Shore & Milwaukee Railroad was an interurban electric railway whose standard multiple-unit passenger cars were equipped with four Westinghouse W557R5 dc series wound traction motors geared 25:52 to 36 in. diameter wheels. The cars weighed 104 400 lb. Balancing speeds on level tangent track in still air of these 55 ft long cars were:

Line Voltage, V	1 Car Train, mph	4 Car Train, mph
650	70.3	82.0
600	66.8	77.5
575	65.0	75.0
550	63.1	72.6
525	61.0	70.0
500	59.1	68.0

This equipment was used in trains up to 8 cars in length for Chicago-Milwaukee service. Determine the balancing speed of a 6 car train at 600 V (see Fig. 9.16).

Our first step is to state as an equality the fact that the accelerating force on the train is the difference between the tractive effort (propelling force of wheels) and the train resistance. At the balancing speed there is no acceleration and this difference is zero.

$$\text{(train resistance)} - \text{(train tractive effort)} = 0 \qquad (9.10)$$

Train resistance on level tangent track is composed of bearing, rolling, and air resistances. These resistances were extensively investigated by W. J. Davis, Jr., and his equations are

$$r, \text{ lbf/T} = 1.3 + 29/w + 0.03V + 0.0024AV^2/(wn) \quad \text{lead car}$$

$$= 1.3 + 29/w + 0.03V + 0.000\ 34AV^2/(wn) \quad \text{trailing car}$$

where: w = axle loading in T/axle
n = numbers of axles per car
A = frontal area of car, ft^2
V = train speed, mph

The first three terms of his equations are generally valid, but the coefficients 0.0024 and 0.000 34 are somewhat equipment specific. For some order of magnitude calculations, let us evaluate r at 70 mph for a lead car:

$$r = 1.3 + 29/13.1 + 0.03(70) + 0.0024(100)(70)^2/[(13.1)4]$$

$$= 1.3 + 2.21 + 2.1 + 22.4 = 28 \text{ lbf/T}$$

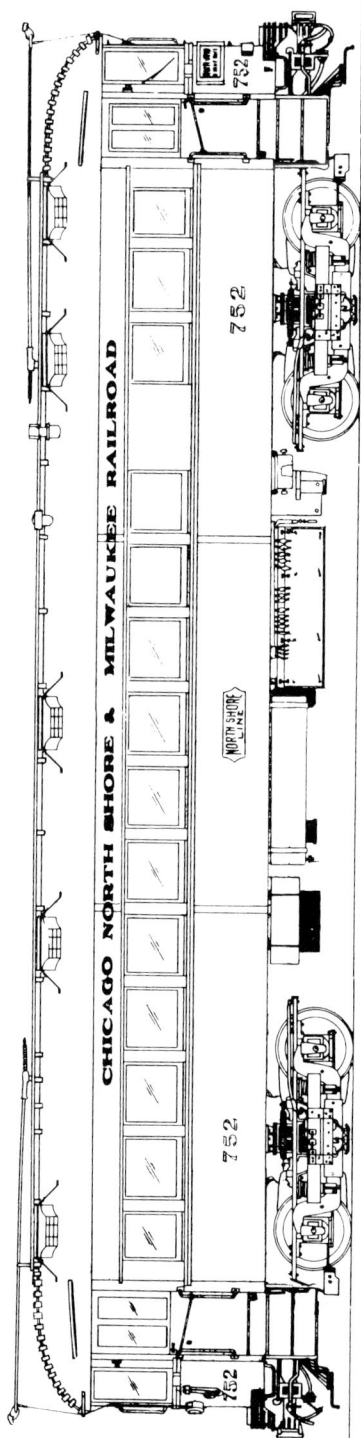
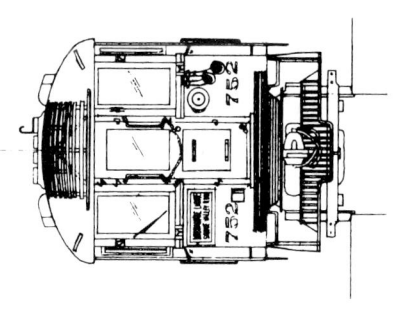
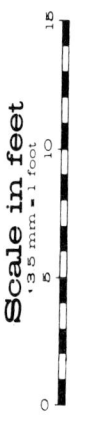

Fig. 9.16. Interurban electric railway coaches of the Chicago, North Shore & Milwaukee Railroad, series 752-776 as originally built. (Copyright 1963 by Central Electric Railfans' Association, Bull. 107, p. 145. Reproduced by permission.)

The bulk of the resistance is sensitive to V^2. For a trailing car,

$$r = 1.3 + 29/13.1 + 0.03(70) + 0.000\ 34(100)70^2/[(13.1)4]$$
$$= 1.3 + 2.21 + 2.1 + 3.17 = 8.78\ \text{lbf/T}$$

These differences are due to the head-end resistance of the lead car, not present in a trailing car. Let N = number of cars in train and W = weight of car in tons. Then Davis's equation becomes

$$R,\ \text{lbf} = rwnN = 1.3wnN + 29/w\ wnN + 0.03VwnN + 0.0024 + 0.000\ 34(N-1)AV^2wn/(wn)$$
$$= 1.3WN + 29nN + 0.03VWN + (0.002\ 06 + 0.000\ 34N)AV^2 \quad (9.11)$$

The constants 1.3, 29, and 0.03 will be quite stable among railway passenger cars generally, but the particular North Shore architecture imposed by necessity of the Chicago elevated operation (cars somewhat narrower, lower, and shorter than conventional passenger cars) require that the numbers 0.002 06 and 0.000 34 be determined for this particular equipment. We will rewrite Davis's equation, incorporating constants α and β as follows:

$$R,\ \text{lbf} = 1.3WN + 29nN + 0.03VWN + (\alpha + \beta N)AV^2 \quad (9.12)$$

where α and β are yet to be determined.

Tractive effort: Westinghouse W557R5 motors are rated at 140 hp at 600 V at 910 rpm. The shaft torque, in.·lbf, is given by

$$T = 63025\ \text{hp/rpm} = 63025(140)/910 = 9700\ \text{in.·lbf}$$

We need to note that 910 rpm is 95.3 rad/s. The running characteristic of a series dc motor [see Fig. 9.17(a)] is given by

$$T\omega^a = K$$

where: T = torque, in.·lbf
  ω = armature angular velocity, rad/s
  a = constant, 3
  K = constant

Figure 9.17(b) shows how a traction motor is connected to the truck axle by a wheelbarrow mount and constrained to the truck frame. Figure 9.17(c) shows the geometric relationship between pinion, gear, and car wheel.

Taking a = 3

$$K = T\omega^3 = (9700)(95.3)^3 = 8.4(10^9)\ \text{in.·lbf/s}^3$$

Referring to Fig. 9.17(d), we can ascertain the relationship between motor torque and tractive effort. Summing moments about the wheel center,

$$(T.E.)D/2 = TG/P$$

where: D = wheel diameter, in.
  G = number of teeth on gear
  P = number of teeth on pinion
  v = train speed, ft/s
  V = train speed, mph
  ω_w = angular velocity of wheels, rad/s

Solving for tractive effort T.E.,

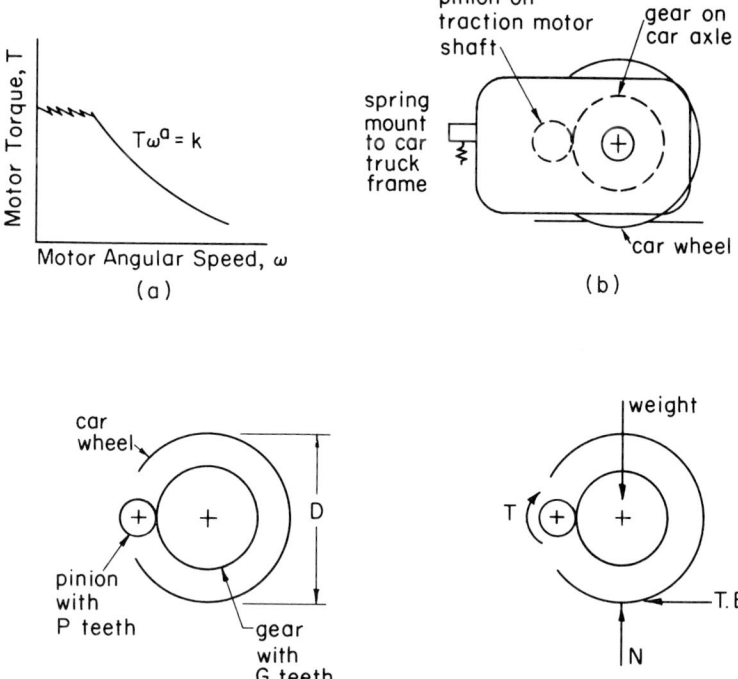

Fig. 9.17. (a) Traction motor torque-speed characteristic; (b) wheelbarrow mount of electric traction motor to car axle; (c) nomenclature; (d) free body of wheel and gears as a system.

$$\text{T.E.} = \frac{2GT}{DP} = \frac{2G}{DP} \frac{K}{[(G/P)\omega_w]^3}$$

Substituting for motor torque,

$$\text{T.E.} = \frac{2}{D}\left(\frac{P}{G}\right)^2 \frac{K}{\omega_w^3} = \frac{2}{D}\left(\frac{P}{G}\right)^2 \frac{KD^3}{8v^3(1728)}$$

Since 88 ft/s is 60 mph,

$$\text{T.E.} = \left(\frac{PD}{G}\right)^2 \frac{K}{[4(88)/60]^3 v(1728)} = 0.386(10^6)\left(\frac{DP}{G}\right)^2 \frac{1}{v^3}$$

If there are m traction motors,

$$\text{T.E.} = m\left(\frac{PD}{G}\right)^2 \frac{0.386(10^6)}{v^3} \tag{9.13}$$

We are now in a position to determine coefficients α and β for North Shore equipment. At the beginning of the section we were given that the 600 V balancing speed of a single four-motor passenger car is 66.8 mph. Now (train resistance) - (train tractive effort) = 0; therefore

$$1.3(52.2)(1) + 29(4)(1) + 0.03(66.8)(52.2)(1)$$
$$+ [\alpha + \beta(1)]100(66.8)^2 - 4\left(\frac{25}{52}\ 36\right)^2 \frac{0.386(10^6)}{(68.8)^3} = 0$$

from which $\alpha + \beta = 0.002\ 84$. Four interurban cars balance at 77.5 mph,

$$1.3(52.2)(4) + 29(4)(4) + 0.03(77.5)(52.2)(4)$$
$$+ (\alpha + 4\beta)100(77.5)^2 - 16\left(\frac{25}{52}\ 36\right)^2 \frac{0.386(10^6)}{(77.5)^3} = 0$$

from which $\alpha + 4\beta = 0.004\ 63$. The simultaneous solution of

$$\alpha + \beta = 0.002\ 84$$
$$\alpha + 4\beta = 0.000\ 463$$

is $\alpha = 0.002\ 243$, $\beta = 0.000\ 597$. Thus the Davis equation, modified for North Shore equipment, is

$$R,\ \text{lbf} = 1.3WN + 29nN + 0.03VWN + (0.002\ 243 + 0.000\ 597N)AV^2 \qquad (9.14)$$

and the tractive effort is

$$T.E. = m\left(\frac{PD}{G}\right)^2 \frac{0.386(10^6)}{V^3}\ \text{lbf}$$

Our task is to determine the balancing speed of a 6 car interurban train at 600 V.

(train resistance of 6 cars) - (tractive effort of 6 cars) = 0

$$1.3(52.2)(6) + 29(4)(6) + (0.03)V(52.2)(6)$$
$$+ [0.002\ 243 + 0.000\ 597(6)]100V^2 - (6)(4)\left(\frac{25}{52}\ 36\right)^2 \frac{0.386(10^6)}{V^3} = 0$$

from which

$$f\left(\frac{V}{100}\right) = 1103.16\left(\frac{V}{100}\right)^3 + 939.6\left(\frac{V}{100}\right)^4 + 5825\left(\frac{V}{100}\right)^5 - 2775.1 = 0$$

The zero place of f can be ascertained using Newton-Raphson first order root-finding strategy. We will need $f'(V/100)$ also.

$$f'\left(\frac{V}{100}\right) = 3309.48\left(\frac{V}{100}\right)^2 + 3758.4\left(\frac{V}{100}\right)^3 + 29125\left(\frac{V}{100}\right)^4$$

Recalling that $x_{i+1} = [x - f(x)/f'(x)]_i$, we write

$$\left(\frac{V}{100}\right)_{i+1} = \left[\left(\frac{V}{100}\right) - \frac{1103.16\left(\frac{V}{100}\right)^3 + 939.6\left(\frac{V}{100}\right)^4 + 5825\left(\frac{V}{100}\right)^5 - 2775.1}{3309.48\left(\frac{V}{100}\right)^2 + 3758.4\left(\frac{V}{100}\right)^3 + 29125\left(\frac{V}{100}\right)^4}\right]_i$$

The initial guess is $(V/100) = 0.8$. Successive values are 0.800, 0.798,

Decisions through Model Building

0.795, and 0.795, and we report 79.5 mph as the balancing speed of a 6 car train at 600 V.

North Shore suburban cars differ only in weight (100 200 lbf) and gearing (23:54 on 36 in. wheels). For a 6 car suburban train,

$$1.3(50.1)(6) + 29(4)(6) + 0.03V(50.1)(6)$$

$$+ [0.002\ 243 + 0.000\ 597(6)]100V^2 - 6(4)\left(\frac{23}{54}\ 36\right)^2 \frac{0.386(10)^6}{V^3} = 0$$

$$f\left(\frac{V}{100}\right) = 996.6\left(\frac{V}{100}\right)^3 + 901.8\left(\frac{V}{100}\right)^4 + 5825\left(\frac{V}{100}\right)^5 = 2178 - 0$$

$$f'\left(\frac{V}{100}\right) = 2989.8\left(\frac{V}{100}\right)^2 + 3607.2\left(\frac{V}{100}\right)^3 + 29125\left(\frac{V}{100}\right)^4$$

The iteration equation becomes

$$\left(\frac{V}{100}\right)_{i+1} = \left[\left(\frac{V}{100}\right) - \frac{996.6\left(\frac{V}{100}\right)^3 + 901.8\left(\frac{V}{100}\right)^4 + 5825\left(\frac{V}{100}\right)^5 - 2178}{2989.8\left(\frac{V}{100}\right)^2 + 3607.2\left(\frac{V}{100}\right)^3 + 29125\left(\frac{V}{100}\right)^4}\right]_i$$

Choosing $(V/100)_0 = 0.8$. Successive values are 0.800, 0.761, 0.757, and 0.757, and we report 75.7 mph as the balancing speed of a 6 car suburban train at 600 V.

9.8 DECISION ON SPEED OF A WIRE WINDER

An arrangement for the winding of resistance wire on a flat form that rotates at constant speed is depicted in Fig. 9.18. A brake is applied to the spool to prevent backlash. The tolerable tension in the 0.010 in. diameter wire cannot exceed 5 lbf. A full spool weighs 5 lbf and an empty spool weighs 0.5 lbf. The radius of gyration of the empty spool is 1.1 in. If the brake setting (brake torque) remains constant, what is the proper brake adjustment for the largest possible rpm of the form? What is the maximum angular speed of the form?

The wire is accelerated, causing tensile force therein (which cannot exceed 5 lbf from position A in Fig. 9.19 to some intermediate position between A and B). In the last part of the half revolution of the form, the wire decelerates and the brake prevents the spool from overrunning the wire. The least tension in the wire is 0 lbf.

The first move toward a solution is to determine the acceleration of the wire at its two extremes: A, when the form first touches the new wind of wire, and B, when the form completes wrapping the wire about its end. From Fig. 9.19 our problem is to determine the linear acceleration of the wire, which is the second time derivative of x, at $\phi = 0$ and $\phi = \pi$. The intuitive thought that this acceleration is the same as for points A and B considered to be on the form and moving in a circle is not correct as we shall see. The law of cosines for the triangle requires

$$x^2 = b^2 + r^2 - 2br\cos\phi$$

Consequently, $x = (b^2 + r^2 - 2br\cos\phi)^{1/2}$. Denoting dx/dt as $\dot{x}$ and noting that $d\phi/dt = \omega$, we write

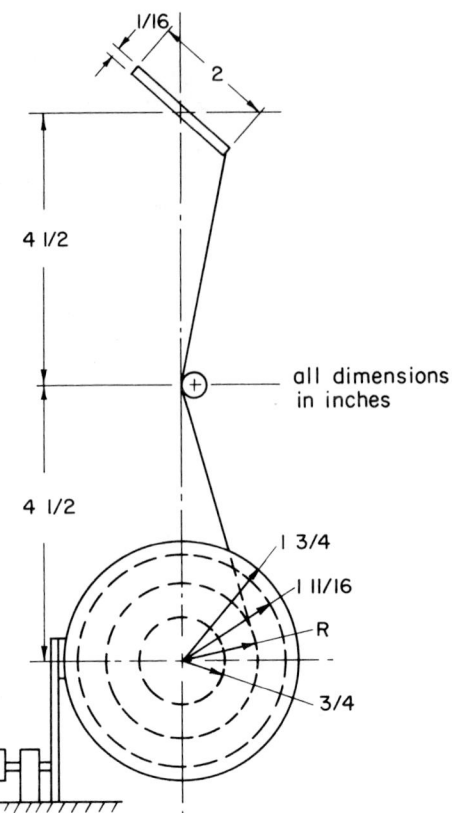

Fig. 9.18. Diagram of a wire winder depicting form, reel, and brake.

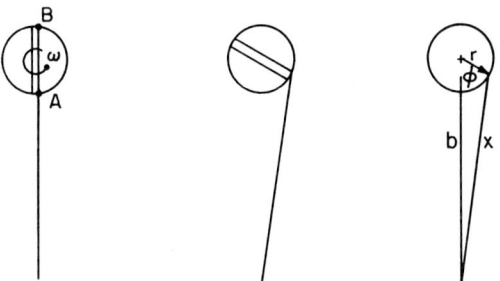

Fig. 9.19. Geometry of the wire-form interaction.

$\dot{x} = (1/2)(b^2 + r^2 - 2br \cos \phi)^{-1/2}(2br\omega \sin \phi) = br\omega \sin \phi (b^2 + r^2 - 2br \cos \phi)^{-1/2}$

The second derivative of x with respect to time is denoted as $\ddot{x}$ and

$\ddot{x} = (br\omega \sin \phi)(-1/2)(b^2 + r^2 - 2br \cos \phi)^{-3/2}(2br\omega \sin \phi)$
$\quad + br\omega^2 \cos \phi (b^2 + r^2 - 2br \cos \phi)^{-1/2}$

Decisions through Model Building 335

This simplifies to

$$\ddot{x} = \frac{br\omega^2}{(b^2 + r^2 - 2br\cos\phi)^{1/2}}\left(\cos\phi - \frac{br\sin^2\phi}{b^2 + r^2 - 2br\cos\phi}\right) \quad (9.15)$$

$$\ddot{x}\Big|_{\phi=0} = \frac{br\omega^2}{(b^2 + r^2 - 2br)^{1/2}} = \frac{r\omega^2}{1 - r/b} = \frac{(1/12)\omega^2}{1 - 1/4.5} = 0.1067\omega^2 \text{ ft/s}^2$$

$$\ddot{x}\Big|_{\phi=\pi} = \frac{-br\omega^2}{(b^2 + r^2 + 2br)^{1/2}} = \frac{-r\omega^2}{1 + r/b} = \frac{-(1/12)\omega^2}{1 + 1/4.5} = -0.0681\omega^2 \text{ ft/s}^2$$

Note that only as b approaches infinity do the accelerations approach $r\omega^2$ and $-r\omega^2$, respectively.

We need an expression for the rotary inertia of the wire on the spool. If the inside radius of a hollow right circular cylinder is a and the outside radius is b, then the inertia is

$$I = \int_a^b (2\pi r l/g) r^2 \, dr = [\pi\gamma l/(2g)](b^2 - a^2)(b^2 + a^2)$$

The weight of the wire on the spool is $W = \pi(b^2 - a^2)l\gamma$; consequently

$$I = [W/(2g)](b^2 + a^2)$$

For a full spool

$$I_f = [(4.5)/2(32.2)][(27/16)^2 + (12/16)^2](1/144) = 0.001\,66 \text{ slug}\cdot\text{ft}^2$$

The rotary inertia of an empty spool, I_s, is W/g multiplied by (radius of gyration)^2,

$$I_s = Wk^2/g = 0.5(1.1/12)^2/32.2 = 0.000\,13 \text{ slug}\cdot\text{ft}^2$$

The total rotary inertia of the full spool is $I_t = I_f + I_s = 0.001\,66 + 0.000\,13 = 0.001\,79$ slug·ft^2. The maximum linear acceleration of the wire occurs with the maximum spool-tangential-acceleration component of the particle of wire just leaving the coil; the magnitude of this acceleration is $a_t = 0.1067\omega^2$. The minimum $a_t = -0.0681\omega^2$ for same reason.

The spool is isolated from the remainder of the universe as a system; see Fig. 9.20. The significant influences are the surface traction of the brake on the spool, which manifests itself as a braking moment B, and the surface tractions of the adjacent wire, which manifest themselves as a normal traction F. The system is rotating about an axis of rotation N fixed in inertial space (the earth), and Euler's equation for this circumstance simplifies to $\Sigma M = I\alpha$. Here ΣM is the sum of all the tractive moments and α is the angular acceleration of the system about its axis of rotation, which can be expressed as the tangential acceleration a_t divided by the radius of reeled wire R. The torque is the wire tension F times the reeled wire radius R. We write

$$FR - B = I\alpha = Ia_t/R$$

When the spool is full of wire,

$$(FR)_{max} = FR_f = B + 0.1067I_t\omega^2/R_f = (5)(27)/[(12)(16)] = 0.702 \text{ lbf}\cdot\text{ft}$$

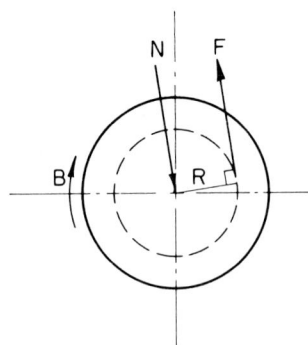

Fig. 9.20. The reel (spool) as a system.

$$(FR)_{min} = B - 0.0681 I_t \omega^2 / R_f = 0$$

When the spool is empty of wire,

$$(FR)_{max} = FR_s = B + 0.1067 I_s \omega^2 / R_s = (5)(3)/[(12)(4)] = 0.312 \text{ lbf·ft}$$

$$(FR)_{min} = B - 0.0681 I_s \omega^2 / R_s = 0$$

Thus for a full spool,

$$B + 0.001\ 31\omega^2 = 0.702 \qquad B - 0.000\ 84\omega^2 = 0$$

and for an empty spool,

$$B + 0.000\ 222\omega^2 = 0.312 \qquad B - 0.000\ 142\omega^2 = 0$$

We have four conditions and two unknowns. The controlling equations must be sought. Solving the full spool equations yields

$$\omega^2 = (0.702)/0.000\ 021\ 5 = 326$$

$$\omega = 18.0 \text{ rad/s} = 173 \text{ rpm} \qquad B = 0.000\ 84(326) = 0.274 \text{ lbf·ft}$$

Substituting $\omega = 18.0$ and $B = 0.274$ into all four equations results in

$$(FR_f)_{max} = B + 0.001\ 31\omega^2 = 0.274 + 0.001\ 31(326) = 0.702$$

$$(FR_f)_{min} = B - 0.000\ 84\omega^2 = 0.274 - 0.000\ 84(326) = 0$$

$$(FR_s)_{max} = B + 0.000\ 222\omega^2 = 0.274 + 0.000\ 222(326) = 0.367 \quad \text{(too big)}$$

$$(FR_s)_{min} = B - 0.000\ 142\omega^2 = 0.274 - 0.000\ 142(326) = 0.229$$

The corresponding wire tensions are 5, 0, 5.88, and 3.93 lbf. The third one is greater than 5 lbf and the speed must be lowered. The controlling equations are

$$B + 0.000\ 222\omega^2 = 0.312 \qquad B - 0.000\ 84\omega^2 = 0$$

Decisions through Model Building

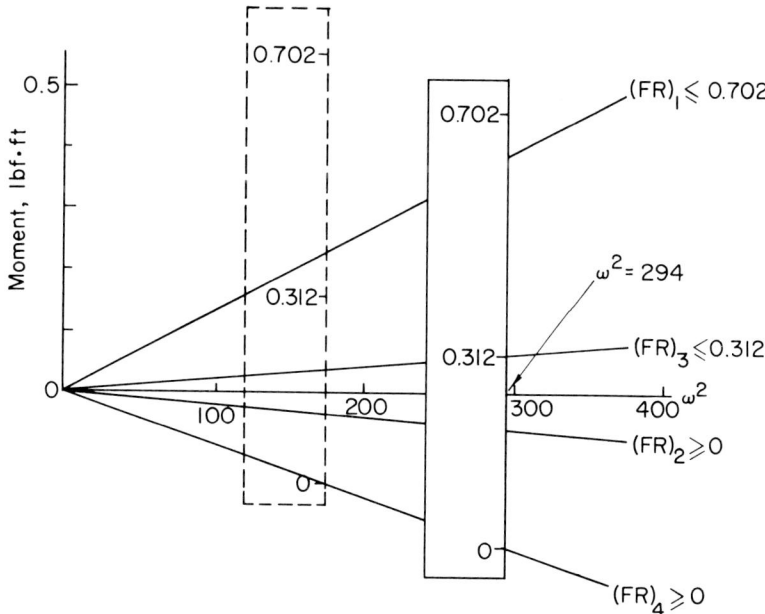

Fig. 9.21. Wire torque versus ω^2 plotted with all loci emanating from an arbitrary point B = 0.

from which $\omega = 17.1$ rad/s = 163.5 rpm and B = 0.247. Another check indicates

$(FR_f)_{max} = 0.247 + 0.001\ 31(294) = 0.632$

$(FR_f)_{min} = 0.247 - 0.000\ 84(294) = 0$

$(FR_s)_{max} = 0.247 + 0.000\ 222(294) = 0.312$

$(FR_s)_{min} = 0.247 - 0.000\ 142(294) = 0.205$

The corresponding wire tensions are 4.48, 0, 5, and 3.28 lbf. A check is in order. The torque on the spool due to the wire tension is

$(FR)_1 = B + 0.001\ 31\omega^2 \qquad (FR)_3 = B + 0.000\ 222\omega^2$

$(FR)_2 = B - 0.000\ 84\omega^2 \qquad (FR)_4 = B - 0.000\ 142\omega^2$

where $(FR)_1 \leq 0.702$, $(FR)_2 \geq 0$, $(FR)_3 \leq 0.312$, and $(FR)_4 \geq 0$. On a plot of ω^2 versus wire torque, all these lines emanate from the same ordinate intercept, the brake torque B; two have positive slopes, two have negative slopes, and all are rectilinear, since the choice of coordinates rectifies the loci. The ordinate erected at the place of largest possible ω^2 would touch one of the positive constraints and one of the zero constraints. Figure 9.21 shows the loci plotted from arbitrary point B = 0. From the ordinate scale, mark a piece of paper with the constraint values 0.702, 0.312 and 0. Place the paper in a vertical orientation with the origin on the $(FR)_4$ curve; translate to the right, keeping the origin on $(FR)_4$, until either the $(FR)_1$ curve touches 0.702 or the $(FR)_3$ curve touches the 0.312, whichever happens first. Translation to

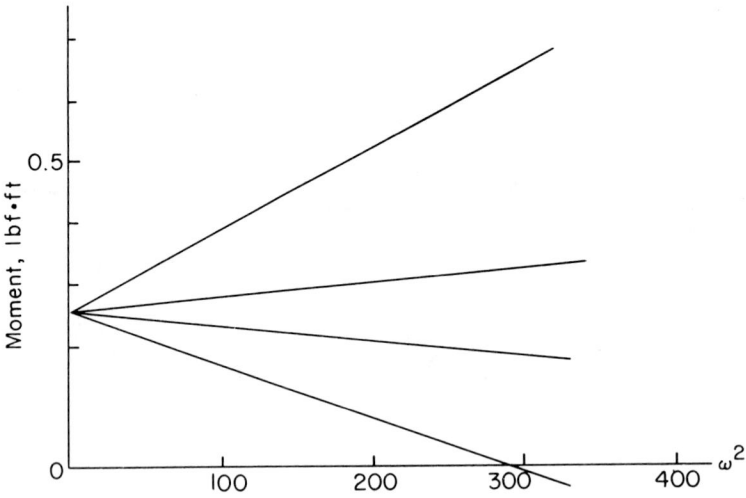

Fig. 9.22. The final plot with the brake torque B established.

the right is stopped at $\omega^2 = 294$ where the $(FR)_3$ line touches the 0.312 as depicted in Fig. 9.21. This confirms the answer previously obtained. Now that B is known, the final plot is displayed in Fig. 9.22.

9.9 DECISION ON A STEP-DOWN GEAR RATIO

Electric motors that are operated at various speeds and geared to rotating loads work against not only the load shaft torque but also inertia torques provided by the resistance of rotating bodies to angular acceleration. The task of analysis is complicated by the fact that some of the inertia is present on the armature shaft in the form of the armature itself and the pinion and present on the load shaft in the form of the gear and load inertias. Fig-9.23(a) depicts a motor shaft connected to a load through a single reduction gear mesh. Separation of the two shafts and their gears into two systems is depicted in Fig. 9.23(b) and (c). The tangential force F at the gear pitch cylinder radius is the manifestation of the influence of the surroundings on the system *for inertia purposes*. For the body in Fig. 9.23(b), the sum of all the torques is the product of the moment of inertia of the body about its rotational axis and the angular acceleration of the body about its rotational axis.

$$\Sigma T = I\dot{\omega}$$

$$T_1(\omega_1) - Fr_1 = (I_a + I_p)\dot{\omega}_1 \tag{9.16}$$

For the body in Fig. 9.23(c),

$$\Sigma T = I\dot{\omega}$$

$$T_2(\omega_2) - Fr_2 = (I_g + I_l)\dot{\omega}_2 \tag{9.17}$$

Equations (9.16) and (9.17) may be combined, eliminating the force F:

$$F = \frac{T_1(\omega_1) - (I_a + I_p)\dot{\omega}_1}{r_1} = \frac{T_2(\omega_2) - (I_g + I_l)\dot{\omega}_2}{r_2} \tag{9.18}$$

Decisions through Model Building 339

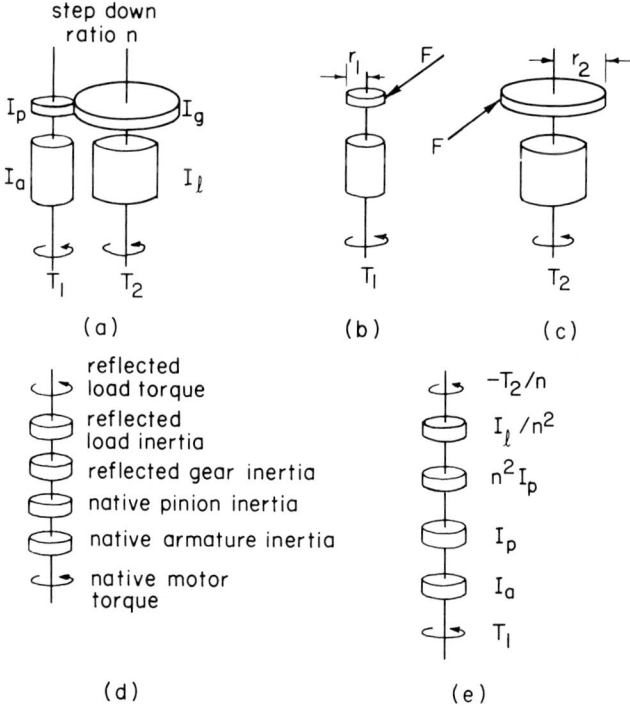

Fig. 9.23. (a) A motor armature and load geared together; (b) and (c) free bodies of both shafts. Shish kebab inertia equivalent on motor shaft: (c) (qualitative) and (d) (quantitative).

Noting that the step-down ratio n is given by

$$n = r_2/r_1 = -\omega_1/\omega_2$$

and $\omega_2 = -\omega_1/n$ and $\dot{\omega}_2 = -\dot{\omega}_1/n$, we can rearrange Eq. (9.18) to read

$$T_1(\omega_1) - T_2(\omega_2)/n = [I_a + I_p + (I_g/n^2) + (I_l/n^2)]\dot{\omega}_1$$

Noting also that the inertia of a disk pinion is (a constant times its pitch cylinder radius)^4 and similarly for a disk gear, we conclude that $I_g = n^4 I_p$ and substitute for I_g in the above relation to give

$$T_1(\omega_1) - T_2(\omega_2)/n = (I_a + I_p + n^2 I_p + I_l/n^2]\dot{\omega}_1 \qquad (9.19)$$

Inspection of Eq. (9.19), viewing the entire gear train problem from the motor shaft and as an instance of $\Sigma T = I\dot{\omega}$, suggests three rules:

1. The inertia on a second shaft is reflected to the first shaft as the native inertia divided by (step-down ratio)^2, or $I_{eq.} = I_l/n^2$. The load inertia I_l in Eq. (9.19) is an example of such an inertia on a second shaft.
2. The inertia of a disk gear on a second shaft *in mesh* with a disk pinion on the first shaft is reflected to the pinion shaft as the *pinion* inertia multiplied by (step-down ratio)^2, or $I_{eq.} = n^2 I_p$. The gear inertia appears as $n^2 I_p$ in Eq. (9.19).

3. A torque on a second shaft geared to a first shaft is reflected to the first shaft as the negative of the original torque divided by the step-down ratio, or $T_{eq.} = -T_2/n$.

Figure 9.23(e) depicts an equivalent rigid system on one shaft, formulated by rules 1, 2, and 3.

The equivalent inertia may be viewed as the inertia of the two-shaft system depicted as a shish kebab on the motor shaft. For the system in Fig. 9.23(a), represented as a shish kebab equivalent in Fig. 9.23(d), the equivalent inertia is

$$I_{eq} = I_a + I_p + n^2 I_p + I_l/n^2$$

The inertia may be reduced by attenuating I_a, I_p, or I_l. Attempts to reduce inertia by increasing the step-down ratio increases gear inertia $n^2 I_p$ while reducing load inertia I_l/n^2. In many machinery applications, gear inertia $n^2 I_p$ is small compared to load inertia I_l/n^2; consequently an increase in step-down ratio usually reduces I_{eq}. For instrument drives and others with gear and load inertias of the same order of magnitude, I_{eq} is minimized when $n = (I_l/I_p)^{1/4}$.

In some circumstances the inertia of the second shaft may be significant. Using rule 1,

$$I'_{eq} = n^2 I_{eq} = n^2 I_a + n^2 I_p + n^4 I_p + I_l$$

According to rule 3, the motor torque transfers as $-T_1(\omega_1)/n$ as in Fig. 9.23(e). The Newtonian equation is

$$T_2(\omega_2) - T_1(\omega_1)/n = (n^2 I_a + n^2 I_p + n^4 I_p + I_l)\dot{\omega}_2$$

If $T_2 = 0$ and T_1 is constant,

$$\dot{\omega}_2 = -T_1(nI_a + nI_p + n^3 I_p + I_l/n)^{-1}$$

Maximizing the absolute value of $\dot{\omega}_2$ is tantamount to minimizing $nI_a + nI_p + n^3 I_p + I_l/n$. The optimal step-down ratio is no longer $(I_l/I_p)^{1/4}$.

The rules just developed are useful in analysis and synthesis wherein torques and inertias are coupled through gearing. As an example, let us consider the problem of making a decision on the gear ratio between a motor and a rotary tipple to be used to dump unit train coal cars. In this application the cars are dumped one by one without uncoupling. The time required to dump 100 cars includes the sum of individual dumping times. We will simplify the problem by

--considering as significant the time to rotate dumping cradle and car through
  π radians
--ignoring the displacement of the center of mass of car and load from rotational axis through couplers
--ignoring the change in inertia as the load leaves the car
--ignoring friction

Under these conditions the resistance of the car and load inertia to the rotational acceleration is the principal resistance to motion. A motor supplies the necessary work through a reduction gear box (or boxes). The traction mo-

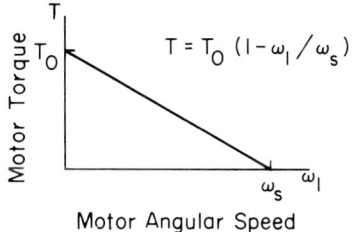

Fig. 9.24. Motor-angular speed characteristic.

tor characteristics are approximated by the linearization in Fig. 9.24. The car dumper is supported by four rollers and driven by the motor through a gear reduction. The dumper with its car and load is taken as the system; the significant influences of the system are the roller reactions N_1, N_2, N_3, and N_4 and the weight of the dumper, car, and load (see Fig. 9.25). The line of action of W is taken through the axis of rotation of the dumper. The surroundings exert a torsional moment on the dumper of magnitude nT (rule 3).

Using Newton's equation relating the sum of external torques and the product of the moment of inertia and angular acceleration, we write

$$\Sigma T = I\dot{\omega} \quad nT = I\dot{\omega} \quad nT_0(1 - \omega_1/\omega_s) = I\dot{\omega}$$

Noting that the speed of the motor armature, ω_1, is n times the cradle angular speed ω, we write

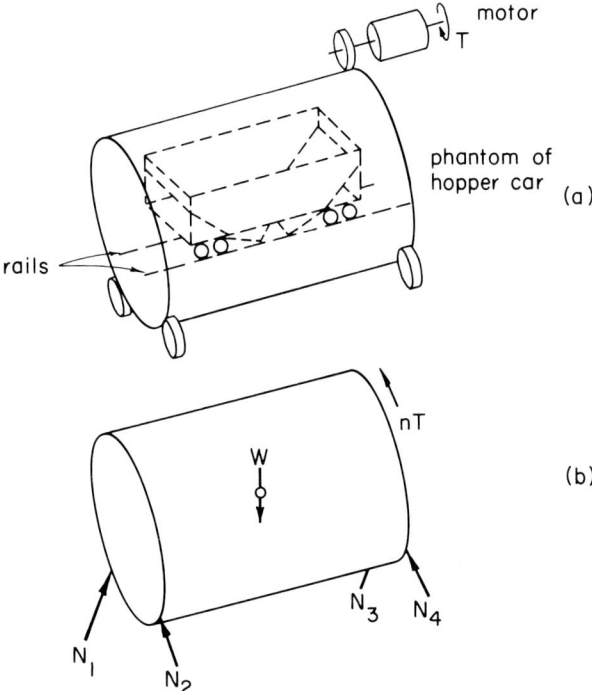

Fig. 9.25. (a) Rotary car dumper; (b) dumper as a free body.

$$nT_0(1 - n\omega/\omega_s) = I\dot{\omega}$$

Rearranging,

$$\dot{\omega} + \frac{n^2 T_0}{I\omega_s}\omega = \frac{nT_0}{I} \tag{9.20}$$

The solution to this first order differential equation is the sum of a complementary solution and a particular integral:

$$\omega = c\,\exp\left(-\frac{n^2 T_0 t}{I\omega_s}\right) + \frac{\omega_s}{n}$$

The constant c can be established by the initial condition of $\omega(0) = 0$ from which $c = -\omega_s/n$; consequently,

$$\omega = \frac{\omega_s}{n} - \frac{\omega_s}{n}\exp\left(-\frac{n^2 T_0 t}{I\omega_s}\right)$$

Now

$$\theta = \int_0^t \omega\,dt = \frac{\omega_s}{n}\int_0^t dt - \frac{\omega_s}{n}\int_0^t \exp\left(-\frac{n^2 T_0 t}{I\omega_s}\right)dt$$

$$\theta = \frac{\omega_s t}{n} + \frac{I\omega_s^2}{T_0 n^3}\left[\exp\left(-\frac{n^2 T_0 t}{I\omega_s}\right) - 1\right] \tag{9.21}$$

Equation 9.21 is a mathematical model relating dumper angle θ to all other problem parameters. If we take as the criterion function the time of the dumper rotating π radians, we discover we cannot solve Eq. (9.21) explicitly for t. Even if we could we would have a transcendental equation. Finding any significant stationary point is even more difficult. Possible approaches to the problem of finding the gear ratio that minimizes time to invert the cradle include:

--Solve Eq. (9.21) on the analog computer by patching a circuit that will draw a curve of θ versus t for a particular n. By varying n and observing solutions, the optimal step-down ratio can be found. An interval-halving or golden section search strategy can be used to minimize the number of graphs to be drawn.
--Program the digital computer to find the root of

$$\pi - \frac{\omega_s t_r}{n} + \frac{I\omega_s^2}{T_0 n^3}\left[\exp\left(-\frac{n^2 T_0 t_r}{I\omega_s}\right) - 1\right] = 0$$

for a given n. Minimizing t_r with respect to n reveals the solution. Superpose a golden section strategy on a subprogram that returns the negative of the time to invert, given a gear ratio n. The subprogram calls a root-finding subroutine that finds t_r corresponding to a given n.
--Exploit the asymptotic nature of Eq. (9.21) if the parameters will allow it.

Decisions through Model Building

We will examine the third alternative. The locus of θ as a function of t may approach its asymptote rapidly. Our physical insight suggests that we expect the dumper to be moving at steady speed by the time the car is inverted. For large t, the exponential term in Eq. (9.21) vanishes and

$$\theta = \frac{\omega_s t}{n} - \frac{I\omega_s^2}{T_0 n^3} \tag{9.22}$$

Solving for t yields

$$t = \frac{\pi n}{\omega_s} + \frac{I\omega_s}{T_0 n^2} \tag{9.23}$$

If the minimum of t occurs at a stationary point,

$$\frac{dt}{dn} = \frac{\pi}{\omega_s} - \frac{2I\omega_s}{T_0 n^3} = 0$$

from which

$$n = \left[\frac{2I\omega_s^2}{\pi T_0}\right]^{1/3} \tag{9.24}$$

If

$$I = 1000 \text{ kg} \cdot \text{m}^2 \quad \omega_s = 10 \text{ rad/s} \quad T_0 = 10 \text{ N} \cdot \text{m}$$

then

$$n = \left[\frac{2(1000)(10^2)}{\pi(10)}\right]^{1/3} = 18.53$$

and the time to invert the car is

$$t = \frac{\pi n}{\omega_s} + \frac{I\omega_s}{T_0 n^2} = \frac{\pi(18.53)}{10} + \frac{1000(10)}{10(18.53^2)} = 8.73 \text{ s}$$

Figure 9.26 shows the loci of θ as a function of t.

We must now check to see if the asymptotes are operative. If Eq. (9.21) is used,

$$\theta = \frac{10(8.73)}{18.53} + \frac{1000(10^2)}{10(18.53^3)} \left\{\exp\left[-\frac{18.53^2(10)(8.73)}{1000(10)}\right] - 1\right\} = 3.168 \text{ rad}$$

which is very close to π. If the motor selected is tenfold larger, i.e., $T_0 = 100$ N·m, then

$$n = \left[\frac{2I\omega_s^2}{\pi T_0}\right]^{1/3} = \left[\frac{2(1000)(10^2)}{\pi(100)}\right]^{1/3} = 8.60$$

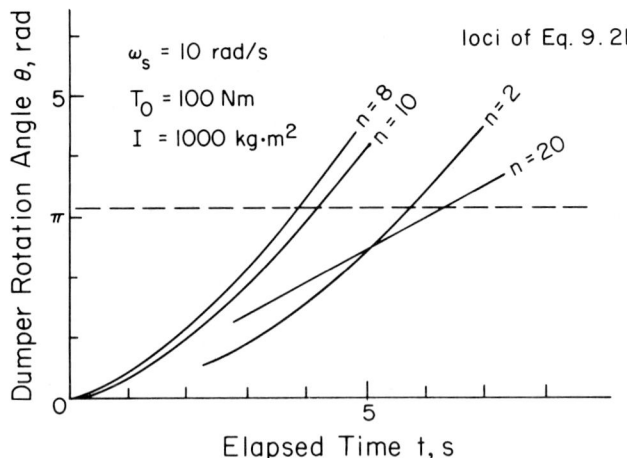

Fig. 9.26. Loci of Eq. (9.21).

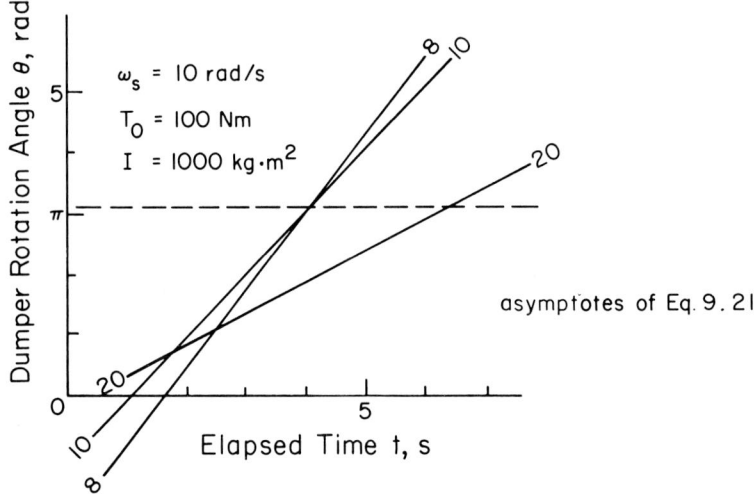

Fig. 9.27. Asymptotes of Eq. (9.21).

$$t = \frac{\pi(8.60)}{10} + \frac{1000(10)}{100(8.60^2)} = 4.05 \text{ s}$$

Checking,

$$\theta = \frac{10(4.05)}{8.60} + \frac{1000(10^2)}{100(8.60^3)} \left\{ \exp -\left[\frac{8.60^2(100)(4.05)}{1000(10)}\right] - 1 \right\} = 3.22 \text{ rad}$$

and the approximation is fading in its utility. Figure 9.27 shows the asymptotes of θ as a function of t.

9.10 DECISION ON A MEASUREMENT

An engineer is confronted with a temperature-time recording (Fig. 9.28) in which the temperature indication must assured within ±5°C. The question to

Decisions through Model Building

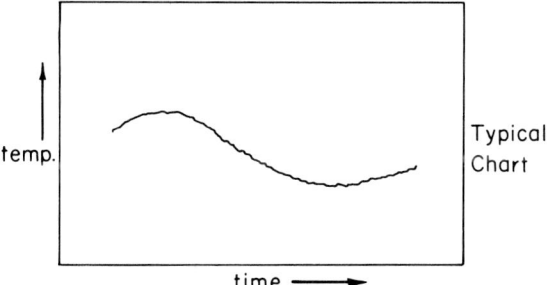

Fig. 9.28. Typical temperature-time chart.

be answered is, Is the indicated temperature within 5°C of the actual temperature? The chart results from a temperature-sensing device immersed in a fluid stream (oil) whose output signal is amplified and used to drive a recording chart. Just what confidence can be placed in the temperature indications of the chart? The engineer must know the thermal sensor's time constant under the conditions of use before proceeding to evaluate the fidelity of the chart recording.

The sensor is a small, metal-encased, bimetallic strip whose mechanical movement turns a miniature variable resistance potentiometer that is an element in a bridge circuit. Examination of the apparatus convinces the engineer that the principal error is in the thermal lag of the sensor. The sensing bulb has good thermal contact with the oil in which it is immersed. The temperature of the bulb should be substantially uniform due to its largely metallic nature, and the sensor will differ in temperature from the fluid because of the time lag in transferring the necessary energy as heat. First, isolate a portion of the apparatus and define the skin of the sensor bulb as the system (see Fig. 9.29). Next, identify the significant influences of the surroundings on the system. Negligible thermal conduction from the bulb exists along the mechanical connection (judgment). There is also a negligible work effect in delivering the bimetallic distortion motion to the surroundings (judgment). The only significant communication of the system with its surroundings is the Newtonian convection effect, qualified as in Fig. 9.30. The first law of thermodynamics, $\dot{Q} = \dot{E} + \dot{W}$, is the basis for an equality in this case; the Newtonian convection is $\dot{Q}$ and

$$-hA(T - T_\infty) = V\rho c \dot{T} + 0$$

$$\dot{T} + \frac{hA}{V\rho c} T = \frac{hAT_\infty}{V\rho c}$$

Designating $hA/(V\rho c) = 1/\tau$, where τ is the time constant of the first order system, we can simplify the differential equation mathematical model of bulb behavior to read

$$\tau \dot{T} + T = T_\infty \tag{9.25}$$

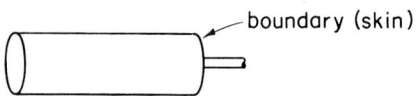

Fig. 9.29. The sensor bulb as system.

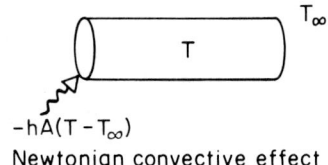

Fig. 9.30. Qualifying the significant influences on the system.

The solution to this nonhomogeneous first order differential equation is composed of two additive parts, the complementary solution and the particular integral:

$$T = Ce^{-t/\tau} + T_\infty$$

The constant C may be evaluated from an initial condition,

$$T(0) = T_0$$

Consequently $C = T_0 - T_\infty$ and the solution is

$$\frac{T - T_\infty}{T_0 - T_\infty} = e^{-t/\tau}$$

Solving for the time t explicitly yields

$$t = \tau \left[- \ln \frac{T - T_\infty}{T_0 - T_\infty} \right] \tag{9.26}$$

This relationship indicates that if the sensor temperature T were observed with the corresponding time t, the data will be rectified by applying the rather complicated logarithmic operator to the temperature observations and the identity operator to the time observations. The data will fall along a straight line and the time constant τ can be inferred from the rectified line's slope. Furthermore, the fidelity of the rectification can be taken as evidence of the first orderedness of the sensor complex and as vindication of judgments made during the mathematical model building process.

A thermometer calibration bath was filled with the oil that surrounds the bulb in actual use. The bath temperature was 82.2°C. Room temperature was 31.1°C. The recorder was started with the bulb in ambient air. The chart exhibited a steady horizontal trace. Sudden immersion of the bulb resulted in a jump in the chart trace such as depicted in Fig. 9.31(a). A difficulty is the lack of good definition of the instant of immersion. If a decision were made on a single trace and points scaled from this (judged) zero, a consistent bias of unknown magnitude would be present. The decision was made to make six charts, scaling one point from each chart, in an attempt to introduce randomness in the time observation (as opposed to a consistent but unknown bias). Figure 9.31(b) is a chart typical of the six.

The principal error is in t with a small error in the bulb temperature T. This was confirmed while the oil bath was heating with sensing bulb immersed. The chart temperature agreed with the bath thermometer. We choose a regression of the form Y = bX corresponding to

$$t = \tau \left[- \ln \frac{T - T_\infty}{T_0 - T_\infty} \right]$$

Decisions through Model Building

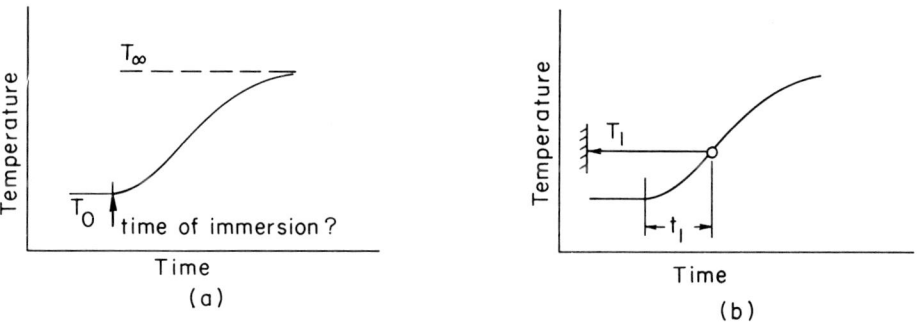

Fig. 9.31. (a) Temperature-time chart with sudden sensor immersion. (b) Chart typical of six.

Data rectification occurs (we hope) when $g(T) \equiv -\ln[(T - T_\infty)/(T_0 - T_\infty)]$ and $h(t) \equiv 1$. See Fig. 9.32(a), which is a plot of the following data.

$X = g(T)$	$Y = t$
0.140	1.0
0.274	1.6
0.427	1.9
0.610	2.7
0.833	3.7
1.121	4.8

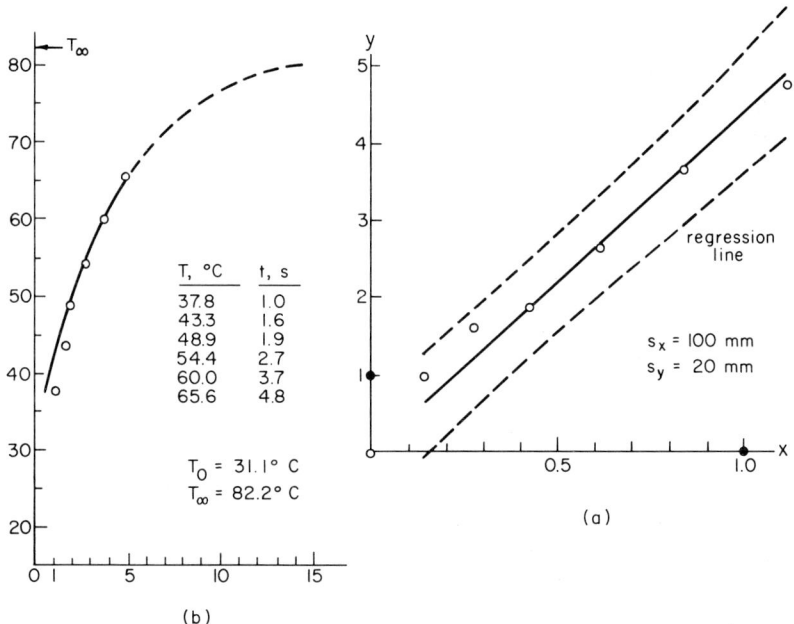

Fig. 9.32. Regression line on (a) rectified coordinates and (b) Cartesian coordinates.

Figure 9.32(a) shows the plotted data and its trend is satisfying to the eye (for six observations with the most error corresponding to the smallest time interval). An eyeball-best-line through the data points is shown in the figure. The abscissa scale factor $s_x = X/x = 100/1 = 100$ mm, $s_y = Y/y = 20/1 = 20$ mm, $\Delta Y/\Delta X = 89/100$ and

$$\tau = (s_x/s_y)(\Delta Y/\Delta X) = (100/20)(0.89) = 4.45 \text{ s}$$

While this estimate of the mean of τ is satisfactory, we need the standard deviation of τ. For this we turn to the statistics of regression and use IOWA CADET subroutine ME0229. We obtain the following information (the details of this step are beyond the scope of this book):

$\hat{\mu}_\tau$, slope of regression line = 4.425 s   (correlation coefficient 0.983)

95 percent upper bound = 4.830 s   95 percent lower bound = 4.019 s

$\hat{\mu}_\tau$, standard deviation of slope = 0.158 s

Figure 9.32(b) is a plot of the sensor bulb temperature as a function of time. From Eq. (9.25) the fluid temperature $T_\infty = T + \tau \dot{T}$, which suggests

$$\mu_{\tau\dot{T}} = \mu_\tau \mu_{\dot{T}} \quad \sigma_{\tau\dot{T}} = \mu_{\tau\dot{T}} \left[\left(\frac{\sigma_\tau}{\mu_\tau}\right)^2 + \left(\frac{\sigma_{\dot{T}}}{\mu_{\dot{T}}}\right)^2 + \left(\frac{\sigma_\tau}{\mu_\tau}\frac{\sigma_{\dot{T}}}{\mu_{\dot{T}}}\right)^2 \right]^{1/2}$$

and T_∞ is the sum of a nearly deterministic T and a probabilistic $\tau\dot{T}$. The equations

$$\mu_{x+y} = \mu_x + \mu_y \quad \sigma_{x+y} = (\sigma_x^2 + \sigma_y^2)^{1/2}$$

suggest

$$\mu_{T_\infty} = \mu_T + \mu_{\tau\dot{T}} \quad \sigma_{T_\infty} = (\sigma_T^2 + \sigma_{\tau\dot{T}}^2)^{1/2}$$

For the circumstances of point A, Fig. 9.33,

$s_y = 10$ mm $\hat{\mu}_\tau = 4.425$ s   (by test)

$s_x = 5$ mm $\hat{\sigma}_\tau = 0.158$ s   (by test)

$\Delta X = -50$ mm $\hat{\mu}_{\Delta Y/\Delta X} = \frac{\Delta Y}{\Delta X} = \frac{+34}{-50} = -0.680$ (by construction)

$\Delta Y = +34$ mm $\hat{\sigma}_{\Delta Y/\Delta X} = |0.058 \mu_{\Delta Y/\Delta X}| = 0.058(0.680)$

$T = 36.4°C$ $= 0.0392$ (by experience)

Now

$$\mu_{\dot{T}} = (s_x/s_y)\mu_{\Delta Y/\Delta X} = (5/10)(-0.680) = 0.340°C/s$$

$$\sigma_{\dot{T}} = (s_x/s_y)\sigma_{\Delta Y/\Delta X} = (5/10)(0.0392) = 0.0196°C/s$$

$$\mu_{\tau\dot{T}} = \mu_\tau \mu_{\dot{T}} = (4.425)(-0.340) = -1.505°C$$

$$\sigma_{\tau\dot{T}} = |\mu_{\tau\dot{T}}| \left[\left(\frac{\sigma_\tau}{\mu_\tau}\right)^2 + \left(\frac{\sigma_{\dot{T}}}{\mu_{\dot{T}}}\right)^2 + \left(\frac{\sigma_\tau}{\mu_\tau}\frac{\sigma_{\dot{T}}}{\mu_{\dot{T}}}\right)^2 \right]^{1/2}$$

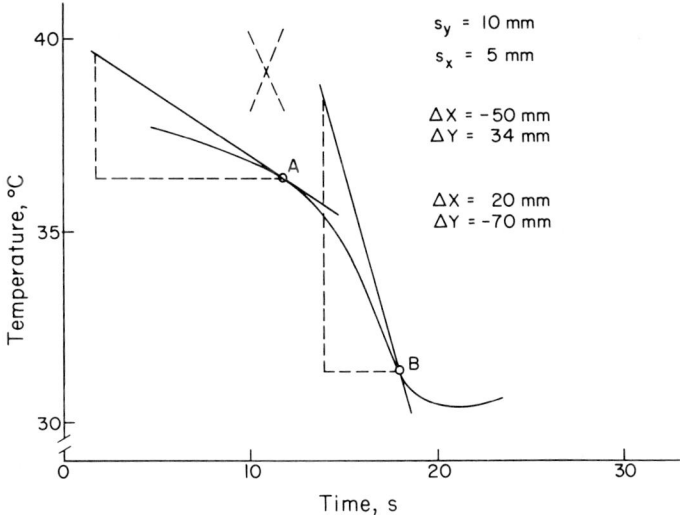

Fig. 9.33. Temperature-time charts and significant stylus slopes.

$$= 1.505\left[\left(\frac{0.158}{4.425}\right)^2 + \left(\frac{0.0196}{0.340}\right)^2 + \left(\frac{0.158}{4.425} \cdot \frac{0.0196}{0.340}\right)^2\right]^{1/2} = 0.102°C$$

The temperature T_∞ of the fluid comes from a value of T read from the chart, τ is from experiment, and $\dot{T}$ is from construction and measurement of the physical slope $\Delta Y/\Delta X$. No tangible error is expected in T taken from the chart. From experience, graphical slope can be established ±10% suggesting a uniform random error in $\Delta Y/\Delta X$:

$$\sigma_{\Delta Y/\Delta X} = (\text{range})/2\sqrt{3} = (0.2\mu_{\Delta Y/\Delta X})/2\sqrt{3} = 0.058\mu_{\Delta Y/\Delta X}$$

The time rate of change of the bulb temperature $\dot{T}$ is the mathematical slope of the charted locus.

$$\dot{T} = (s_x/s_y)(\Delta Y/\Delta X)$$

Since scale factors are known with high precision, s_x/s_y can be taken as a deterministic constant. The above equation is of the form

$$\mu_{ax} = a\mu_x \qquad \sigma_{ax} = a\sigma_x$$

$$\mu_{\dot{T}} = (s_x/s_y)\mu_{\Delta Y/\Delta X} \qquad \sigma_{\dot{T}} = (s_x/s_y)\sigma_{\Delta Y/\Delta X}$$

Also,

$$\sigma_{\dot{T}}/\mu_{\dot{T}} = \sigma_{\Delta Y/\Delta X}/\mu_{\Delta Y/\Delta X}$$

The expression $\tau\dot{T}$, the product of two uncorrelated random variables τ and $\dot{T}$, is of the form

$$\mu_{xy} = \mu_x\mu_y \qquad \sigma_{xy} = \mu_{xy}\left[\left(\frac{\sigma_x}{\mu_x}\right)^2 + \left(\frac{\sigma_y}{\mu_y}\right)^2 + \left(\frac{\sigma_x}{\mu_x}\frac{\sigma_y}{\mu_y}\right)^2\right]^{1/2}$$

and

$$\mu_{T_\infty} = \mu_T + \mu_{\tau\dot{T}} = 36.4 + (-1.505) = 34.9°C$$

$$\sigma_{T_\infty} = \sigma_{\tau\dot{T}} = 0.102°C$$

The mean $T_\infty \pm 2\sigma_{T_\infty} = 34.9 \pm 0.2°C$ represents (approximately) the 95 percent confidence interval. The error of the chart is $\leq 1.5°C$ with a 0.95 confidence.

For point B of Fig. 9.33,

$s_y = 10$ mm $\quad \hat{\mu}_\tau = 4.425$ s $\quad$ (as before)

$s_x = 5$ mm $\quad \hat{\sigma}_\tau = 0.158$ s $\quad$ (as before)

$\Delta X = +20$ mm $\quad \mu_{\Delta Y/\Delta X} = -70/20 = -3.5$

$\Delta Y = -70$ mm $\quad \sigma_{\Delta Y/\Delta X} = 0.058(-3.5) = 0.202$

$T = 31.5°C$

Proceeding similarly,

$$\mu_{\dot{T}} = (s_x/s_y)\mu_{\Delta Y/\Delta X} = (5/10)(-3.5) = -1.75°C/s$$

$$\sigma_{\dot{T}} = (s_x/s_y)\sigma_{\Delta Y/\Delta X} = (5/10)(0.202) = 0.101°C/s$$

$$\mu_{\tau\dot{T}} = \mu_\tau \mu_{\dot{T}} = (4.425)(-1.75) = -7.744°C$$

$$\sigma_{\tau\dot{T}} = \mu_{\tau\dot{T}}\left[\left(\frac{\sigma_\tau}{\mu_\tau}\right)^2 + \left(\frac{\sigma_{\dot{T}}}{\mu_{\dot{T}}}\right)^2 + \left(\frac{\sigma_\tau}{\mu_\tau}\frac{\sigma_{\dot{T}}}{\mu_{\dot{T}}}\right)^2\right]^{1/2}$$

$$= |-7.744|\left[\left(\frac{0.158}{4.425}\right)^2 + \left(\frac{0.101}{1.75}\right)^2 + \left(\frac{0.158}{4.425}\frac{0.101}{1.75}\right)^2\right]^{1/2} = 0.526°C$$

$$\mu_{T_\infty} = \mu_T + \mu_{\tau\dot{T}} = 31.5 + (-7.744) = 23.8°C$$

$$\sigma_{T_\infty} = \sigma_{\tau\dot{T}} = 0.526°C$$

Thus $\mu_{T_\infty} \pm 2\sigma_{T_\infty} = 23.8 \pm 1.05°C$ and the probability is 0.95 that the value of the fluid temperature is in this range. Compare this with the charted temperature of 31.5°C at point B. The chart does not indicate to within 5 percent in the neighborhood of B. The decision is not to believe the chart as an indicator of the fluid temperature.

The charted temperature (bulb temperature) is within $\pm 5°C$ of the fluid temperature T when $\pm 5 = \mu_T - \mu_{T_\infty}$ or

$$|\mu_T - \mu_{T_\infty}| = 5 \quad |\mu_{\tau\dot{T}}| = 5 \quad \mu_\tau|\mu_{\dot{T}}| = 5$$

$$\mu_\tau|(s_x/s_y)\mu_{\Delta Y/\Delta X}| = 5$$

or

$$\mu_{\Delta Y/\Delta X} = \pm 5 s_y/(\mu_\tau s_x) = \pm 5(10)/[4.425(5)] = \pm 2.26$$

This simple relation is only as valid as $\hat{\mu}_\tau$ and the mathematical model employed to understand it. Almost all the preceding work was necessary to have confidence in its meaning.

Draw a pair of lines on the chart with a physical slope of +2.26 and -2.26 (plotted on Fig. 9.32 as dashed lines). If the inked temperature locus is ever steeper than these lines, the error exceeds $\pm 5°C$. These lines are

Decisions through Model Building

significant provided the sensor is *immersed in the oil of the test*.

While the statistical aspects of regression are properly the subject of a measurements and instrumentation course (or statistics), it is pertinent to indicate that when more understanding is available, the appropriate approach is to

- choose as a regression model a form, such as $y = a + bx$, that contains within it the mathematical model $y = bx$
- take sufficient data to allow a regression of $y = a + bx$ to substantiate a significant correlation through the correlation coefficient
- show by statistical hypothesis testing that a is (statistically) zero and b is (statistically) nonzero, allowing the data to permit a rejection of a in the regression that validates the mathematical model

Then

- choose a regression model of $y = bx$
- substantiate a significant regression of $y = bx$ through the correlation coefficient
- use statistical information $\hat{\mu}_b$, $\hat{\sigma}_b$, and $\hat{\sigma}_{y/x}$ as appropriate

To proceed otherwise is folly. We use many mathematical models because of previous validation. The surprises occur when we stretch the assumptions. By forcing statistical information gathered by the measurement process to validate the mathematical model through which we interpret the measurement, we save ourselves the embarrassment of being wrong (but only at the β confidence level).

9.11 THE ENGINEER'S NOTEBOOK

Engineers keep notebooks that are master records of events and conclusions. Test engineers keep records that are somewhat different from those of design engineers. They show records of technical planning (whether executed or aborted), analyses, experimental data, references, and a running "thinking out loud" narrative. A notebook becomes, by the end of a project, the primary record of understanding as it unfolded.

At convenient junctions, say at the completion of steps, analysis and evaluation of project progress is the subject of entries in the notebook. At that point the next step is formulated in detail. If backtracking is made necessary by the unfolding of events or the growth of understanding, it is treated as it occurs.

In experimental situations, engineers record names, qualifications, and assignments of personnel observing and identify all apparatus, instrumentation, and specimen numbers so that steps can be retraced and recalibrations made. Places, times, environmental conditions, observations, and graphs should all be entered in a way that is understandable at a later date by the engineer or someone else.

In a design situation any detail drawing should reflect decisions documented in the notebook. Entries should be legible, clear, and complete.

Careful freehand sketches, graphs, and drawings are helpful. When precision graphing is needed, paste-in displays are acceptable. The witnessing of entries by competent individuals who see *and* comprehend (understand) is an element in patent development.

When product liability suits are instituted against the corporation (or the engineer), it is far more convincing to refer to primary documents to show that safety and foreseeable use were systematically considered and decisions were made to provide reasonable or mandated safety than to rely on memory.

EPILOGUE

    Congratulations! You have seen it through, all the way to the end. Now where is the pot of gold?
    It is resident in your mind. You will have the relatively rare opportunity to study your basic engineering subjects with an a priori appreciation of mathematical modeling and its role in analysis--something that is an integral part of your instructor's exposition and essential to the comprehension of the disciplinary subject matter at hand. While this background will enrich your comprehension and appreciation of subsequent course work, please do not believe that you have mastered mathematical model building or have even seen a substantial portion of it. You have been introduced to the philosophy, some of the techniques, and a bit of history. In this you are richer than your antecedents in the pursuit of an engineering education, and more will be expected of you.

Competence is, after all, a relative rather than an absolute quality. It is a matter of being able to do what is expected of one. But these expectations change, and sometimes the things one is able to do change too—in either direction. So competence itself is essentially changeable, not fixed, and we can reasonably expect it to fluctuate during the course of any man's career. Nevertheless we have traditionally regarded competence as being rather like a beard; that is, something not acquired until adulthood and never really lost thereafter. Regrettably, it is becoming increasingly clear that competence is more like the hair on top of a man's head, in the sense that it is not necessarily there to stay.

S. W. GELLERMAN

Good judgment comes from experience, the most valuable experience from poor judgment.

T. VON KARMAN

The ability to foresee that some things cannot be foreseen is a very necessary quality.

J. J. ROUSSEAU

APPENDIXES

APPENDIX A   SOME UTILITY ROUTINES FROM CADET

Slider-Crank Position Analysis
ME0004(A1,A2,A3,B1,B2,B3,B4)   Mischke
This subroutine performs a position analysis for a singular configuration of
the mechanism in coordinates depicted on p. 123, Figs. 5-1 and 5-2, C. R.
Mischke, 1963, *Elements of Mechanical Analysis*, Addison-Wesley, Reading, Mass.
(used with permission).

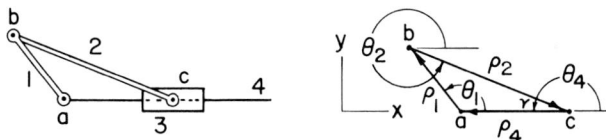

The crank angle must not be larger than 6π rad (three revolutions). The connecting rod must be longer than the crank throw, and both must be greater than zero.

CALLING PROGRAM REQUIREMENTS:

    None

CALL LIST ARGUMENTS:

    A1 = θ_1, abscissa angle of crank vector, rad
    A2 = ρ_1, length of crank vector, in.
    A3 = ρ_2, length of connecting rod vector, in.
    B1 = θ_2, abscissa angle of connecting rod vector, rad
    B2 = γ, angle between connecting rod and piston path, rad
    B3 = ρ_4, distance between main bearing and wrist pin, in.
    B4 = s, distance of wrist pin from top-dead-center position of pin, in.

PREEMPTED NAMES:

    None

SIZE:

    2872 bytes WATFIV compiler

Geneva Mechanism Analysis
ME0011(I1,A2,A3,B1,B2,B3,B4,B5)   Mischke
This subroutine calculates angular position, velocity, and acceleration of a
radial slot Geneva wheel such as depicted in Fig. 7-7(A), p. 185, Mischke, *Elements of Mechanical Analysis* (used with permission).

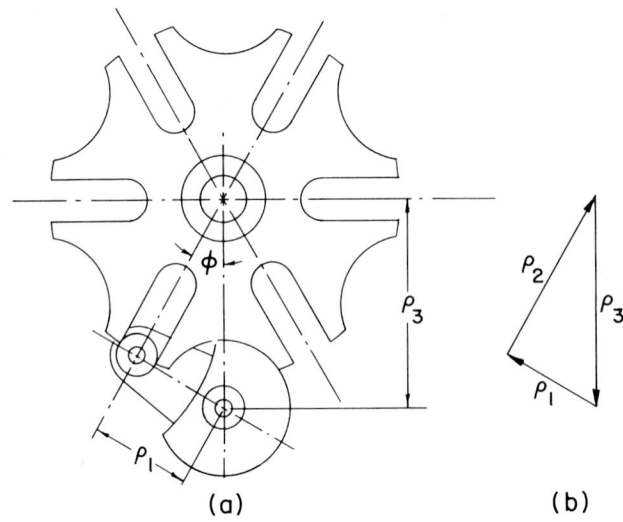

The Geneva wheel must have at least three slots. The crank angle must not be
greater than 6π rad (three revolutions) and the crank must have an angle such
that it is engaged in a slot.

CALLING PROGRAM REQUIREMENTS:

    None

CALL LIST ARGUMENTS:

    I1 = m, number of slots in Geneva wheel, integer
    A2 = γ, abscissa angle of crank vector, rad
    A3 = ω_1, angular velocity of crank, rad/s
    B1 = ρ_3/ρ_1, ratio of center-to-center distance to crank length
    B2 = α, angle between slot and line of centers, rad
    B3 = γ, angle between crank vector and line of centers, rad
    B4 = ω_2, angular velocity of Geneva wheel, rad/s
    B5 = $\dot{\omega}_2$, angular acceleration of Geneva wheel, rad/s^2

PREEMPTED NAMES:

    None

SIZE:

    3064 bytes WATFIV compiler

Appendix A   357

N-Dimensional Gradient Search
GRAD4(I1,I2,I3,A4,A5,A6,A7,A8,B1,B2,J3,B4,B5)   Mischke
This subroutine conducts an N-dimensional gradient search in up to 9 independent variables. The subprogram maps the search domain into and onto a unit hypercube. The method employed is that of steepest ascent with bolder steps and correction of direction at each step.

The search terminates when the logic is unable to improve the merit ordinate when standing on a mapped hypersquare of a size specified by the user.

See Mischke, 1968, *Introduction to Computer-Aided Design*, Prentice-Hall, p. 74.

CALLING PROGRAM REQUIREMENTS:

 Provide the equivalent to the following declaration

   DIMENSION A7(9),A8(9),B2(9),B4(9),B5(9)

 Provide a subroutine MERIT4(X,Y) that, when tendered a column vector of abscissas X, returns a corresponding dependent variable Y.

CALL LIST ARGUMENTS:

 I1 = number of independent variables (up to 9)
 I2 = convergence monitor print every I2-th evaluation
 I3 = 1 start search centrally
   = 2 start search lower corner of search domain
   = 3 start search upper corner of search domain
   = 4 start search at position specified by column vector B2
 A4 = initial exploration step size, fraction of unit hypercube side
   length, such as 0.1
 A5 = step size growth multiplier, suggest a number between 1. and 1.5
 A6 = size length of final hypersquare
 A7 = lower bounds of region of uncertainty, column vector
 A8 = upper bounds of region of uncertainty, column vector

 B1 = extreme ordinate discovered during search, maximum
 B2 = corresponding abscissas of extreme ordinate found, column vector
 J3 = number of merit function evaluations expended during search
 B4 = forward slopes at termination of search, column vector
 B5 = backward slopes at termination of search, column vector

PREEMPTED NAMES:

 Subroutine MERIT4, UNNORM

SIZE:

 9436 bytes WATFIV compiler

Grid Search
GRID1(I1,I2,A3,A4,A5,A6,B1,B2,B3,B4,J5)   Mischke

This subroutine conducts a grid-type search within regional constraints in a hyperspace of up to eight independent variables. The fractional reduction of grid size must be greater than 2/3 but less than 1.

The search pattern alternates between a hypercube and a hyperstar oriented about a central point. For additional details see Mischke, *Introduction to Computer-Aided Design*, p. 89.

CALLING PROGRAM REQUIREMENTS:

    Provide a dimension declaration equivalent to the following

        DIMENSION A3(9),A4(9),B2(9),B3(9),B4(9)

    Provide a subprogram MERIT1(X,Y) that will return to GRID1 an ordinate Y corresponding to the column vector of abscissas, X.

CALL LIST ARGUMENTS:

    I1 = number of independent design variables present, integer
    I2 = 0 convergence monitor will not print
       = 1 convergence monitor will print each new grid center
    A3 = column vector of initial lower extremities of interval
    A4 = column vector of initial upper extremities of interval
    A5 = fractional reduction of interval of uncertainty desired
    A6 = fractional reduction in grid size desired (between 2/3 and 1)
    B1 = largest merit ordinate discovered during grid search
    B2 = column vector of abscissas corresponding to largest ordinate
    B3 = column vector of final lower extremity of interval
    B4 = column vector of final upper extremity of interval
    J5 = number of function evaluations expended during search

PREEMPTED NAMES:

    Subroutine MERIT1, NORMAL, UNNORM, HIGH, REGION

SIZE:

    10416 bytes WATFIV compiler

Appendix A 359

One-Dimensional Golden Section Search
GOLD(I1,A2,A3,A4,A5,B1,B2,B3,B4,J5)   Mischke

This subroutine will search over a one-dimensional unimodal function and report the largest ordinate found, its abscissa, final abscissas found in the interval of uncertainty, and the number of function evaluations expended during the search.

The subroutine requires the specification of the present interval of uncertainty, the fractional reduction required in the interval of uncertainty (or bilateral tolerance on abscissa of the extreme), and whether a convergence monitor printout is desired. The necessary number of function evaluations may be predicted from

$$N = \left[1 + 2.08 \, \ln \frac{1}{F}\right]_+$$

when the fractional reduction in interval of uncertainty F is given, or

$$N = \left[2.08 \, \ln \frac{x_r - x_l}{a}\right]_+$$

when the bilateral tolerance a on abscissa of extreme is given.

See Mischke, *Introduction to Computer-Aided Design*, p. 64.

CALLING PROGRAM REQUIREMENTS:

   Provide a subroutine A5(X,Y) that returns the ordinate Y when the abscissa X is tendered.

   Provide the equivalent of the following statement: EXTERNAL A5

CALL LIST ARGUMENTS:

        I1 = 0, convergence monitor will not print
           = 1, convergence monitor will print
        A2 = x_l, original left-hand abscissa of interval of uncertainty
        A3 = x_r, original right-hand abscissa of interval of uncertainty
        A4 = fractional reduction in interval of uncertainty desired, F, entered positive or bilateral tolerance on abscissa at extreme, a, entered negative
        A5 = name of one-dimensional unimodal function SUBROUTINE
        B1 = y^*, extreme ordinate discovered during search (maximum)
        B2 = x^*, abscissa of extreme ordinate
        B3 = x_1, final left-hand abscissa of interval of uncertainty
        B4 = x_2, final right-hand abscissa of interval of uncertainty
        J5 = N, number of function evaluations expended during search

PREEMPTED NAMES:

        None

SIZE:

        4264 bytes WATFIV compiler

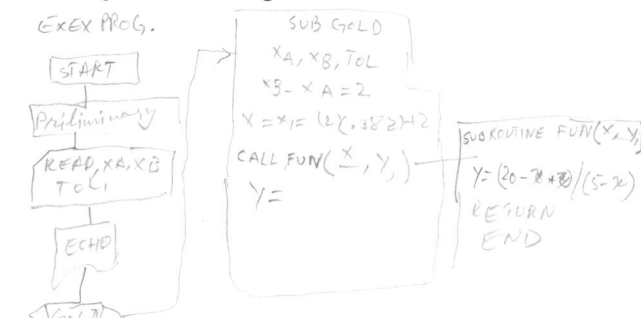

Lagrange Interpolation
INTER(I1,A2,A3,A4,B1,J2)   G. R. Wagner
This subroutine conducts a three-point Lagrangian interpolation, given the
column vectors of point pairs (A2,A3) and the abscissa of interest, A4.

The method is described by Carl Erik Froberg, 1969, *Introduction to Numerical Analysis*, Addison-Wesley, pp. 164-68.

CALLING PROGRAM REQUIREMENTS:

    Provide the equivalent to the declaration

        DIMENSION A2(I1), A3(I1)

CALL LIST VARIABLES:

    I1 = number of data points (must be $\geq$ 3)
    A2 = column vector of abscissa values of data points (must be in ascending order)
    A3 = column vector of ordinate values corresponding to abscissa values in A2
    A4 = abscissa of interest

    B1 = interpolated ordinate corresponding to abscissa A4
    J2 = 0 interpolation occurred
       = 1 extrapolation occurred

PREEMPTED NAMES:

    None

SIZE:

    1592 bytes WATFIV compiler

Appendix A

Newton-Raphson Root-Finder of First Order
NEWRA1(I1,A2,B1,B2)   Mischke
This subroutine implements the first order Newton-Raphson root-finding iteration formula

$$x_{i+1} = [x - f(x)/f'(x)]_i$$

The method requires more iterations as the slope in the neighborhood of the root gets smaller and the method is defeated by a zero first derivative at the root. The user must supply an estimate of the root, the closer the better.

CALLING PROGRAM REQUIREMENTS:

The user provides two FUNCTION subprograms F(X) and DF(X) that provide NEWRA1 the value of the function $f(x)$ and its first derivative $f'(x)$ at any x the Newton-Raphson strategy requires.

```
FUNCTION F(X)              FUNCTION DF(X)
F=(expression in x)        DF=(expression in x)
RETURN                     RETURN
END                        END
```

CALL LIST ARGUMENTS:

I1 = 0, convergence monitor will not print
   = 1, convergence monitor will print
A2 = x_0, the initial estimate of abscissa that is root

B1 = x_r, Newton-Raphson abscissa of zero place of the function
B2 = y_r, value of ordinate at x_r

PREEMPTED NAMES:

None

SIZE:

1624 bytes WATFIV compiler

Pattern Search
PATTRN(I1,I2,I3,A4,A5,A6,I7,A8,A9,A10,A11,B1,B2,J3)   G. R. Wagner

This subroutine conducts a pattern search within regional constraints in a hyperspace of up to 30 independent variables. Whenever the logic is unable to improve the merit ordinate the step size is decreased. The search terminates when the step size falls below a preselected minimum.

CALLING PROGRAM REQUIREMENTS:

   Provide the equivalent of the following declaration

          DIMENSION A8(I7), A9(I7), B1(I7), A10(I7)

   Provide a subroutine A11(X,Y) that returns the ordinate Y when the column vector of abscissas X is tendered.

   Provide the equivalent of the following statement

          EXTERNAL A11

CALL LIST ARGUMENTS:

   I1 = 0 convergence monitor does not print
      = 1 convergence monitor prints
   I2 = 0 best nonridge-following technique
      = 1 best ridge-following technique
   I3 = maximum number of function evaluations allowable
   A4 = initial step size for the surface mapped onto unit square, such as 0.1
   A5 = step size on unit hypersquare at termination
   A6 = reduction in step size when logic is unable to improve the merit ordinate. Suggest 0.1 for I2 = 0 and 0.25 for I2 = 1
   I7 = number of independent variables
   A8 = column vector of starting abscissas
   A9 = column vector of lower bounds
   A10 = column vector of upper bounds
   A11 = name of merit subroutine
   B1 = column vector of abscissas of extreme ordinate found
   B2 = extreme ordinate found, maximum
   J3 = number of function evaluations expended

PREEMPTED NAMES:

   None

SIZE:

   11824 bytes WATFIV compiler

Appendix A 363

Random Variable Algebra
RANDOM   Mischke

The subroutine RANDOM has no call list. It is called in order to load a battery of FUNCTION subprograms that aid the user in executing the algebra of random variables.

Programming is based on Eqs. (18)-(25) given by C. R. Mischke, 1974, in "A Rationale for Mechanical Design to a Reliability Specification," *Proceedings of First Design Technology Transfer Conference of ASME*, p. 221.

When the exponent n is fractional, the mean of x must be positive.

Real Function	Algebraic Operation	Value of Function	Eq.
MEAN1($\mu_x, \sigma_x, \mu_y, \sigma_y, \rho$)	$z = x + y$	μ_z	(18)
SIGMA1($\mu_x, \sigma_x, \mu_y, \sigma_y, \rho$)		σ_z	(19)
MEAN2($\mu_x, \sigma_x, \mu_y, \sigma_y, \rho$)	$z = x - y$	μ_z	(20)
SIGMA2($\mu_x, \sigma_x, \mu_y, \sigma_y, \rho$)		σ_z	(21)
MEAN3($\mu_x, \sigma_x, \mu_y, \sigma_y, \rho$)	$z = xy$	μ_z	(22)
SIGMA3($\mu_x, \sigma_x, \mu_y, \sigma_y, \rho$)		σ_z	(23)
MEAN4($\mu_x, \sigma_x, \mu_y, \sigma_y, \rho$)	$z = x/y$	μ_z	(24)
SIGMA4($\mu_x, \sigma_x, \mu_y, \sigma_y, \rho$)		σ_z	(25)
MEAN5($\mu_x, \sigma_x, \mu_y, \sigma_y$)	$z = x^y$	μ_z	
SIGMA5($\mu_x, \sigma_x, \mu_y, \sigma_y$)		σ_z	

See ME0233 for more complicated functions.

CALLING PROGRAM REQUIREMENTS:

   Declaration: REAL MEAN1, MEAN2, MEAN3, MEAN4, MEAN5

CALL LIST ARGUMENTS:

   μ_x = mean of variate x, real FORTRAN variable
   σ_x = standard deviation of variate x, real FORTRAN variable
   μ_y = mean of variate y, real FORTRAN variable
   σ_y = standard deviation of variate y, real FORTRAN variable
   ρ = correlation coefficient, real FORTRAN variable

PREEMPTED NAMES:

   None

SIZE:

   12068 bytes WATFIV compiler

Simpson's Rule Integration
SIMPSN(A1,I2,I3,A4,A5,A6,B1,B2,B3)   G. R. Wagner
This subroutine determines the value of the definite integral of a function by means of Simpson's rule. I3 must be an odd integer greater than 4 and of the form 4*N+1 where N is a positive integer. If I3 is of the form 4*N+3, B2 and B3 cannot be evaluated and consequently will be undefined.

CALLING PROGRAM REQUIREMENTS:

    Provide the equivalent of the following declaration

        DIMENSION A4(I3)

    Provide the equivalent of the following statement

        EXTERNAL A1

    Provide a function subprogram A1(X) that will be called by SIMPSN. The subprogram returns an ordinate A1(X) for a tendered abscissa X. If I2 = 1, the user must supply a dummy function subprogram A1(X).

CALL LIST ARGUMENTS:

    A1 = name of function subprogram containing integrand
    I2 = 0 generate ordinates using function subprogram A1
       = 1 use I3 equally spaced ordinates contained in column vector A4
    I3 = number of ordinates to be generated or supplied in column vector A4
    A4 = column vector of I3 ordinates
    A5 = lower limit of integral
    A6 = upper limit of integral

    B1 = value of integral using I3 ordinates of column vector A4
    B2 = value of integral using alternate ordinate of column vector A4
    B3 = estimated error in B1

PREEMPTED NAMES:

    None

SIZE:

    2080 bytes WATFIV compiler

Appendix A   365

Fine Root Finder
VARSEC(A1,A2,I3,B1)   G. R. Wagner

This subroutine determines the zero place of a function by the Regula Falsi
(variable secant) method. The subroutine requires an initial estimate for the
location of the root. If the root is not found within 50 iterations the subroutine stops iterating and returns its present estimate of the location of
the root to the calling program.

A convergence monitor can be printed, if desired.

CALLING PROGRAM REQUIREMENTS:

   Provide a function subprogram A1(X) that will be called by VARSEC; the
subprogram returns an ordinate A1 for a tendered abscissa X.

   Provide an external statement in the calling program to pass the name of
the function subprogram to subroutine VARSEC.

CALL LIST ARGUMENTS:

   A1 = name of function subprogram
   A2 = initial estimate for location of root
   I3 = 0 convergence monitor does not print
      = 1 convergence monitor will print
   B1 = location of root of function A1 correct to 6 digits

PREEMPTED NAMES:

   None

SIZE:

   2104 bytes WATFIV compiler

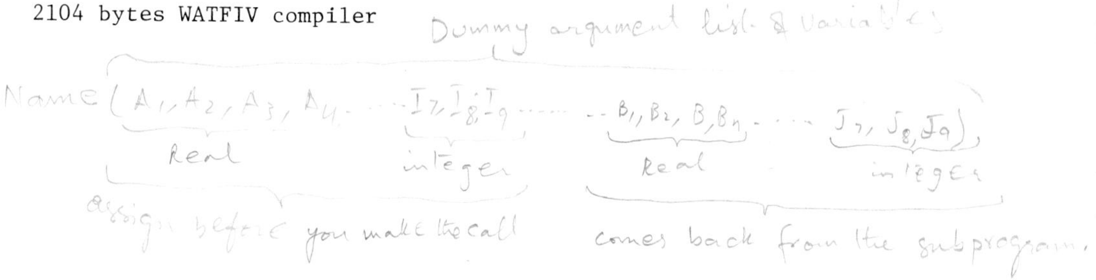

[Handwritten annotations:]

where the value of function is zero.

Dummy argument list & variables

Name $(A_1, A_2, A_3, A_4, \ldots I_7, I_8, I_9, \ldots B_1, B_2, B_3, B_4, \ldots J_7, J_8, J_9)$
         Real   integer   Real   integer

assign before you make the call   comes back from the subprogram.

Regula Falsi method. $x_n = x_{n-1} - \dfrac{(x_{n-1} - x_{n-2}) f(x_{n-1})}{f(x_{n-1}) - f(x_{n-2})}$

Calling program requirement.

Executive program
```
I3 = 1
XI = 1.0
CALL VARSEC(FOX, XI, I3, XR)
WRITE XR
```

VARSEC(A_1, A_2, I_3, B_1)

Function subprogram
```
FUNCTION FOX(X)
FOX = X^3 + 2X^2 + 10X - 20
RETURN
END
```

$f(X) = FOX$

user supplies this

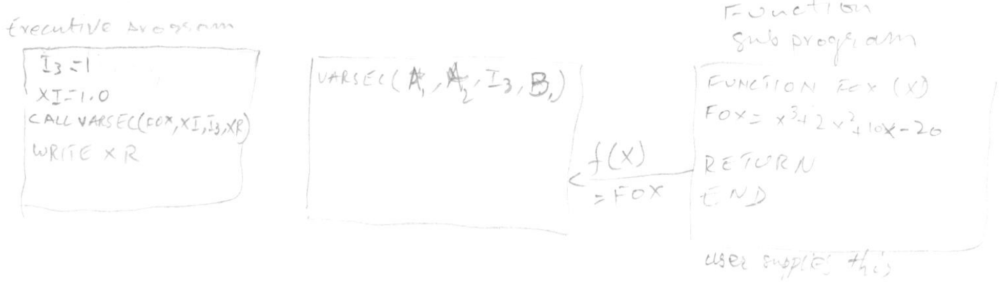

APPENDIX B INTERNATIONAL SYSTEM OF UNITS
Information on the SI Measurement System (Copyright 1975 by Deere & Company. Reproduced by special permission)

The SI system embodies more than the metric system. Several distinct features are as follows:

1. As an international system, strict symbology promotes common worldwide communication capability.

2. The introduction of a new unit (newton) in the system helps clarify the distinction between force and mass.

3. Energy units are now common to electrical and mechanical systems.

4. In general, unit prefixes are encouraged to keep numbers between 0.1 and 1000.

In going to the new SI system, even Europeans familiar with the old metric system will have to make changes. A review of the metric system as taught in high school or college does not suffice to be proficient with the SI system.

This grouping of information is intended to help mechanical engineering personnel learn the necessary fundamentals of the new system.

The material is presented in the following outline:

1. Development of SI units.

2. Use of SI units, prefixes, abbreviations, etc.

3. Rules for conversion and rounding.

4. Examples of engineering calculations.

5. Review problems.

6. Appendix:

Since 1790, when France designed the first metric system, the U.S. and rest of the world has studied its adoption. The International Conference on Weights and Measures has met at least every six years to review and update the system. In 1960, the modernized metric system was named the Système International d'unités - the International System of Units.

The International Standard consists of two parts; the first part deals with the actual units, the second, selected decimal multiples and sub-multiples for general use.

Appendix B

I. DEVELOPMENT OF SI UNITS

The new system has three classes of units. These are:

1. Base Units
2. Supplementary Units
3. Derived Units

1. **Base Units** There are seven base units in the SI system of measurement. These "base" units are defined in such a way that they may be reproduced anywhere in the world by a quality laboratory (except kilogram).

Base Units

Quantity	Name of base SI Unit	Symbol
Length	metre	m
Mass	kilogram	kg
Time	second	s
Electric Current	ampere	A
Thermodynamic Temperature	kelvin	K
Amount of Substance	mole	mol
Luminous Intensity	candela	cd

a. *Length (metre)* The metre is defined in terms of a certain number of wavelengths of krypton-86 atom.

b. *Mass (kilogram)* This unit is obtained by comparison with the international prototype of the kilogram.

c. *Time (second)* The unit of time is the second. It is defined in terms of the frequency of the Cesium-133 atom.

d. *Electric Current (ampere)* The unit of current is the ampere (A) and is defined in terms of the force between two electrical conductors.

e. *Temperature (kelvin)* The unit of temperature is kelvin (K) with 0 kelvin being absolute zero and 273.16 kelvin the triple point of water. The size of the unit is the same as the familiar Centigrade. For common usage, the old Centigrade may be used but renamed Celsius. For example, the boiling point of water is 100°C (see Page 6 for use of temperature unit).

f. *Amount of Substance (mole)* The mole is the amount of a substance of a system which contains as many elementary entities as there are atoms in 0.012 kg of Carbon 12.

g. *Luminous Intensity (candela)* The base unit candela relates to the intensity of light from platinum at a certain temperature.

2. **Supplementary Units**

Quantity	Name of Supplementary SI Unit	Symbol
plane angle	radian	rad
solid angle	steradian	sr

The unit for the plane angle is radian, a term familiar to engineers. The base unit for the solid angle is steradian, rarely used in our engineering activities.

3. **Derived Units** Derived units are obtained by the combination of base and supplementary units. For mechanical engineering applications, the first five are of principal importance.

Quantity	Name of derived SI Unit	Symbol	Expressed in terms of base or supplementary SI units or in terms of other derived SI units
frequency	hertz	Hz	$1 \text{ Hz} = 1 \text{ s}^{-1}$
force	newton	N	$1 \text{ N} = 1 \text{ kg} \cdot \text{m/s}^2$
pressure, stress	pascal	Pa	$1 \text{ Pa} = 1 \text{ N/m}^2$
energy, work, quantity of heat	joule	J	$1 \text{ J} = 1 \text{ N} \cdot \text{m}$
power	watt	W	$1 \text{ W} = 1 \text{ J/s}$
electric charge, quantity of electricity	coulomb	C	$1 \text{ C} = 1 \text{ A} \cdot \text{s}$
electric potential, potential difference, tension, electromotive force	volt	V	$1 \text{ V} = 1 \text{ J/C}$
electric capacitance	farad	F	$1 \text{ F} = 1 \text{ C/V}$
electric resistance	ohm	Ω	$1 \Omega = 1 \text{ V/A}$
electric conductance	siemens	S	$1 \text{ S} = 1 \Omega^{-1}$
flux of magnetic induction, magnetic flux	weber	Wb	$1 \text{ Wb} = 1 \text{ V} \cdot \text{s}$
magnetic flux density, magnetic induction	tesla	T	$1 \text{ T} = 1 \text{ Wb/m}^2$
inductance	henry	H	$1 \text{ H} = 1 \text{ Wb/A}$
luminous flux	lumen	lm	$1 \text{ lm} = 1 \text{ cd} \cdot \text{sr}$
illuminance	lux	lx	$1 \text{ lx} = 1 \text{ lm/m}^2$

a. *Frequency (hertz)* Here the unit is the old "cycles per sec." but now called hertz (Hz). In terms of base units, it is s^{-1}.

b. *Force (newton)* The name newton (N) has been given to the term of force. In terms of base units, the newton is the force required to give a mass of 1 kg an acceleration of 1 m/s^2.

$$F = ma$$
$$1 \text{N} = 1 \text{ kg} \times 1 \text{ m/s}^2$$

Appendix B

This unit will definitely alter some of our calculating procedures. Forces will be expressed in newtons. The difference in calculations is well illustrated by calculating potential and kinetic energy.

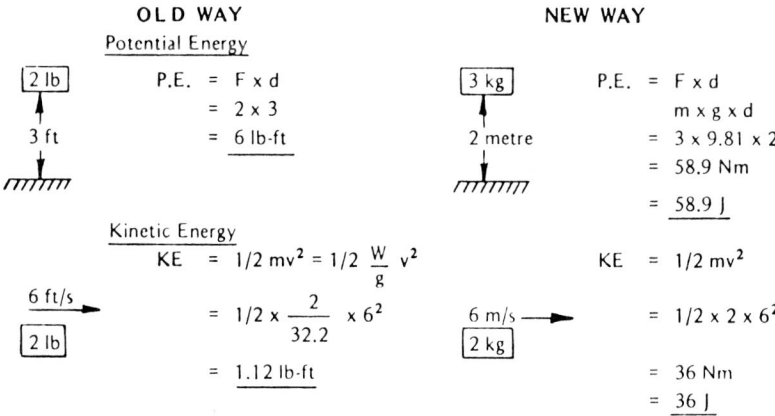

	OLD WAY	NEW WAY
	Potential Energy	
	P.E. = F x d = 2 x 3 = 6 lb-ft	P.E. = F x d m x g x d = 3 x 9.81 x 2 = 58.9 Nm = 58.9 J
	Kinetic Energy	
	KE = $1/2\, mv^2 = 1/2\, \frac{W}{g} v^2$ = $1/2 \times \frac{2}{32.2} \times 6^2$ = 1.12 lb-ft	KE = $1/2\, mv^2$ = $1/2 \times 2 \times 6^2$ = 36 Nm = 36 J

You will note more complexity for calculating potential energy, less for kinetic energy. This will generally be the case. Static problems will have to use gravity ($9.81\ m/s^2$) more frequently, dynamic problems will be simpler.

c. *Pressure*   In the SI system, the unit of pressure is pascal (Pa). It is defined as the per unit pressure developed when a force of 1 newton acts against a surface area of 1 square metre.

This is an exceedingly small unit that necessitates the use of prefixes. The term kPa and MPa are more suitable units. Another unit that is preferred in fluid power work is the "bar." It is 100 000 (10^5) pascal. Its desirability stems from its being within 2% of the previous kg/cm^2, familiar to former metric users. Stresses of materials will be expressed in N/mm^2.

d. *Energy, Work, Quantity of Heat*   In the SI system, the unit for energy is joule (J), defined as a newton-meter (Nm). By expressing energy in joules, we distinguish between energy and torque moment.

e. *Power*   The unit for power is watt (W). This is the same "power" as we have had in electrical expressions for many years. Whenever we discuss the rate of energy use, electrical, mechanical or thermal, it will be in watts or kilowatts. It is defined as 1 W = 1 J/s.

II. USE OF SI UNITS, PREFIXES, ABBREVIATIONS, ETC.

A. **Use of SI Units** An SI recommended procedure is to use prefixes and multiples so that the numerical value of the measurement is between 0.1 and 1000. Engineers can make better judgement when working in this range. For example, if someone says the distance between two cities is 2 000 000 inches, the normal reaction is to divide by 12 or 10 to get it in ft. (about 200 000) - still too large, so divide by 5000 to get into miles -- 40. This is a number that is easier to grasp and use. The 0.1 to 1000 rule should be used with judgement - if values are to be compared or are in a table, they should have the same prefix. Engineering drawings are to be in mm only.

Factor by which the unit is multiplied	Prefix Name	Symbol
10^{12}	tera	T
10^{9}	giga	G
10^{6}	*mega	M
10^{3}	*kilo	k
10^{2}	hecto	h
10	deca	da
10^{-1}	deci	d
10^{-2}	centi	c
10^{-3}	*milli	m
10^{-6}	*micro	μ
10^{-9}	nano	n
10^{-12}	pico	p
10^{-15}	femto	f
10^{-18}	atto	a

*Preferred

The above prefixes are all officially recognized but not all are preferred. For dimensions and weights, it is preferred that prefixes be in increments of 1 000. Thus, all distances will be in μm, mm, m, km. Mass in mg, g, kg and Mg. When working with areas or volumes, the smaller prefix increments will have to be used.

The cubic decimeter commonly called litre will continue to be a popular unit. Compound prefixes should not be used; for example, milligram is mg, not μkg.

Engineers are encouraged to emphasize the use of powers of 10 in their calculations as a means of reducing errors.

B. **Symbology & Punctuation** To be successful as an international language, we must use common symbology and punctuation as follows:

1. The names of all units are in small letters if written out. Where symbols are used, it is capitalized only if the symbol is derived from a proper name.

    Examples:
    - m   metre
    - s   second
    - N   newton
    - J   joule

2. Except for very large numbers (M, G & T), numerical prefixes are not capitalized.

3. No periods are used after symbols unless it is the end of a sentence.

4. Symbols do not have "s" added to indicate plural. SI Units written out may have "s" for plural.

Appendix B

5. Compound units may be indicated as N·m, N.m, or Nm.

6. Due to the practice of European countries using commas as decimal points, commas will no longer be used to separate numbers in groups of 3, a space will be used.

   Example:
   $$3{,}146.60491 \text{ (wrong)}$$
   $$3\ 146.604\ 91 \text{ (right)}$$

7. Either a comma or period may serve as a decimal point, however, English speaking countries will use the period

8. With decimal numbers less than one, a zero will precede the decimal point:

   Example:
   $$0.146$$

9. Caution should be used in sequence of units; for example, Nm is newton metres. If reversed, mN could be interpreted as milli newtons. The multiplication period can clarify this m·N or N·m.

10. When a compound unit is formed by dividing one unit by another, it is acceptable to show the unit either as m/s or m(s^{-1}).

11. No more than one "/" should be used to express a compound unit.

12. In a limited number of cases, component units with other than base units are permitted; for example: km/h.

13. Since computers do not have small case letters, a suggested substitute set of abbreviations is found in Appendix E.

14. The use of Celsius in temperature units violates several previously mentioned rules, following are examples of correct and incorrect usage:

    Correct Usage
    240 K   24 degrees Celsius
    240 kelvin   24°C

    Incorrect Usage
    240°K   24 Celsius
    240 Kelvin   24 C

Appendix B

Examples of right and wrong usage of SI abbrevations, rules, etc.;

Wrong	Right	Wrong	Right
.491	0.491	50 kw	50 kW
34,620	34 620	65 meters	65 metres*
64 N.	64 N	50 cc	50 cm^3 or 50 x 10^{-6} m^3
64 j	64 J	g/kW/hr	g/(kWh)
95 Newton	95 newtons	m·kg/s^3/A	m·kg/(s^3·A) or M·kg·s^{-3}·A^{-1}
1 ton	1 megagram (Mg)**	1.71mm	1.71 mm
1 hr	1 h	64l	64 l
4 liters	4 litres*	distance of	
164°K	164 K	69 kms	69 km
95C	95°C	74 kilometre	74 kilometres
95 celsius	95 degrees Celsius		
300 Kelvin	300 kelvin		

*This spelling goes along with the international standard. Presently, both spellings are used.

**In many applications, it is expected that metric tonne (t) will be used rather than Mg.

Appendix B 373

III. ENGINEERING DRAWINGS; DIMENSIONS AND ROUND-OFF

The millimetre is the standard linear dimension for engineering drawings. As an interim practice, the drawing will be provided with a conversion table to show the inch conversion of the millimetre dimension. Rules for conversion are as follows:

A. One more digit following the decimal point in inch dimensions as compared with millimetre, with minimum to two digits.

B. Rounding will be "simple," .5 and larger to next higher digit, below .5, number is truncated.

Caution should be exercised in totalizing dimensions in both systems and expecting final dimensions to check. Totalizing should be done in millimetres only and final results converted to inch. An even metre dimension will not have a decimal point and following zero (es) unless greater than standard precision is required.

IV. CALCULATING WITH SI UNITS

The working of problems is the fastest way to become familiar with SI units. Following are general statements which may help you get oriented in the general approach to solving problems.

A. Fundamental relationships still hold; for example, $F = ma$, $T = I\alpha$, $KE = 1/2\, mv^2$ etc.

B. Proficiency with dimensional analysis will be very helpful. For example: ability to recognize

$$\text{Watt} = \frac{J}{s} = \frac{Nm}{s} = \frac{N}{m^2} \times \frac{m^3}{s} = (\text{Pressure} \times \text{Flow})$$

And other similar relationships will facilitate the use of SI units.

C. Generally speaking, adjust all variables to base units (kg, metre, etc.) for simplification of calculations.

D. For stress calculations, use of mm results in more conveniently sized numbers.

E. Develop a proficiency for exponential notation of numbers to be used on conjunction with unit prefixes.

F. Use units which are consistent with corporate selections in Design Manual 30

V. REVIEW PROBLEMS

1. A bar of steel has a length of 600 mm at room temperature (20°C). To what temperature must it be heated to increase its length by one mm? Coefficient of linear expansion is 12×10^{-6}/K.

   (158.9°C)

2. If a block of cast iron has a density of 7.2 at 20°C, what is the density at 100°C? Assume the coefficient of linear expansion of cast iron is 12×10^{-6}/K.

   (7170 kg/m³)

3. An oil burner has an efficiency of 80%. If it burns 10 litres per day of a fuel with a calorific value of 40 MJ/kg, what is the power output in kW? Assume specific gravity of fuel as 0.8.

   (2.96 kW)

4. The rectangular wall of a house is 15 m × 4 m with a thickness of 200 mm. What is the cost per day of the heat loss through the wall if the inside temperature is 20°C and the outside is −20°C. Assume the wall has a thermal conductivity of 1.2 W/(m K) and electricity costs two cents per kWh.

   ($6.91/d)

5. A one litre measure is filled with mercury at 20°C. How much mercury will overflow if the temperature of container and mercury is raised to 100°C. Assume the coefficient of expansion for the metal container is 12×10^{-6}/K and the value for mercury is 180×10^{-6}/K.

   (13.44 cm³)

6. A 5 kW electric furnace operates at 75% efficiency. It is used to heat a 10 kg mass of steel from 20°C to 800°C. The specific heat capacity of steel is 444 J/(kgK).
   a. How much energy in joules is required? *(3.46 MJ)*
   b. How long does it take? *(15.4 min)*
   c. What is cost at two cents/kWh? *(2.6 cents)*

7. A ladle contains 100 kg of tin at its melting temperature of 230°C. A block of steel with mass of 20 kg at a temperature of 20°C is dropped into the tin. If the specific heat of steel is 0.444 kJ/(kgK) and heat of fusion for tin is 60 kJ/kg, determine the grams of tin solidified by the time the steel and tin reach equilibrium.

   (31.1 kg)

8. A steel gear of mass 2 kg and specific heat capacity of 444 J/(kg K) is heated to 800°C and plunged into oil of specific heat capacity of 2000 J/(kg K) at 20°C. If one litre has a mass of 0.8 kg, what is the minimum amount of oil (litres) required so that the temperature rise does not exceed 10°C?

   (42.7 litres)

9. A steam boiler which is 75% efficient overall burns coal with a calorific value of 40 MJ/kg. Feed water is supplied at 70°C and the boiler generates 800 kg/h of steam superheated to 160°C at atmospheric pressure. The latent heat of vaporization is 2 250 000 J/kg, specific heat of water is 4200 J/(kgK) and superheated steam is 2000 J/(kgK). What mass of coal is consumed per hour?

   (66.6 kg/h)

Appendix B $PE = mgh$
         $HE = cmT$ Physics   375

10. A waterfall is 200 m high. If all the potential energy is converted into heat energy, find the difference in temperature between the top and bottom of the fall (g = 9.81 and specific heat of water = 4200 J/(kgK).

(0.47°C)

11. A 3 kW electric kettle contains 1 kg of water. The heat capacity of the kettle is 300 J/K. Find the time to raise the water from 20°C to the boiling point.

(2 min)

$TE = KE + PE$ Physics

12. What is the total energy possessed by a 5000 kg airplane flying at 500 km/h at an altitude of 1200 m?

(107 MJ)

13. A lead bullet travelling at 300 m/s impacts a steel target. If 75% of its energy is converted into heating the bullet, what is the rise in temperature? (Specific heat of lead = 140 J (kgK)$^{-1}$)

(241°C)

14. A motor car is running on a level road at a constant speed of 100 km/h. If the total frictional resistance is 200 newtons, what is the engine power?

(5.56 kW)

15. A steam engine cylinder is 0.4 m in diameter, the mean pressure during the stroke is 35 bar and the length of stroke is 0.5 metre. What is work done in one stroke and, if at 120 strokes per minute, what is the power?

(219.9 kJ)
(440 kW)

16. A tapered pipe 200 m long lays on a hill such that its centerline is at a gradient of sin^{-1} 0.15. At the upper end, it has a diameter of 1 m and at lower end a diameter of 0.5 m. When running full, the pressure at the upper end is 4 bar and the flow velocity is 3 m/s. Calculate for water:
   a. The flow in litres/sec.   *(2356 l/s)*
   b. The pressure at the lower end.   *(6.26 bar)*

17. An 8 metric tonne flywheel has a radius of gyration of 1 m. While running at 600 r/min, it is suddenly connected to a second flywheel having a mass of 2 tonnes and radius of gyration of 0.6 m. Determine:
   a. Common angular velocity, in rev/min, after connection.   *(550.5 r/min)*
   b. The loss of energy, in joules.   *(1.304 MJ)*
   c. The time it would take, in seconds, to bring the system to rest, if a retarding torque of 2 Nm were applied.
                                                                 (251 323 s)

18. A solid disc flywheel has an outside diameter of 0.5 m and a width of 0.1 metre. It is made from a material with specific density of 7.2. If the flywheel is to drive a ram which provides a constant force of 100 kN through a distance of 5 mm, at what initial speed should the flywheel be rotating to just complete the job?

(143.7 r/min)

19. A symmetrical "I" beam with web vertical spans a 4 meter space. The depth of the beam is 250 mm, the width is 200 mm, the flange thickness is 25 mm and web thickness is 20 mm. Neglecting the weight of the beam, what load at the center of the beam will give a bending stress of 120 N/mm²?

(134.8 kN)

20. How much will a 30 metre steel tape, 1 cm wide and 0.5 mm thick stretch under a pull of 300 newton, if E, Young's modulus for steel is 210 kN/mm².

(8.6 mm)

21. A small generator is driven from a large flywheel via a steel shaft 10 mm in diameter and 0.4 m long. If I for the generator rotor is 3.0×10^{-3} kgm², determine the torsional natural frequency (shear modulus = 80 kN/mm²).

(40.7 Hz)

22. A volume of 10 litres of air at an absolute pressure of 1 bar and a temperature of 20°C is compressed until the volume is 6 litres and absolute pressure is 2 bars. What is final temperature in degrees Celsius?

(78.6°C)

23. Calculate the mass of 1 cubic metre of air at an absolute pressure of 6 bars and a temperature of 27°C. The gas constant R for air is 288 J/(kg·K).

(6.96 kg)

24. The compressed air used for starting a diesel engine is stored in a 500 litre cylinder. During a start, the pressure dropped from 30 to 25 bar, the temperature dropped from 17°C to 7°C. Using R = 288 J/(kg·K), what is mass of air required to start the engine?

(2.46 kg)

25. A small diesel engine drives a dc generator which is 80% efficient. The diesel engine consumes 2 litre of fuel (specific gravity = 0.8) per hour with heat value of 46 MJ/kg and is 30% efficient. The electrical power is used to heat a room which is a suspended 5 metre cube surrounded by 0°C air. The walls are 0.15 metre thick and have a thermal conductivity of 1.5 W/(mK). What is the steady state temperature inside the room?

(3.27°C)

26. A 7 litre 6-cylinder diesel engine with 150 mm bore cylinders produces 100 kW at 2400 r/min. What is the BMEP?

(7.143 bar)

27. A turbo-air compressor draws 3 m³ of air per second at 1 bar absolute and 27°C. The air is delivered at 5 bar absolute and a temperature of 77°C. The area of the suction pipe is 0.1 m² and of the discharge pipe 0.02 m² and the discharge pipe is 10 m above the suction inlet. A 6 kg mass of water enters at 27°C and leaves the water jacket at 52°C each second. Assuming no other losses, what is the power required to drive the compressor?

(491 kW)

Appendix B

Standard Representations for SI Units Using All Upper Case Characters

AMPERE	A	KELVIN	K	POINT	PNT
YEAR	ANN	KILOGRAM	KG	RADIAN	RAD
ARE	ARE	LITRE	L	SECOND (TIME)	S
BAR	BAR	LUMEN	LM	SECOND (ANGLE)	SEC
COULOMB	C	LUX	LX	SIEMENS	SIE
CANDELA	CD	METRE	M	STERADIAN	SR
DEGREE CELSIUS	CEL	MINUTE (TIME)	MIN	STOKES	ST
DAY	D	MINUTE (ANGLE)	MNT	TESLA	T
DEGREE (ANGLE)	DEG	MONTH	MO	METRIC TÔNNE	TNE
ELECTRONVOLT	EV	MOLE	MOL	TWSP	TWP
FARAD	F	NEWTON	N	ATOMIC MASS UNIT	U
GRAM	G	NAUTICAL MILE	NMI	VOLT	V
HENRY	H	OHM	OHM	WATT	W
HOUR	HR	POISE	P	WEBER	WB
HERTZ	HZ	PASCAL	PA	WEEK	WK
JOULE	J	PICA	PC		

Representation of Prefixes

Prefix	Multiplier	International Symbol	Single Case Upper
tera	10^{12}	T	T
giga	10^{9}	G	G
mega	10^{6}	M	MA
kilo	10^{3}	k	K
hecto	10^{2}	h	H
deca	10^{1}	da	DA
deci	10^{-1}	d	D
centi	10^{-2}	c	C
milli	10^{-3}	m	M
micro	10^{-6}	μ	U
nano	10^{-9}	n	N
pico	10^{-12}	p	P
femto	10^{-15}	f	F
atto	10^{-18}	a	A

Dimensionally Correct Units for Rotation

Our changeover to the metric system affords us the opportunity to eliminate several practices that over the years have been troublesome and confusing, particularly to engineers and students in engineering. The discontinuance of the use of the mass unit to express both mass and force is an example of taking advantage of such an opportunity; accordingly, there is general agreement that the kilogram-force unit is obsolete and unacceptable for use internationally.

The confusion that results from the use of newton meter as the unit of both work and torque can and should similarly be eradicated during the current changeover to SI. The purpose of this brief note is to explain how this laudable objective can be attained.

The root of the problem becomes apparent when dimensional analysis is applied to well-known formulas that involve work, torque, moment of inertia, and angular momentum. In Table 1, well-known quantity equations are shown in the left column. To the right of each equation is given the corresponding unit relationship. Starting with the first equation the units for work (N·m) and angular displacement (rad) are indisputably SI and are shown without brackets. To make the unit equation dimensionally correct, the unit for torque must then be [N·m/rad] and this sought-after, dimensionally correct unit is shown in brackets. A similar procedure applies to the second and third equations, resulting in the dimensionally correct units for moment of inertia [kg·m^2/rad^2] and the moment of momentum [kg·m^2/(rad·s)] instead of the customarily used units kg·m^2 and kg·m^2/s, respectively.

TABLE 1

Formula	Unit Relationship
$W = T\theta$	N·m = [N·m/rad] × rad = [kg·m^2/(rad·s^2)] × rad
$T = I\alpha$	kg·m^2/(rad·s^2) = [kg·m^2/rad^2] × rad/s^2
$L = I\omega$	[kg·m^2/(rad·s)] = kg·m^2/rad^2 × rad/s

In Table 2 are listed the dimensionally incorrect units given in ISO Standard R31/III (column 2) and the proposed dimensionally correct units (column 3). The use of the dimensionally correct units would not only eliminate the difficulty that results from using the same unit for work and torque; it would also eliminate the confusion that arises from the lack of dimensional integrity of equations that describe phenomena of rotation.

TABLE 2

Quantity	ISO/R31/III	Proposed
T	N·m	N·m/rad
I	kg·m^2	kg·m^2/rad^2
L	kg·m^2/s	kg·m^2/(rad·s)

A memorandum for members of the SAE Metric Advisory Committee from L. E. Barbrow, National Bureau of Standards (May 17, 1976).

Table 3 explains how the proposed units for T, I, and L may be derived directly. As often explained (as, for example, in 3.4.4.2 of E380-76), if vectors are used, the distinction between energy and torque becomes obvious. However, there is no means for carrying vector notation over into the symbols. In Table 3 a simple substitute for vector notation is used. In column 1 of Table 3, the subscripts t and r designate the tangential direction and radial direction, respectively, of the vector quantities to which they are affixed.

TABLE 3

Formula	Unit
$T = F_t d_r$	$[N \cdot m/rad]$
$I = m d_r^2$	$[kg \cdot m^2/rad^2]$
$L = m v_t d_r$	$[kg(m/s)(m/rad)] = [kg \cdot m^2/(rad \cdot s)]$

Thus, in Figure 1, d_r represents a radial direction from the point of rotation 0 to point A, and Δd_t represents a tangential increment of distance from point A to point B. Since $d_r = \Delta d_t / \Delta \theta$, the SI unit of measurement of d_r is that of $\Delta d_t / \Delta \theta$; namely, m/rad. Thus when substituting units for the quantities on the right side of the equations listed in column 1 of Table 3, insert m/rad for d_r and m for d_t; since $v_t = d_t/s$, its unit is m/s. These substitutions result in the units for T, I, and L listed in column 2 of Table 3, which are the same as the units derived in Table 1.

Illuminating engineers will find a similar approach advantageous in what has to them been a troublesome area. In the well-known formula $E = I/d^2$ (where E is illuminance on an area A, I is luminous intensity of a source, and d is distance between the source and area A), substituting traditional units for E, I, and d results in the dimensional relationship

$$lm/m^2 = cd/m^2$$

This relationship indicates that lm and cd are dimensionally the same, which is obviously untrue. Since d is perpendicular to A, it should preferably be represented by the symbol d_r. Then for d_r^2, the unit that should be used is $m^2/(sr)$. Making this substitution results in the dimensional relationship

$$lm/m^2 = cd/(m^2/sr)$$

$$lm = cd \cdot sr$$

which is recognized as a valid relationship.

The rationale behind the use of $m^2/(sr)$ as the proper unit in which d_r^2 should be expressed is shown in Figure 2 in which d_r represents the radial distance from 0 to the spherical increment of area ΔA and $\Delta \omega$ represents the increment of solid angle subtended at 0 by ΔA. Since $d_r^2 = \Delta A / \Delta \omega$, the SI unit of measurement of d_r^2 is that of $\Delta A / \Delta \omega$; namely $m^2/(sr)$.

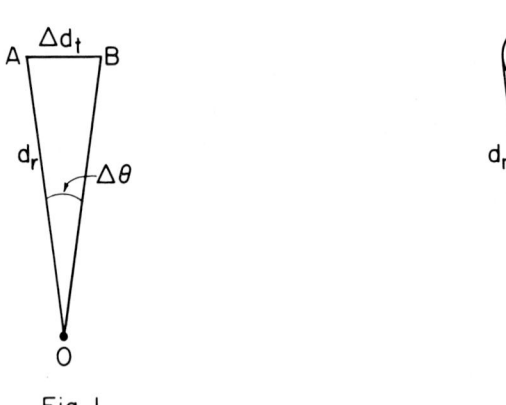

Fig. 1   Fig. 2

Implementation of the above proposal is recommended simultaneously at various levels by:

(a) standards writing bodies, both international and national
(b) educators
(c) authors of books
(d) engineers and scientists, particularly in problems that involve rotation

APPENDIX C   THEOREMS OF RANDOM VARIABLE ALGEBRA

A body of theorems can be assembled that rests on the algebraic properties listed at the beginning of Sec. 7.11. This body of theorems constitutes an algebra of *expectation*. The fundamental definitions are

$$\mu_{\phi(x)} = E[\phi(x)] = \int_{-\infty}^{\infty} \phi(x)f(x)dx$$

$$\sigma_{\phi(x)} = \{\int_{-\infty}^{\infty} [\phi(x) - \mu_{\phi(x)}]^2 f(x)dx\}^{1/2}$$

where x is a random variate, $f(x)$ is its probability density, and ϕ is some function of x. It follows that

$$\mu_a = a \quad (1) \qquad \sigma_{x+a} = \sigma_x \quad (5)$$

$$\mu_{x+a} = \mu_x + a \quad (2) \qquad \sigma_{ax} = a\sigma_x \quad (6)$$

$$\mu_{ax} = a\mu_x \quad (3) \qquad \sigma_{x^2} = (4\mu_x^2 \sigma_x^2 + 2\sigma_x^4)^{1/2} \quad (7)$$

$$\mu_{x^2} = \mu_x^2 + \sigma_x^2 \quad (4)$$

uncorrelated sums:

$$\mu_{x+y} = \mu_x + \mu_y \quad (8) \qquad \sigma_{x+y} = (\sigma_x^2 + \sigma_y^2)^{1/2} \quad (9)$$

uncorrelated differences:

$$\mu_{x-y} = \mu_x - \mu_y \quad (10) \qquad \sigma_{x-y} = (\sigma_x^2 + \sigma_y^2)^{1/2} \quad (11)$$

uncorrelated products:

$$\mu_{xy} = \mu_x \mu_y \quad (12) \qquad \frac{\sigma_{xy}}{\mu_{xy}} = \left[\left(\frac{\sigma_x}{\mu_x}\right)^2 + \left(\frac{\sigma_y}{\mu_y}\right)^2 + \left(\frac{\sigma_x \sigma_y}{\mu_x \mu_y}\right)^2\right]^{1/2} \quad (13)$$

uncorrelated quotients:

$$\mu_{x/y} \simeq \frac{\mu_x}{\mu_y} \quad (14) \qquad \frac{\sigma_{x/y}}{\mu_{x/y}} \simeq \left[\left(\frac{\sigma_x}{\mu_x}\right)^2 + \left(\frac{\sigma_y}{\mu_y}\right)^2\right]^{1/2} \quad (15)$$

If $\phi = \phi(x_1, x_2, \ldots, x_n)$, propagation of error techniques give good approximations to (μ_ϕ, σ_ϕ)

$$\mu_\phi \simeq \phi(\mu_{x_1}, \mu_{x_2}, \ldots, \mu_{x_n}) + \frac{1}{2} \sum_{i=1}^{n} \frac{\partial^2 \phi}{\partial x_i^2} \sigma_{x_i}^2 \quad (16)$$

$$\sigma_\phi = \left[\sum_{i=1}^{n} \left(\frac{\partial \phi}{\partial x_i}\right)^2 \sigma_{x_i}^2 + \frac{1}{2} \sum_{i=1}^{n} \left(\frac{\partial^2 \phi}{\partial x_i^2}\right)^2 \sigma_{x_i}^4\right]^{1/2} \quad (17)$$

For cases in which correlation exists between random variates, for correlated sums

$$\mu_{x+y} = \mu_x + \mu_y \quad (18) \qquad \sigma_{x+y} = (\sigma_x^2 + \sigma_y^2 + 2\rho \sigma_x \sigma_y)^{1/2} \quad (19)$$

correlated differences:

$$\mu_{x-y} = \mu_x - \mu_y \quad (20) \qquad \sigma_{x-y} = (\sigma_x^2 + \sigma_y^2 - 2\rho \sigma_x \sigma_y)^{1/2} \quad (21)$$

correlated products:

$$\mu_{xy} = \mu_x \mu_y \left(1 + \rho \frac{\sigma_x}{\mu_x} \frac{\sigma_y}{\mu_y}\right) \tag{22}$$

$$\frac{\sigma_{xy}}{\mu_{xy}} = \left[\left(\frac{\sigma_x}{\mu_x}\right)^2 + \left(\frac{\sigma_y}{\mu_y}\right)^2 + \left(\frac{\sigma_x}{\mu_x}\frac{\sigma_y}{\mu_y}\right)^2 + 2\rho \frac{\sigma_x}{\mu_x}\frac{\sigma_y}{\mu_y} + \rho^2\left(\frac{\sigma_x}{\mu_x}\frac{\sigma_y}{\mu_y}\right)^2\right]^{1/2} \tag{23}$$

correlated quotients:

$$\mu_{x/y} \simeq \frac{\mu_x}{\mu_y}\left[1 + \frac{\sigma_y}{\mu_y}\left(\frac{\sigma_y}{\mu_y} - \rho \frac{\sigma_x}{\mu_x}\right)\left(1 + 3\left(\frac{\sigma_y}{\mu_y}\right)^2 + \ldots\right)\right] \tag{24}$$

$$\frac{\sigma_{x/y}}{\mu_{x/y}} = \left[\left(\frac{\sigma_x}{\mu_x}\right)^2 + \left(\frac{\sigma_y}{\mu_y}\right)^2 - 2\rho \frac{\sigma_x}{\mu_x}\frac{\sigma_y}{\mu_y} + 8\left(\frac{\sigma_y}{\mu_y}\right)^4 - 16\rho \frac{\sigma_x}{\mu_x}\left(\frac{\sigma_y}{\mu_y}\right)^3 \right.$$
$$\left. + 3\left(\frac{\sigma_x}{\mu_x}\right)^2\left(\frac{\sigma_y}{\mu_y}\right)^2 + 5\rho^2\left(\frac{\sigma_x}{\mu_x}\right)^2\left(\frac{\sigma_y}{\mu_y}\right)^2 + \ldots\right]^{1/2} \tag{25}$$

For powers of x, any real value of n,

$$\mu_{x^n} = \mu_x^n\left[1 + \frac{1}{2}n(n-1)\left(\frac{\sigma_x}{\mu_x}\right)^2 + \ldots\right] \tag{26}$$

$$\sigma_{x^n} = |n|\mu_x^{n-1}\sigma_x\left[1 + \frac{1}{4}(n-1)^2\left(\frac{\sigma_x}{\mu_x}\right)^2 + \ldots\right] \tag{27}$$

and for commonly encountered values of n,

function

$$\frac{1}{x^3} \quad \mu_{1/x^3} = \frac{1}{\mu_x^3}\left[1 + 6\left(\frac{\sigma_x}{\mu_x}\right)^2\right] \quad (28) \quad \sigma_{1/x^3} = \frac{3}{\mu_x^3}\frac{\sigma_x}{\mu_x}\left[1 + 4\left(\frac{\sigma_x}{\mu_x}\right)^2\right] \tag{29}$$

$$\frac{1}{x^2} \quad \mu_{1/x^2} = \frac{1}{\mu_x^2}\left[1 + 3\left(\frac{\sigma_x}{\mu_x}\right)^2\right] \quad (30) \quad \sigma_{1/x^2} = \frac{2}{\mu_x^2}\frac{\sigma_x}{\mu_x}\left[1 + \frac{9}{4}\left(\frac{\sigma_x}{\mu_x}\right)^2\right] \tag{31}$$

$$\frac{1}{x} \quad \mu_{1/x} = \frac{1}{\mu_x}\left[1 + \left(\frac{\sigma_x}{\mu_x}\right)^2\right] \quad (32) \quad \sigma_{1/x} = \frac{1}{\mu_x}\frac{\sigma_x}{\mu_x}\left[1 + \left(\frac{\sigma_x}{\mu_x}\right)^2\right] \tag{33}$$

$$x^0 \quad \mu_{x^0} = 1 \quad (34) \quad \sigma_{x^0} = 0 \tag{35}$$

$$\sqrt{x} \quad \mu_{\sqrt{x}} = \sqrt{\mu_x}\left[1 - \frac{1}{8}\left(\frac{\sigma_x}{\mu_x}\right)^2\right] \quad (36) \quad \sigma_{\sqrt{x}} = \frac{\sqrt{\mu_x}}{2}\frac{\sigma_x}{\mu_x}\left[1 + \frac{1}{16}\left(\frac{\sigma_x}{\mu_x}\right)^2\right] \tag{37}$$

$$x \quad \mu_x = \mu_x \quad (38) \quad \sigma_x = \sigma_x \tag{39}$$

Appendix C

$$x^2 \quad \mu_{x2} = \mu_x^2\left[1 + \left(\frac{\sigma_x}{\mu_x}\right)^2\right] \quad (40) \quad \sigma_{x2} = 2\mu_x^2 \frac{\sigma_x}{\mu_x}\left[1 + \frac{1}{4}\left(\frac{\sigma_x}{\mu_x}\right)^2\right] \quad (41)$$

$$x^3 \quad \mu_{x3} = \mu_x^3\left[1 + 3\left(\frac{\sigma_x}{\mu_x}\right)^2\right] \quad (42) \quad \sigma_{x3} = 3\mu_x^3 \frac{\sigma_x}{\mu_x}\left[1 + \left(\frac{\sigma_x}{\mu_x}\right)^2\right] \quad (43)$$

$$x^4 \quad \mu_{x4} = \mu_x^4\left[1 + 6\left(\frac{\sigma_x}{\mu_x}\right)^2\right] \quad (44) \quad \sigma_{x4} = 4\mu_x^4 \frac{\sigma_x}{\mu_x}\left[1 + \frac{9}{4}\left(\frac{\sigma_x}{\mu_x}\right)^2\right] \quad (45)$$

CLOSURE

If random variates $x \sim N(\mu_x, \sigma_x)$ and $y \sim N(\mu_y, \sigma_y)$, then for these Gaussian variates summation is *closed* since for $z = x + y$, $z \sim N(\mu_z, \sigma_z)$. If random variate $x \sim N(\mu_x, \sigma_x)$ and $y \sim N(\mu_y, \sigma_y)$ are multiplied to form $z = xy$, the product is not closed since the distribution of the product of Gaussian variables is not Gaussian. However, if the errors involved in presuming closure are small, the presumption of a Gaussian distribution of a product is said to be *robust*. If random variables $x \sim LN(\mu_x, \sigma_x)$ and $y \sim LN(\mu_y, \sigma_y)$ are log normally distributed, their product $z = xy$ is closed since $z \sim LN(\mu_z, \sigma_z)$.

SELECTED PROBLEM ANSWERS

CHAPTER 2
1. $v^* = 21.05$ mph
2. $x^* = 2.76$ $y^* = 5.53$ hundreds of tires
3. $x^* = (c - b - t)/(2e)$, unit price $p = (b + c + t)/2$
4. 2340, $0.251(10^{-6})$ \$/in.·lbf
5. 2340, $0.100(10^{-5})$ in.$^{-1}$
6. Still the same, 2340.
7. See p. 99 et al.
8. $M = -nPx \dfrac{\rho}{S}\left[\dfrac{\cos\beta}{\sin(\alpha+\beta)\cos\alpha} + \dfrac{\cos\alpha}{\sin(\alpha+\beta)\cos\beta}\right]$
   $\alpha^* = \beta^* = 45°$
9. $M = -\rho \dfrac{x}{\cos\theta}\left[\dfrac{2nP\cos(\theta-\xi)}{S\sin 2\theta}\right]$
   For $\xi = 30°$, $36.2° < \theta^* < 36.4°$.

CHAPTER 3
1. See Sec. 9.6.
6. $M = bh^2 = b(4r^2 - b^2)$, $(b/h)^* = 1/\sqrt{2}$
7. $M^2 = b^2h^6 = 4r^2h^6 - h^8$, $(b/h)^* = 1/\sqrt{3}$
8. Optimal wire size decreases with increasing number of strands.
9. $p = 0.014$
10. $7/2$
12. $\hat{\mu} = 0.109$, $\hat{\sigma} = 0.0117$ lbf
13. $\hat{\mu} = 0.109$, $\hat{\sigma} = 0.0096$ lbf
14. $E(x^n) = 1/(n+1)$

CHAPTER 4
1. $\int_0^\pi \sin^2 x\, dx = \pi/2$
2. 0.0, 4.730, 7.853, 10.996, 14.137, 17.279
3. 1.875, 4.694, 7.855, 10.996, 14.137, 17.279

CHAPTER 5
1. An experiment is a *planned* event designed to reveal valuable information in an efficient manner.
4. $S_e = (1/2)S_u$ $50 < S_u < 200$ kpsi, steel

$S_e = 100$ $S_u > 200$ kpsi, steel
$S_e = 0.4 S_u$ $75 < S_u < 170$ kpsi, cast iron

6. $S_f' = 36$ kpsi for 1020 hot rolled

   $S_f' = 169$ kpsi for 2340 Q & T at 400°F

7. $S_u = 1277 + 529$ BHN psi
   $S_u = 8.51 + 3.79$ BHN N/mm^2

8. $\mu = 19\,800/\#^{1.06}$

9. $P_u = 4799/e^{0.222\#} = 4799/10^{0.096\#}$

10. $R = 90.56 + 0.3875t$ Ω

11. $\Delta W = -0.0054 + 0.0858t$ mg/cm^2

12. $GPM = 36\sqrt{h}$

13. $v = 332.78/p^{0.937}$ ft^3/lbm

14. $Q = 1875 \, D^{0.907}$ Btu/ft/day

15. $S = 1.672(10^7)\exp(-0.0079t)$ psi

16. $S = 2.90 \exp(0.0354t)$

17. $\pi_1 = sr^3/M$, $\pi_2 = h/r$, $\pi_3 = A/r^2$, $\pi_4 = e/r$

18. $\pi_1 = \rho/I^{1/4}$, $\pi_2 = M/(EI^{3/4})$

19. $\pi_1 = U/(A^{3/2}E)$, $\pi_2 = s/E$, $\pi_3 = l/\sqrt{A}$

20. $\pi_1 = \mu$, $\pi_2 = T/(Fl)$, $\pi_3 = d_m/l$

21. $\pi_1 = \tau d^2/F$, $\pi_2 = D/d$

22. $\pi_1 = f$, $\pi_2 = p/(\mu N)$, $\pi_3 = r/c$

26. $\pi_1 = \tau J^{3/4}/T$, $\pi_2 = r/J^{1/4}$

27. $\pi_1 = s/p$, $\pi_2 = D/t$

28. $\pi_1 = U/(GJ^{3/4})$, $\pi_2 = T/(GJ^{3/4})$, $\pi_3 = l/J^{1/4}$

29. $\pi_1 = K/(DG)$, $\pi_2 = N$, $\pi_3 = d/D$

30. $\pi_1 = T/dF$, $\pi_2 = f$, $\pi_3 = D/d$

31. $y = 13.506 + 0.7902x$

CHAPTER 7

1. $0 = (m + M)V - m\dot{x}$

2. $V = \dfrac{m_1}{m_1 + M} \dot{x}_1$

3. $f = \dot{w}[V + v_1 \sin\theta_2 - u(\cos\theta_1)(\cos\theta_2)] + W \sin\theta_2$

4. $f_z = \dot{w}V$, $f_x = -\dot{w}u$

5. $\dot{w} = 43.06$ slug/s, $f_x = 3789$ lbf, hp = 606

6. $F = \dot{w}V = 22.5$ lbf or 100 N

8. $F = \dot{w}(V - u)$

19. $R_x = -p_1 A_1 + p_2 A_2 \cos\theta + \rho_1 V_1 A_1 (V_2 \cos\theta - V_1) = 44\,451$ N

    $R_y = W + p_2 A_2 (\sin\theta) + \rho_1 V_1 A_1 (V_2 \sin\theta) = 35\,125$ N

22. $p = 0.032\,78$ psi, 0.395 psi (negative, meaningless)

23. $p = 0.252$ psf (exponential)

24. $p = 15.2315$ psi, 15.241 psi, 15.239 psi

26. Sixth day.

29. $\mu_F = 120$ lbf/in.^{2}, $\sigma_F = 7.21$ lbf/in.^{2}

30. $\mu_{\sin x} = 0.9274$, $\sigma_{\sin x} = 0.0868$

31. $\mu = 15.62$, $\sigma = 1.30$

Selected Problem Answers   387

37. Exact $\mu_y = 1/2$, approximate $\mu_y = 0.494$ for $n = 2$.
38. For $n = 2$, $\mu_y = 7/3$, $\mu_{z2} = 7/3$.

CHAPTER 8
1. Maximum.
2. Maximum.
5. $x^* = 2.76$ is max, $x^* = 7.24$ is min
6. $(4, -2)$ min
7. $x^* = \$0.89$, $p^* = \$2829.49$
   $y^* = \$0.99$
8. Thirds.
9. W is minimum at $z = 0$ for any x and y.
10. $(5/2, 3/2, 1)$
11. $x_1^* = 15/14$, $x_2^* = 9/7$
12. $x_1^* = 15/14$, $x_2^* = 9/7$
13. $x_1^* = 2$, $x_2^* = 0$
14. $x^* \pm 1 = 43.36$ cm
15. $x^* \pm 2 = 68.75$ cm
16. $x^* \pm 1/16 = 0.65$ in.
17. $x^* \pm 0.5 = 4.38$ cm
18. $x^* \pm 0.1 = 1.6875$ Ω
19. $x^* \pm 0.1 = 2.0625$ Ω
20. $x^* \pm 0.05 = 1.97$ hundreds of grade A tires
21. $x^* \pm 0.05 = 2.75$ hundreds of grade A tires
22. $x^* \pm 1 = 43.16$ cm
23. $x^* \pm 2 = 70.03$ cm
24. $x^* \pm 1/16 = 0.66$ in.
25. $x^* \pm 0.3 = 4.16$ cm
26. $x^* \pm 0.05 = 1.695$ Ω
27. $x^* \pm 0.05 = 2.04$ Ω
28. $x^* \pm 0.05 = 1.96$ hundreds of grade A tires
29. $x^* \pm 0.05 = 2.76$ hundreds of grade A tires
30. $\overline{\Delta x_1 = 7.41°}$
    $\overline{\Delta x_2 = 6.72°}$
33. (C) $N = 20$
34. (C) $N = 17$
35. $x^* \pm 0.001 = 3.873$
36. $x_1^* = 0.24$
    $x_2^* = 1.62$
    $y^* = 0.722$
    $\Delta y^* = -0.078$
37. $x_1^* = 1.7889$, $x_2^* = 0.8944$, $y^* = 6.111$
38. $x_1^* = -1/2$, $x_2^* = -\sqrt{3}/2$; also $x_1^* = -1/2$, $x_2^* = \sqrt{3}/2$
39. $D^* = 1.37$ ft, $L^* = 10.2$, $\$^* = 3174.40$ dollars
    $\Delta y^* = 55.80$ dollars
40. $D^* = 0.977$ ft
    $L^* = 20$ ft
    $\$^* = 3275$ dollars
41. $b/a = 1.618\ 034 = 1 + \eta$
42. $x = 0.618\ 034, -1.618\ 034$; η is one such number.

INDEX

Abbreviations, SI, 368, 370, 371
   computer, 377
Abstraction, levels of, 110
Acceptability, 5
Accountability, 195
   equation for continuous countables, 208, 210
   equation for discrete countables, 201
Accuracy, 112, 189
Adiabatic system, 215
Algebra of expectation, 247
Algorithm, 73
   diagrammed, 74
   IOWA CADET, 78
   root-finding, 86, 89
   for square root, 73–74, 75
   as steps, 76
   as stratagem, 76
Alternatives, 5
   best, 29
Amount of substance (mole), 367
Ampere (A), 367
Analysis, 21
Angle (rad, sr), 368
Approximate number, 256
Archimedes' buoyant effect, 186
Argument
   of FORTRAN subprogram, 81
   list, CADET convention, 87
Assembly operations, uncertainties, 44
Assertive deductive model, 228
   of epidemic, 240
Asymptote exploitation, 243

Balancing speed, 65, 328
Ballistic effects, 189
Biot, Jean Baptiste, 184
Body, 218

Boltzmann, Ludwig, 185
Boundary
   adiabatic, 183
   diathermic, 183
   of system, 179
Boyle, Robert, 105
Brainstorming, 15
Bridgeman, Percy Williams, 123
Buckingham, E., 140
Buoyant effect, 186

CADET, IOWA, 78
   adding programs to, 84
   documentation, 81
   error messages, 83
   fine root-finder, 365
   Geneva mechanism analysis, 356
   golden section search, 359
   grid search, 358
   Lagrangian interpolation, 360
   n-dimensional gradient search, 357
   Newton-Raphson root-finder, 361
   pattern search, 362
   random variable algebra, 363
   Simpson's rule, 364
   slider-crank position analysis, 355
   structure, 80
Calculator, hand-held
   epidemic model program, 244
   keystroke sequence, 75
   program, 75
   root-finding program, 87, 90, 91
Calorimetric heat effect, 214
Cam, circular disk, 316
Candela (cd), 367
Can proportions, 325
Capacitive effect, 187
Capillarity, 186
Car dumper, rotary, 339
Celsius scale, 122

Centroid of data, 127
Charles, Jacques, 105
Chemical effects, 188
Chemical squib, 248
Commas in numbers, SI, 371
Complete set, 143
  of dimensionless variables, 143
  equivalent, 147
Computer
  capabilities, 77
  in engineering design, 76
  library, 84
Concept of
  change, 109
  force, 109, 110
  length, 109, 110
  mass, 109, 110
  simultaneity, 108
  succession, 108
  time, 108-9
Conduction of heat, 184
Confidence bands, 121
Conservation
  of electric charge, 211
  of energy, 213
  of matter, mass, 212
Constraints
  equality, 27, 40
  functional, 27, 40
  inequality, 27, 40
  legal code, 44
  prior experience, 44
  problem-solving, 22, 31
  regional, 27
  slack, 270
  solution, 22, 31
  tight, 270
Continuum of matter, 105
Control region, space, or volume, 180
Control region delineation, 222
Control surface, 180
  thermodynamics of, 120
Convection of heat, 184
Convergence monitor, 82
  of program NEWRA1, 98, 100
Conveyor, sand, 218
Coordinate of graph, 112, 117
Cost
  expected, 48
  and figure of merit, 47
  incremental, 48, 50
Cost function, 274
Coulomb (C), 368

Coulomb's charge attraction effect, 188
Countables, 196
  conservation of, 197
  destruction of, 198
  species of, 199, 200
  students as, 206
Couple, measurement, 256
  statistical, 189
Critical population, susceptibles, 247
Curve fitting, 125

Dantzig, George, 277
Data
  centroid of, 127
  rectification of, 129
  transformation of, 129
Davis, W. J., Jr., 332
Davis equation, 328, 332
Decimal points, in SI, 371
Decision making, 5, 239
Deduction, 21
Deductive mathematical model, 228, 247
Definition, 107
  of entity, 108, 112
  of measurable attributes, 108, 112
  operational, 108, 110, 111, 112
Design, 8
  material selection, 33
  operations, 12, 13
  preliminary, 104
  set, 26
  variable, geometric, 33
  variables, 26, 29
Design factor
  as figure of merit, 43
  need for, 44
Design imperative, 22, 31, 263
Deterministic deductive model, 228, 241
Deterministic statements, 112
Dimensional homogeneity, 141, 239
Dimensions, 141
Discrepancy, in accounting, 200
Documentation, 81
  structure of, 82

Edelbaum's sufficiency conditions, 268
Effect, 183
  ballistic, 189
  Biot, 184
  buoyant, 186
  calorimetric, 214
  chemical, 188
  conduction, 184
  convection, 184

Index 391

electric charge, 187
heat, 183
Hookian, 186
impossible, 189
magnetic, 188
motor, 188
radiative, 185
work, 183
Electric current (ampere), 367
Eleven Toes, 197
Empiricism, achievements of, 104-5
Endurance limit, 123
Energy, 7, 110, 214, 369
  internal, 215
  kinetic, 204
Engineering, 21
  skills, 9, 10
Engineers, attitudes and abilities, 14
Entropy, 110, 111
Epidemic model, 240
Error message, IOWA CADET, 83
Expected value, 51-52, 247
Extreme. *See* Maxima; Minima
Eyeball-best-line, 127

Factor of safety. *See* Design factor
Fahrenheit scale, 123
False minimum, 265
Farad (F), 368
Feasibility, 5
Fence-walking search, 297
Figure of merit, 39, 79
  cost as, 47
  discrete-valued, 34
  formal, 41
  penalized, 43
  qualitative factors in, 68
  reliability as, 59
  tactical, 42, 46
  time as, 64
  weight as, 32
Finger-Toe law, 198
First law of thermodynamics, 111, 213, 214
First principles, 211
Flowchart, logic, 74
Flowchart, universal, 85
Force, 109, 141, 368
FORTRAN
  declaration statement, 77
  IOWA CADET, 78
  subprogram, 79
Frequency, 368

Function
  cost, 274
  criterion, 26, 39, 274
  cumulative density, 55
  FORTRAN, 79
  merit, 26, 274
  objective, 26, 274
  probability density, 52, 167
  trade-off, 66
  unimodal, 278
  utility, 39
Fundamental dimensions, 141
  in CADET documentation, 83

Gear ratio, 66, 338
Golden section search, 284
Goodness-of-fit, 126
Gradient-sensitive searches, 290
Graph paper, log-log, semilog, 130
Graphs, 112
Gravitation, Newton's law of, 221

Hazard rate, 56, 57
Heat, 184, 185
  quantity of, 369
Heater, hen house, 29
Hegel, George William, 123
Henry (H), 368
Hertz (Hz), 368
Homogeneity of dimensions, 141
Hooke, Robert, 185
Hooke's equation, 23-24

Impossible effects, 189
Incomplete number, 256
Incremental cost, 48
Inductive effect, 187
Inertia equivalents, shish kebab, 339
Infant mortality interval, 57
Influences
  external, 11
  internal, 11
  significant, 224
Inspection procedure, 49
Interest, 48
Internal energy, 215
International Conference on Weights and Measures, 366
Interval
  infant mortality, 57
  random failure, 57
  of uncertainty, 278
  useful life, 57
  wear-out failure, 57
Interval-halving search, 281, 282

Inverse analysis, 21
IOWA CADET. *See* CADET, IOWA

Joule (J), 368

Kelvin (K), 367
Kelvin scale, 122
Kilogram (kg), 367
Kinetic energy, 204

Lagrange, Compte Joseph Louis, 268
Lagrange's method, 268
Lagrangian, 269
Lagrangian multipliers, 269
Laws, 211
  Finger-Toe, 198
  Newton's, of gravitation, 211
  Newton's, of motion, 217-18
  of thermodynamics, 111, 213, 214
Least squares line, 121
Length, 109, 141, 367
"l'Hospital law," 199
Linear impulse, 203
Linear programming, 277
Line graphs, 112
  construction, 115
  pair of, to form graph, 114
Loads, uncertainty due to, 44
Logic error, 82
Lumen (lm), 368
Luminous intensity, 367
Lux (lx), 368

Magnetic effect, 188
Manufacturing process, uncertainty due to, 44
Mass, 109, 141, 367
Materials properties, uncertainty in, 44
Mathematical curve fitting, 125
Mathematical model. *See* Model, mathematical
Matrix
  of dimensions, 143
  rank of, 145
  of solutions, 144
Matter
  continuum hypothesis, 105
  particulate viewpoint, 105
  phase of, 180
  property of, 180
Maxima, 265
Maxwell, James Clerk, 105
Mean, 52

Measurement couple, 256
  decisions concerning, 344
Measurements, 111
Merit. *See* Figure of merit
Meter. *See* Metre
Methods, numerical
  of integration, 92
  of root-finding, 86
  of square root, 73-75
Metre, 41, 367
  operational definition, 111
Metric convention, 165
Metric system, 165
Miller problem, 196
Mine hoist, 44
Minima, 265
Minimum, false, 265
Model, mathematical
  assertive deductive, 228
  deterministic deductive, 228
  probabilistic deductive, 228, 247
  probabilistic simulative, 252
  steps in development, 228
  uncertainty due to, 44
Model, physical, 160
Modulus of elasticity, rigidity, 185
Mole (mol), 367
Monitor, convergence, 82
  of program NEWRA1, 98, 100
Monte Carlo simulation, 253
Motor effect, 188
Multidimensional searches, 290

Need, identification of, 13
Newton (N), 368
Newton-Raphson method, 86
Newton's first order method, 86
  flowchart for, 96
  FORTRAN program for, 96
Newton's law, 65, 202, 217-18, 221
  convective heat effect, 184, 345
  viscous shear effect, 106
Newton's second order method, 89
Notebook, engineer's, 351
Numerical methods
  integration, 92
  Newton-Raphson, 86
  Newton's first order, 86
  Newton's second order, 89
  root-finding, 86
  square root, 73-75

Objective function, 8, 39, 265, 274
Ohm (Ω), 368
Operational definition, 108, 110,

Index

123
of temperature, 122
Optima, 266
Optimization, 265
  Lagrange's method, 268
  mathematical programming method, 274, 278-97
  problem statement, 274
Origin, plotting, 113
Overloading a transformer, 309

Partial derivative, 147
Pascal (Pa), 368
Patent development, 351
Path, locus of states, 180
Phase, of matter, 180
Physical model, 160
Pipeline routing, 312
π-terms
  complete set of, 143
  construction of, 141
π-theorem, 147
Plotting, 130
Pocket calculator
  keystroke sequence, 75
  program, 75
  root-finding program, 87-91
Points
  on boundary, 266
  interior stationary, 266
Poisson's ratio, 185
Power, 369
Precision, 111, 189
Prediction, 228
Preliminary design, 104
Pressure, 369
Probabilistic mathematical model.
    See Model, mathematical
Probability density, 51
  pdf function, 52, 167
Problem-solving, 21
  questions during, 308
Product liability, 351
Program
  executive, 101
  FORTRAN, for mine hoist, 99
  FORTRAN, for square root, 75
  pocket calculator, for square root, 75
  test, for NEWRA1, 97
Property, 180
Pure substance, 180

Qualitative factors, 68

Radian (rad), 368
Radiation, 185
Railroad, Chicago, North Shore & Milwaukee
  equipment statistics, 328
  interurban car plan, 329
  motor tractive effort, 330
Random failure interval, 57
Random number generation, 253
Random variable algebra, 247, 348
Rank, of matrix, 145
Rankine scale, 122
Rapid transit car, 64, 328
Rectification, of curves, data strings, 129
Reference, inertial, noninertial, 223
Region, control, 180
Regression line, 347
Relay, inspection, 48, 202
Reliability, 55
  as figure of merit, 59
  systems in parallel, 59
  systems in series, 58
Reproducibility, 189
Resistive effect, 187
Risk, 54
Rock crusher, pan angle of, 322
Root-finding methods, Newton's
  first order, 86
  second order, 89
Rounded (truncated) number, 256
Round-off, in SI, 373

Sampling, random, 53, 54
Scale factors, 112, 113
Search
  direct, 275
  exhaustive, 280
  fence-walking, 297
  fruitless, 29
  golden section, 284
  gradient sensitive, 290
  interval-halving, 281, 282
  multidimensional, 290
  simultaneous, 279
Second (s), 111, 367
Second law of thermodynamics, 111
Sensitivity analysis, 272
Sensitivity coefficients, 273
Set, complete. *See* Complete set
Set, decision, 275
Siemens (S), 368
Significant number, 256
Similarity, 162

Similitude, 160
Simpson's first rule, 91
Simulative model, 252
SI prefixes, 370
SI units, 164-65, 370
  correct usage, 372
  problems, 374
Size, uncertainty due to, 44
Slackness, complementary, 270
Slack variables, 269
Slope of line, 114
Spaces in numbers, SI, 371
Species, of countables, 199
Speed
  balancing, 65
  of wire winder, 333
Spring block system, 216, 228
Standard deviation, 53
  in random variable algebra, 248
Stationary points, 266
Statistical couple, 189
  statements, 112
Steady state, 200
Stefan, Josef, 185
Steradian (sr), 368
Stress concentration, uncertainty due to, 44
Subroutine, FORTRAN, 79
Substance, pure, 180
Substance, thermometric, 122
Suitability, 5
Suitability, feasibility, acceptability test, 6
Surface tension, 187
Surroundings
  of control region, 180
  of system, 179
Survival equation, 58
Synthesis, 13, 21
System, 179-83
  adiabatic, 215
  delineation of, 222
  heterogeneous, 180
  homogeneous, 180
  of matter, 179
  modernized metric, 164
  open, 180
Système International. *See* SI units
Temperature, 111, 141, 367
  scales, 122-23
Template, engineering, 8
Thermodynamics
  first law, 111, 213, 214
  second law, 111

Thermodynamic surface, 120
Thermometer, gas, 117
Ticks, scale, 115, 166
Tie-bolt, 22
Time
  uncertainty due to, 44
  as unit, 109, 367
Time constant, 345
Traction motor, 64
Tractive effects, 185
Trade-off function, 66
Train, 64, 328
  brakes, 65
  resistance equation, 328
Transformation of data, 129
Transformer, overloading of, 309
Twelve Fingers, 197

Uncertainty, 44, 54
  interval of, 278
Unimodal function, 278
Unit prefixes, 377
Units
  base, 367
  derived, 368
  fundamental, 109
  SI. *See* SI units
  supplementary, 368
Useful life, 57
Utility function, 39

Validity, checking for, 238-40
Variability, recognition of, 167
Variable loading, 123
  stress, 124
Variables
  reduction in number of, 139
  slack, 269
Variance, 53
Viscous fluid effect, 186
Volt (V), 368

Watt (W), 368
Wear-out interval, 57
Weber (Wb), 368
Weibull
  random numbers, 255
  survival equation, 255
Weierstrass's theorem, 265
Weight, as figure of merit, 32
Widget machine example, 48, 202
Wire winder, speed of, 333
Work, 369
World, verbal-symbolic, 110